3rd ISESCO International Workshop and Conference On Nanotechnology 2012

(IWCN2012)

AA002799

Journal of Physics: Conference Series Volume 431

Selangor, Malaysia
5-7 December 2012

Editors:

Akrajas Ali Umar
Muhamad Mat Salleh
Muhammad Yahaya

ISBN: 978-1-62748-308-7
ISSN: 1742-6588

Printed from e-media with permission by:

Curran Associates, Inc.
57 Morehouse Lane
Red Hook, NY 12571

CURRAN ASSOCIATES INC. proceedings.com

Some format issues inherent in the e-media version may also appear in this print version.

Printed by Curran Associates, Inc. (2013)

For permission requests, please contact the Institute of Physics at the address below.

Institute of Physics
Dirac House, Temple Back
Bristol BS1 6BE UK

Phone: 44 1 17 929 7481
Fax: 44 1 17 920 0979

techtracking@iop.org

Additional copies of this publication are available from:

Curran Associates, Inc.
57 Morehouse Lane
Red Hook, NY 12571 USA
Phone: 845-758-0400
Fax: 845-758-2634
Email: curran@proceedings.com
Web: www.proceedings.com

TABLE OF CONTENTS

PREFACE

The ISESCO Conference on Nanomaterials and Applications (IWCN2012) is one of a series of nanotechnology seminars organized by ISESCO, Malaysian Solid State Science and Technology Society (MASS), the Universiti Kebangsaan Malaysia. IWCN2012 is the third seminar, following IWCN2007 and IWCN2010, held in the Universiti Kebangsaaan Malaysia, Bangi Malaysia from 5–7 December 2012. The conference was attended by more than 250 participants from 15 countries, including 150 students. The conference and workshop provided a forum for researchers and students, policymakers and other professionals especially from the ISESCO Member States to exchange information, enhance understanding and more importantly to engage in the development of new nanoscience and nanotechnology research in multidisciplinary areas in physics, chemistry and biology. Together with the conference, the third Meeting of the ISESCO Expert Panel on Nanotechnology was held to chart the future activities in Nanoscience and Nanotechnology in ISESCO member countries.

The objective of the conference is to communicate and discuss recent progress in nanoscience and nanotechnology research and its potential applications in future economic growth. The main focus for the present activity is on energy and environment. The conference received 105 papers in total and 50 of them were selected to be considered for publication in *Journal of Physics: Conference Series*. However, finally, after undergoing vigorous and thorough revisions by respected editors and reviewers in the fields, 29 papers were accepted. This volume covers the follwing topics:

- Nanomaterials Synthesis and Characterization
- Nanomaterials for Energy and Catalysis
- Nanoelectronics, Sensors and MEMS Devices
- Nanophotonics

We are indebted to all the keynote speakers, invited speakers, participant and authors for their contribution to this event. Their contribution has led to the success of this conference. We are also very grateful for the time, effort and unwavering support dedicated by all the referees. Their tireless work has helped guarantee the high scientific level of this series. Many thanks are also addressed to the ISESCO, COMSATS, MASS, UKM and FST for their financial support. Logistics support provided by IMEN and MNA is also much appreciated.

We realized that this volume would never have materialized without the hard work of Ms Siti Khatijah Md Saad who helped the editorial secretariat in many ways communicating with the authors and reviewers and checking and typesetting the papers in the final stage. Finally, great appreciation is addressed to all who have worked hard and given support ensuring that the conference was a success.

Editors
Associate Professor Dr Akrajas Ali Umar

Professor Dr Muhamad Mat Salleh
Professor Dato' Dr Muhammad Yahaya

IWCN 2012 Organizing Committee

International Advisory Board
Professor Dato' Dr Muhammad Yahaya (UKM, Malaysia)
Dr Faiq Bilal (ISESCO, Morocco)
Professor Dr Michael Graetzel (Switzerland)
Professor Dr Muhamad Rasat (Malaysia)
Prof Dr Masbah R T Siregar (Indonesia)
Associate Professor Dr Munetaka Oyama (Japan)
Professor Dr Ismat Shah (USA)
Professor Dr Muhamad Mat Salleh (Malaysia)
Chairman
Associate Professor Dr Mohammad Kassim
Co-Chairman
Dr Mohammad Hafizuddin Haji Jumali
Secretary
Dr Lorna Jeffry Minggu
Treasurer
Dr Sharina Abu Hanifah
Committee Members
Associate Professor Dr Mohd Azmi Abd Hamid
Associate Professor Dr Akrajas Ali Umar
Dr Zahari Ibarahim
Dr Rozidawati Awang
Dr Farah Hannan Anuar
Secretariat
Dr Khuzaimah
Mohd Firdauz Ismail
Izura Izzuddin
Nor Huwaida Janil Jamil
Siti Khatijah Md Saad
Noor Razinah Rahmat
Mark Lee Wun Fui
Ng Kim Hang
Lee Thian Khoon
Law Kung Pui
Choong Yan Yi

All papers published in this volume of *Journal of Physics: Conference Series* have been peer reviewed through processes administered by the proceedings Editors. Reviews were conducted by expert referees to the professional and scientific standards expected of a proceedings journal published by IOP Publishing.

Ultrafast Formation of ZnO Nanorods via Seed-Mediated Microwave Assisted Hydrolysis Process

S T Tan[1], A A Umar[2*], M Yahaya[1], C C Yap[1] and M M Salleh[2]

[1]School of Applied Physics, Faculty of Science and Technology, Universiti Kebangsaan Malaysia, 43600 Bangi, Selangor, Malaysia
[2]Institute of Microengineering and Nanoelectronics (IMEN), Universiti Kebangsaan Malaysia, 43600 Bangi, Selangor, Malaysia

E-mail: [*]akrajas@ukm.my

Abstract. One dimensional (1D) zinc oxide, ZnO nanostructures have shown promising results for usage in photodiode and optoelectronic device due to their high surface area. Faster and conventional method for synthesis ZnO nanorods has become an attention for researcher today. In this paper, ZnO nanorods have been successfully synthesized via two-step process, namely alcothermal seeding and seed-mediated microwave hydrolysis process. In typical process, the ZnO nanoseeds were grown in the growth solution that contained equimolar (0.04 M) of zinc nitrate hexahydrate, Zn $(NO_3).6H_2O$ and hexamethylenetetramine, HMT. The growth process was carried inside the inverted microwave within 5- 20 s. The effect of growth parameters (i.e. concentration, microwave power, time reaction) upon the modification of ZnO morphology was studied. ZnO nanostructures were characterized by Field emission scanning electron microscope (FESEM) and X-ray diffraction (XRD). The densities of nanorods were evaluated by the Image J analysis. It was found that the morphology (e.g. shape and size) of nanostructures has changed drastically with the increment of growth solution concentration. The density of ZnO nanorods was proven to increase with the increasing of reaction time and microwave power. We hypothesize that the microwave power might enhance the rate of nucleation and promote the faster nanostructure growth as compared with the normal heating condition due to the superheating phenomenon. This method might promote a new and faster alternative way in nanostructure growth which can be applied in currently existing application.

1. Introduction

Zinc oxide (ZnO) is a common material that is readily used in semiconductor fabrication especially in solar cell and gas sensor due to its special properties. ZnO is a wide energy band gap material (3.37 eV) that exhibits excellent optical, electrical and thermochromism properties. Recently, it has been demonstrated that ZnO readily self-assembles into diversity of nanocrystalline structures such as nanospines, tetrapods, nanorod [1-3].

Many approaches have been done to produce a vertical array of ZnO nanorods such as vapor-solid-liquid method, chemical vapor deposition and electrochemical growth method [4]. However, all of these methods consume a lot of energy and time. In order to synthesize a well-crystalline nanostructure in an economical way, hydrothermal process was introduced. Wang et.al [5] have reported the formation of various ZnO nanostructures via hydrothermal process. The effect of growth conditions (i.e. concentration, pH and growth time) was studied. It was found that an interesting structure called latch-like structure was formed at higher pH: 8.0 and 120 °C for 4 h. Nevertheless, this method was time consuming and cannot produce a uniform homogeneous sample due to the existence of temperature gradient between the vial and growth solution during the convection heating process.

A fast and easier way which is called as microwave hydrothermal process was introduced in recent years. Microwave radiation synthesis is a method that applies the electromagnetic radiation to alter the dipole moment of the molecule and induce formation of nanostructure through the superheating mechanism. This method has been widely used in medical science, organic polymerization and nanomaterial synthesis [6, 7] due to its accelerate synthesis and high yield of the nanostructure. Recently, we have synthesized the ZnO nanoplate via the reflux microwave oven inside microwave transparent vial (i.e. sealed Teflon) [8]. This method has shown a rapid growth of nanostructure in homogeneity distribution within minute. Unfortunately, the special design microwave is costly which retards the mass reproduction. Considerable attention has also been given to reaction time and growth rate of nanostructure in recent research. Mahpeykar et al. [9] have synthesized ZnO nanowire in rapid growth rate of 200 nm min^{-1} by the ammonia assisted growth process. Several reports have suggested the effective increment of growth rate of nanostructure synthesized directly in solution form. However, the coverage of nanostructure spin coated from the as-grown solution on to the substrate is very difficult to control.

Our group have been actively involved in the synthesis of ZnO nanostructure directly on the surface with controllable shape and morphology [10, 11]. Here, we report the growth of ZnO nanostructure on the surface of FTO via ultrafast microwave seed assisted method within 10-20 s. This experiment was carried in two-step synthesis process, namely alcothermal seeding and microwave-assisted hydrolysis process. An equimolar (0.04 M) of zinc nitrate hexahydrate, Zn $(NO_3).6H_2O$ and hexamethylenetetramine, HMT was used as a growth solution. The effect of precursor concentration, microwave power and the reaction time was studied. The optimum and high density of ZnO nanorods was obtained at microwave power of 1100 W in growth solution with concentration of 0.04 M for 20 s. This technique has been shown to be a simple and faster way in producing vertically aligned ZnO nanorods. This result could be potentially used in rapid production of ZnO nanorods.

2. Experimental details

2.1. Synthesis of ZnO nanostructures

One dimensional (1D) hexagonal nanorods were synthesized via two-step route method, namely alcothermal seeding and hydrothermal growth process. Alcothermal seeding [12] was used to prepare uniform nanoseed particles. In typical procedure, 20 mL of ethanolic solution (Sigma Aldrich, Reagent grade) of zinc acetate (0.01 M) was prepared. The solution was then spin coated on a FTO substrate with 3000 rpm for 30 s. After that, the as-coated substrate was heated on a hotplate at 100 °C for 15 min. This process was repeated for 3 times to obtain a high density of nanoseed particles. Finally, the substrate was transferred into the horizontal tube furnace (VT Furnace) and annealed in air for 1 h at 350 °C. Using this process, ZnO nanoseeds with size of 5-10 nm in high density were obtained.

Hexagonal ZnO nanorods were grown from the nanoseeds in a growth solution that contained equimolar (0.04 M) of zinc nitrate hexahydrate, Zn $(NO_3).6H_2O$ and hexamethylenetetramine, HMT (Reagent grade, Sigma Aldrich) via microwave assisted hydrolysis process. A Panasonic inverted home application microwave system (NN-GD577M, 1100 W, 2.45 GHz) was used in this study. In typical process, the seeded substrate was placed inside the glass vial that consisted of 5 mL of growth solution. After that, the glass vial was put into the Teflon container and positioned on the centre of microwave during the growth process. Lastly, the sample was taken out from the system. The as-grown nanorods were cleaned with copious of distilled water and dried with a jet of nitrogen gas. Effect of structural growth of ZnO was studied by varying the growth solution concentration (0.01-0.05 M), microwave power (180-1100 W) and reaction time (5-20 s).

2.2. Characterization

The morphology of the nanostructures was examined using Field emission scanning electron microscope (FESEM, ZEISS Supra 55 VP) that was operated at the pressure of 10^{-5} Pa and acceleration voltage, EHT of 5 kV. The crystallinity and purity of the sample were characterized by using advanced x-ray diffraction spectrometer (XRD, Bruker D8) with CuKα radiation. Diffraction angle of 30-50 ° and scanning rate of 0.002°/s was thus studied.

3. Results and discussion

3.1 Effect of Concentration

Figure 1 shows the FESEM image of ZnO nanorods grown at different growth solution concentrations, namely 0.01 M-0.05 M in 1100 W for 20 s. ZnO nanorods were successfully grown on the surface using this present approach. As seen from Figure 1, the nanorods diameter was found to be dependent on the growth solution concentration. In typical results, the diameter increased from c.a. 30 nm to 80 nm with the increasing concentration up to 0.04 M. But it decreased to c.a. 50 nm once the concentration exceeded 0.05 M. Besides, it was found that the density of nanorods also changed with the concentration. The density of the sample increased exponentially with the concentration. The densities of the sample were 0.54, 0.58, 0.63, 0.74 and 0.91 nanorods/ nm^2, respectively. Furthermore, it was noticed that the present nanorod end showed a rough surface. This phenomenon contradicts with the other reports that nanorods synthesized via usual hydrothermal method only showed a flat hexagonal surface. The existence of rough surface might be caused by the Ostwald ripening which will be discussed later. Figure 1 (G) shows the XRD spectra of the samples grown at different growth solution concentrations, namely 0.01 M-0.05 M. A dominant peak correlated to (002) plane of ZnO hexagonal wurtzite structure (JCPDS No file: 36-1451) was obtained at around 2θ of 34.4° in all the samples. Besides that, it was noticed that three small peaks which denoted the (100), (101) and (102) planes were obtained at 2θ of 31.3°, 36.2° and 47.5°, respectively. The Bragg intensity increased with the concentration due to the increment of diameter and density of nanorods. This result is in good agreement with the FESEM image of nanorods. Apart from ZnO and FTO peaks, there was no other Zn complex peak appeared in the spectrum. From this result, it was confirmed that the pure ZnO nanostructures were successfully grown on the substrate by using the current method.

The effect of precursor concentration on the nanorods diameter and density can be considered as follows. In the microwave-assisted hydrolysis process, the microwave radiation directly heated the whole volume of the solution and induced the hydrolysis process of the Zn complex. In the growth process, the ionic species aligned within the electrical field and nucleated on the nanoseeds surface in an effort to minimise the surface energy. The growth process only occurred when the critical size of nanoseeds was obtained. The nanoseeds then grew until reaching the critical size that promoted to vertically-aligned nanorods growth (refer Figure 2). For the case of concentration of 0.01 to 0.04 M, the nanorods diameter and density increased with the increasing of concentration. This can be understood as the result of quick formation of critical size on most of the nanoseeds on the surface due to the increasing precursors. Amount of the nanoseeds which resulted in nanorods growth increased accordingly, leading to higher density of the sample. Since the nanostructure growth was toward all directions including long and short axis of nanorod, the diameter also increased with the increasing of concentration. However, at relatively high concentration i.e. 0.05 M, the nanorods diameter decreased as compared to those prepared at lower concentration. This could be due to the rapid growth process at such high concentration, where the critical size was expected to be smaller than those in the lower concentration. Hence smaller diameter and higher density nanorods were obtained.

Meanwhile, for the case of concentration effect on the nanorods end surface roughness, Ostwald annealing process can be considered (refer Figure 2). It is agreed that the Ostwald annealing often occurs in the variation of time reaction. But in this experiment, we have applied a high power synthesis, which was 1100 W for the microwave growth process. This high power might increase the rate of nucleation. The precursor might be quickly consumed in a short period and cause a drastic drop in the nucleation rate. This condition may cause the atom at the vertex site of nanorods surface to dissolve into the solution and then nucleate on the other site of the surface. This process is predicted to be more active at high concentration. Thus, the surface roughness of the nanorods end increased with the increasing concentration.

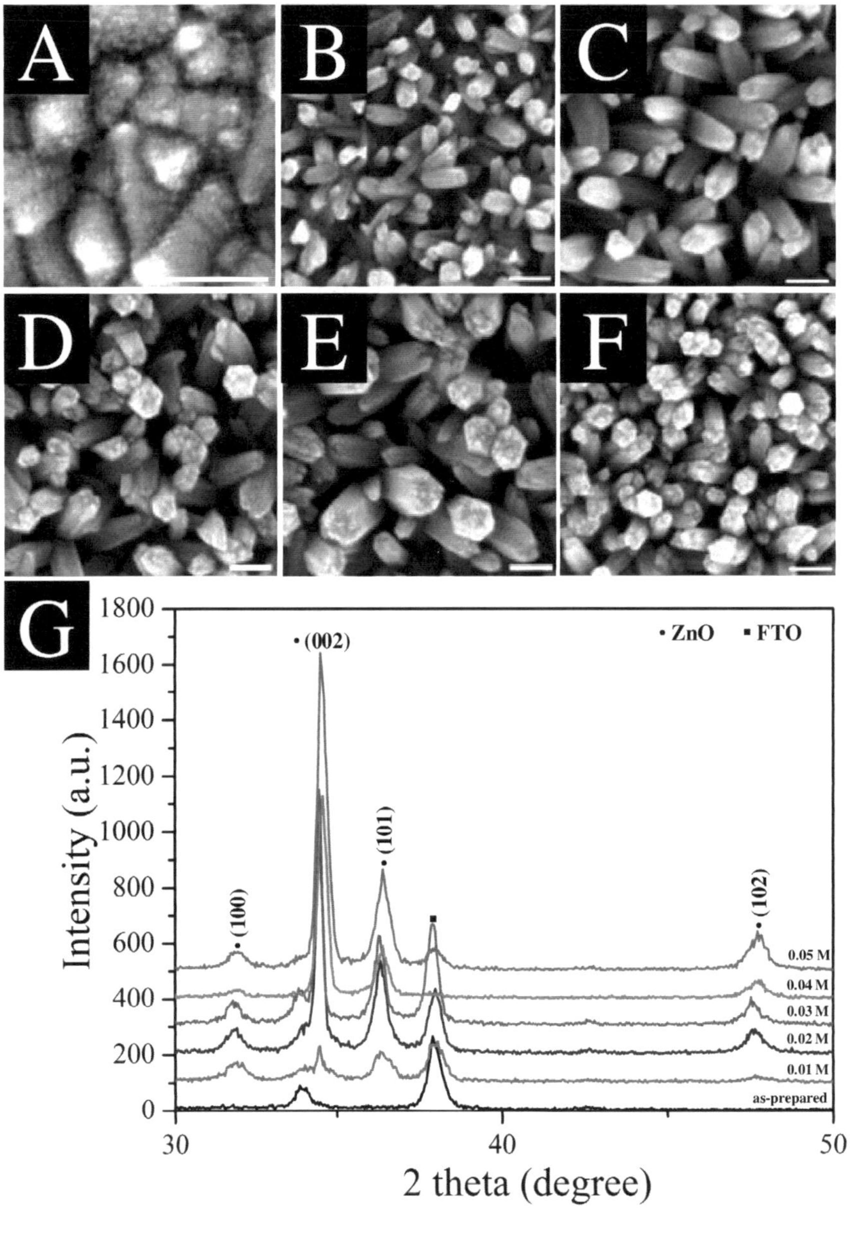

Figure 1. FESEM image of ZnO nanoseeds (A) and nanorods grown in 0.01 M (B), 0.02 M (C), 0.03 M (D), 0.04 M (E), 0.05 M (F) of growth solution at 1100 W for 20 s inside the inverted microwave and their corresponding XRD spectra (G). Scale bars are 100 nm.

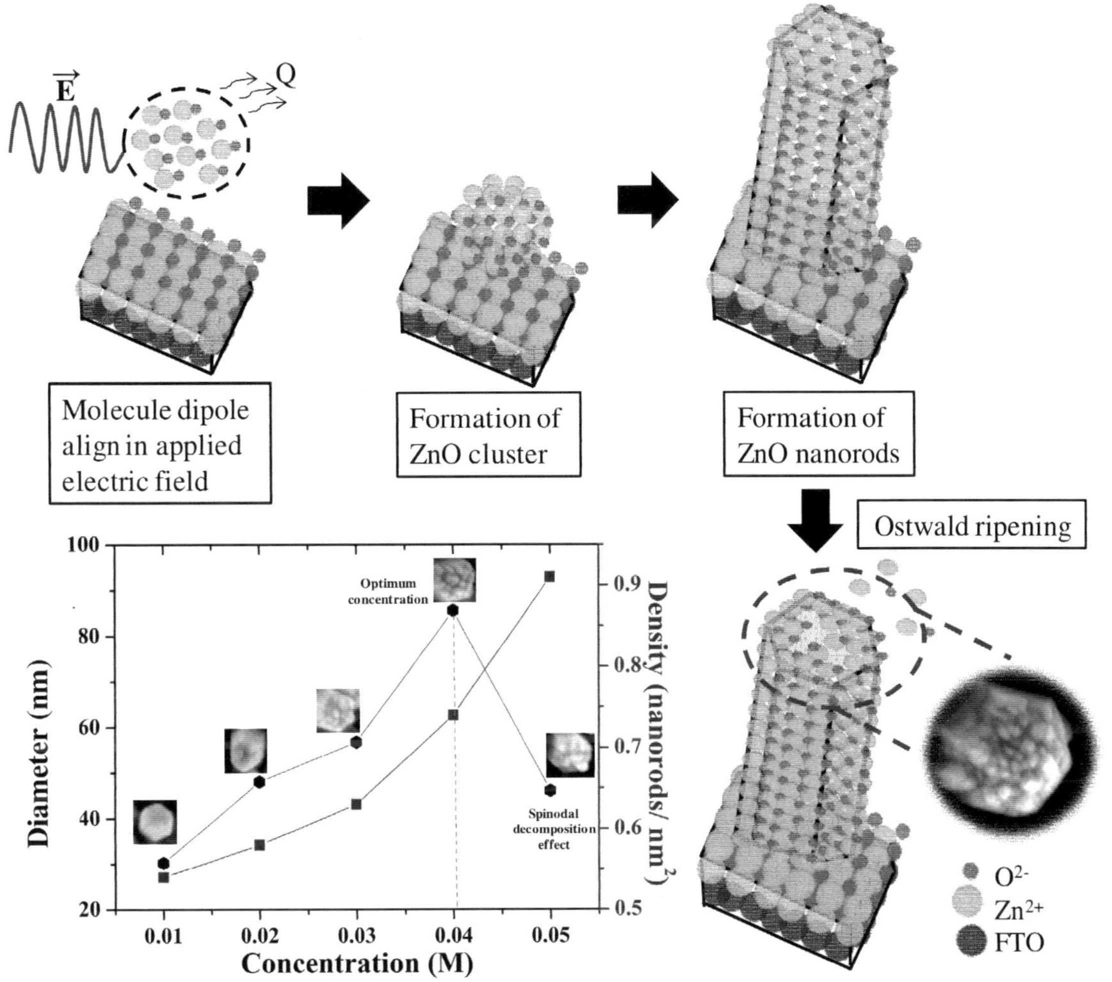

Figure 2. Mechanism of ZnO nucleation process under microwave-assisted hydrolysis method.

3.2 Effect of microwave power

Figure 3 shows the FESEM and XRD results of the corresponding untreated and treated samples which were grown at microwave power of 187-1100 W. As seen from Figure 3(A), there was no nanostructure formed at 187 W. The nanorods formed as the power reached around 300 W. The diameter of the nanorods increased from c.a. 28 nm to 80 nm with the increasing microwave power. The density of nanostructures also increased from 0.49 to 0.74 nanorods/ nm^2 as the synthesis power reached 1100 W. The surface roughness of the sample also increased with the microwave power. Effect of the power on crystalline structure of the sample was also studied here. The result is shown in Figure 3(G). It was found that the XRD profile on the sample prepared by different powers showed similar trend with that observed in Figure 1(G) without any shifting in the peak. This infers that there was no modification in the crystallinity of nanostructure in the application of microwave although different microwave powers were applied.

The effect of microwave power on the diameter, density and surface roughness can be explained as follows. Microwave power played an important role in determining the superheating state of the precursor solution in the nucleation process. The small microwave power was insufficient to generate dielectric heating for the hydrolysis process. As the power increased, the hydrolysis process happened more thoroughly in the solution. Indirectly, the ionic species concentration in the solution would be affected. The phenomenon is similar to the concentration effect. As mentioned earlier, the density and the diameter of the nanorods might change with the critical size. Furthermore, the increment of surface roughness of the ZnO nanorods end might also be caused by the rate of microwave radiation emission besides the Ostwald ripening. As the power increased, the rate of microwave emission increased and caused the ionic species to oscillate in a higher speed. The collision between the dipolar ZnO molecules might also induce the defect on the nanorods end.

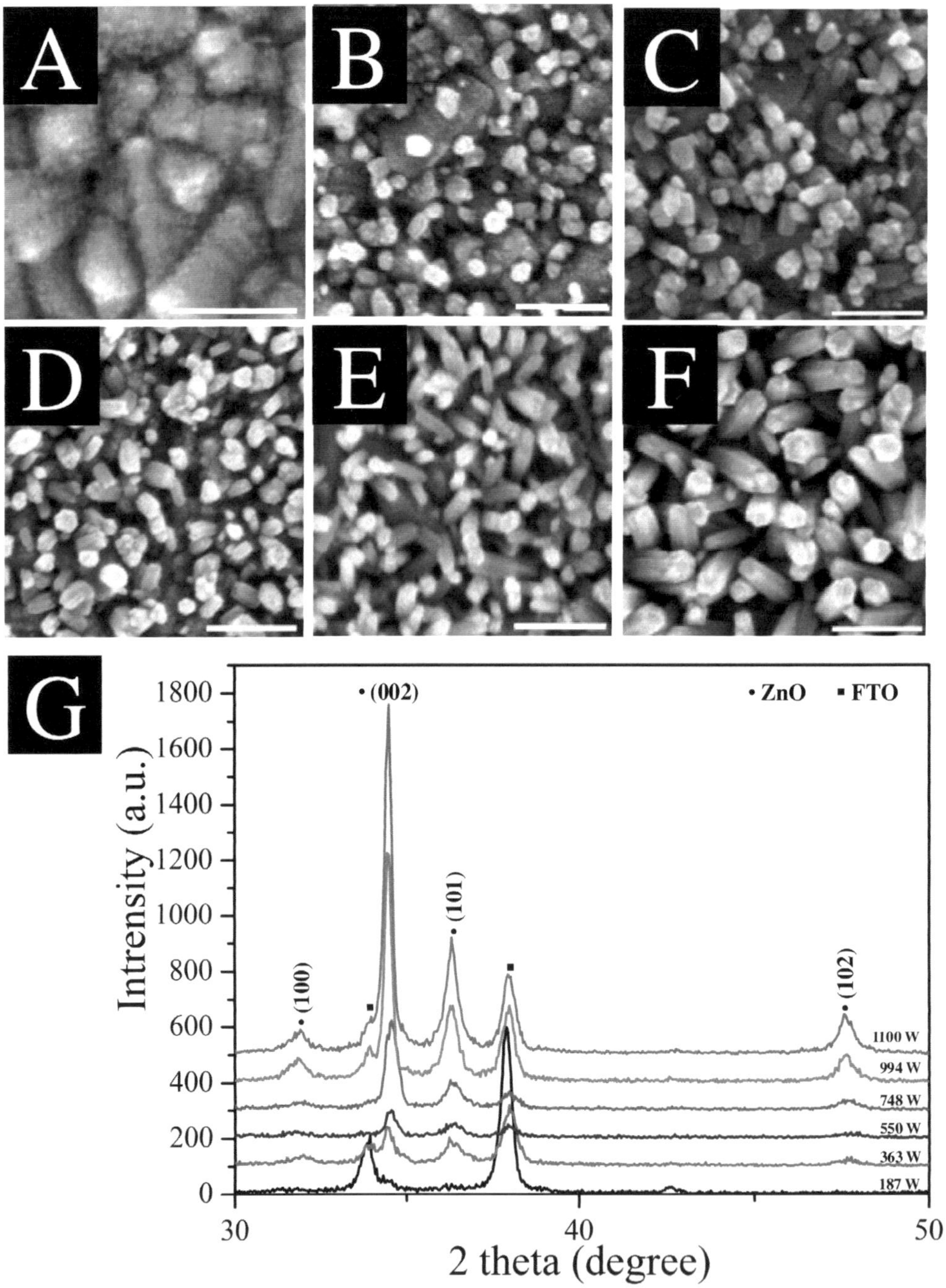

Figure 3. FESEM image of ZnO nanorods grown at 187 W (A), 363 W (B), 550 W (C), 748 W (D), 994 W (E) and 1100 W (F) for 20 s inside the inverted microwave and their corresponding XRD spectra (G). Scale bars are 100 nm.

3.3 Effect of reaction time

Figure 4 shows the FESEM image of ZnO nanostructure grown for different reaction times which were in the range of 5-20 s. As can be seen from Figure 4(A), none of the ZnO nanostructure was obtained on the surface. As the time prolonged to 8 s, the nanorods were found starting to grow and the growth solution turned into milky white colour after the synthesis process. The diameter and density of the ZnO nanorods increased proportionally with the extension of the reaction time, which were from c.a. 27 nm to 80 nm and from 0.44 to 0.74 nanorods/nm^2 for the growth time of 8- 20 s, respectively. This could be related to the thermal energy required for initialisation of hydrolysis process. Thus, sufficient microwave irradiation time is needed. Based on this result, we predict that the hydrolysis could be started earlier than 8 s in the growth process. It was also found that the increasing of time also caused increase in the surface roughness of the nanorods end. This is a normal phenomenon as the Ostwald annealing process increased with the reaction time.

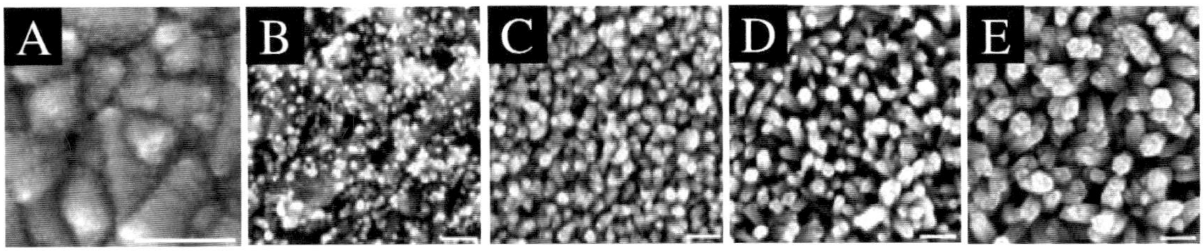

Figure 4. FESEM image of ZnO nanorods grown at 1100 W for 5 s (A), 8 s (B), 10 s (C), 15 s (D) and 20 s (E) inside the inverted microwave. Scale bars are 100 nm.

4. Conclusion

ZnO nanorods have been successfully synthesized using the ultrafast microwave-assisted hydrolysis method in 8-20 s. The diameter, density and surface roughness were found to depend on the concentration, microwave power and the reaction time. The presence approach should become an alternative strategy for rapid preparation of ZnO nanorods with a controlled-density, -diameter and -surface roughness for usage in currently existing applications.

Acknowledgment

We acknowledge the contribution of financial supports by the Malaysian Ministry of Higher Education under research grant ERGS/1/2011/STG/UKM/01/27 and FRGS/1/2012/SG02/UKM/02/3 and En.Mohamad Hasnul Naim B. Abdul Hamid and En. Idris in FESEM characterization.

References

[1] Kumar R S, Sathyamoorthy R, Matheswaran P, Sudhagar P and Kang Y S 2010 Growth of novel ZnO nanostructures by soft chemical routes *J Alloy Compd* **506** 351-5

[2] Tan T, Li Y, Liu Y, Wang B, Song X M, Li E, Wang H and Yan H 2008 Two-step, preparation of Ag/tetrapod-like ZnO with photocatalytic activity by thermal evaporation and spluttering *Mater. Chem. Phys* **111** 305-8

[3] Zhou Z Z, Ding Y, Zu X H and Deng Y L 2011 ZnO spheres and nanorods formation: their dependence on agitation in solution synthesis *J Nanopart Res* **13** 1689-96

[4] Lee H K, Kim M S and Yu J S 2011 Effect of AZO seed layer on electrochemical growth and optical properties of ZnO nanorod arrays on ITO glass *Nanotechnology* **22** 445602

[5] Wang H, Xie J A, Yan K P and Duan M 2011 Growth Mechanism of Different Morphologies of ZnO Crystals Prepared by Hydrothermal Method *J Mater Sci Technol* **27** 153-8

[6] McBurney R T, Portela-Cubillo F and Walton J C 2012 Microwave assisted radical organic syntheses *Rsc Adv* **2** 1264-74

[7] Atkins T M, Louie A Y and Kauzlarich S M 2012 An efficient microwave-assisted synthesis method for the production of water soluble amine-terminated Si nanoparticles

Nanotechnology **23** 294006

[8] Noor J R, Umar A A, Alosfur F, Jumali M H H and Salleh M M 2012 Microwave Assisted Hydrothermal Method for Porous Zinc Oxide Nanostructured Films *J Nanosci Nanotechno* [In Press]

[9] Mahpeykar S M, Koohsorkhi J and Ghafoori-fard H 2012 Ultra-fast microwave-assisted hydrothermal synthesis of long vertically aligned ZnO nanowires for dye-sensitized solar cell application *Nanotechnology* **23** 165602

[10] Tan S T, Umar A A, Yahaya M, Yap C C and Salleh M M 2012 Effect of ZnO Nanoseed Structure o the Growth Orientation of Vertical Array ZnO Nanorods via Hydrothermal Process *Appl Mech Mater* **229-231** 239-42

[11] Tahir S F B M, Ramli N, Izzuddin I, Salleh M M, Yahaya M, Umar A A and Jumali M H H 2012 Effect of ZnO addition on structural properties of ZnO-PANi/carbon black thin films *Sains Malays* **41** 1001-4

[12] Umar A A, Abd Rahman M Y, Taslim R, Salleh M M and Oyama M 2011 A simple route to vertical array of quasi-1D ZnO nanofilms on FTO surfaces: 1D-crystal growth of nanoseeds under ammonia-assisted hydrolysis process *Nanoscale Res Lett* **6** 564

Mechanical and Electrical Properties of p-type $Bi_{0.4}Sb_{1.6}Te_3$ and n-type $Bi_2Se_{0.6}Te_{2.4}$ Bulk Material for Thermoelectric Applications

A Kadhim [a], A Hmood [b] and H A Hassan [c]
School of physics, Universiti Sains Malaysia, 11800, Pulau Penang, Malaysia

Email: [a]areejkadhim@yahoo.com, [b]arshad.phy73@gmail.com, [c]haslan@usm.my

Abstract. This study reports on the fabrication of a chalcogen-based thermoelectric power generation (TEG) device using p-type $Bi_{0.4}Sb_{1.6}Te_3$ and n-type $Bi_2Se_{0.6}Te_{2.4}$ bulk thermoelectric materials. The microstructure of the samples was characterized by field emission scanning electron microscopy (FESEM). The phase composition of the powders was characterized by X-ray diffraction (XRD), revealing a rhombohedral structure. The thermoelectric (TE) properties such as Seebeck coefficient (S) and the electrical conductivity (σ) of the resulting alloys were studied in the temperature range of 300 K to 523 K. The power factor (P_{factor}) for a $Bi_{0.4}Sb_{1.6}Te_3$ as p-type sample was found to be 4.96 mW/mK^2 at 373 K, whereas 2.22 mW/mK^2 was obtained at 383 K for a $Bi_2Se_{0.6}Te_{2.4}$ as n-type sample. Electrical power generation characteristics such as high open circuit voltage (V_{oc}) and maximum output power (P_{max}) were monitored by changing the temperature conditions required to generate maximum power. The significance of the resistances including the internal resistance (R_{in}) and contact resistance (R_C) between legs and electrodes, are discussed. The maximum output power obtained with the 9 p-n couples device was 39.4 mW under the thermal condition of T_H=523 K hot side temperature and ΔT = 184 K temperature difference.

1. Introduction

A thermoelectric power generation (TEG) device produces voltage when there is a temperature difference (ΔT) between the hot and the cold sides as a result of Seebeck thermoelectric effect (TE) [1]. The following are the advantages of these generators: no moving parts, small and lightweight, maintenance-free, acoustically silent and electrically "quiet", and environmentally friendly [2]. Bi_2Te_3-based chalcogenides have attracted great interest due to their promising TE properties [3]. Yet, enhancing in the TE properties is strongly required for empirical applications. One approach to enhance the TE properties is optimization of the doping effective route. For example, recent devices use Bi_2Te_3, a semiconductor, which when alloyed with antimony (Sb) or selenium (Se) becomes an efficient TE material for power generation as a result of the variations in the carrier concentration and carrier mobility [4, 5]. As an alternate approach, researchers have attempted to improve the efficiency of materials based on Bi_2Te_3 by creating structures with one or more reduced dimensions [6]. In one case, an n-type Bi_2Te_3 has been shown to have improved Seebeck coefficient (S). However, S and electrical conductivity (σ) have a trade-off; a higher S results in decreased carrier concentrations and decreased σ [7]. The fundamental physical parameters of Bi-Sb-Te and Bi-Se-Te TE materials such as σ, S, and power factor (P_{factor}) are referred in our previous papers [5,8], which reported that the highest P_{factor} have been obtained for the compositions of $Bi_{0.4}Sb_{1.6}Te_3$ and $Bi_2Se_{0.6}Te_{2.4}$. For $Bi_2Se_{0.6}Te_{2.4}$ as n-type materials, the carrier concentration is adjusted by doping with tellurium (Te) which is done in the present paper. Te atoms exhibit a donor action because they replace Se atoms in the lattice and each contributes one electron to the conduction band [9]. It appears that the ionization energy of a Te atom is very low and therefore this atom in $Bi_2Se_{0.6}Te_{2.4}$ is almost fully ionized. As an extension of our previous work, we fabricated in this paper 9 couples of TEG device using solid-state microwave synthesis. We focused on the thermoelectric properties of the TEG device that use $Bi_{0.4}Sb_{1.6}Te_3$ as p-type and $Bi_2Se_{0.6}Te_{2.4}$ as n-type.

2. Experimental

Bi, Sb, Se, and Te that were used in this study were highly pure powders (99.999%). The typical element ratio for the preparation of p-type $Bi_{0.4}Sb_{1.6}Te_3$ is as follows: 0.2528g Bi, 0.5892 g Sb, and 1.1580 g Te. The n-type $Bi_2Se_{0.6}Te_{2.4}$ with excess Te was prepared by mixing 1.0834g Bi, 0.1228g Se, and 0.7938g Te. The p-type and n-type ingots were grown using a solid-state microwave synthesis that was described in a previous literature [5, 8]. After grinding, the samples were then characterized to determine their crystallization via X-ray diffraction (XRD, PANalytical X'Pert PRO MRD PW3040, Almelo, The Netherlands). Both types of powders were pressed into disk shapes (5 mm diameter and 3.5 mm thickness) through cold pressing at 10 tons. Selected regions of the samples were imaged using field emission scanning electron microscopy (FESEM) (Leo-Supra 50VP, Carl Zeiss, Germany). σ as a function of temperature was measured via the standard 4-terminal DC method under a vacuum of 10^{-3} mbar, whereas, S was determined by the slope of the linear relationship between the thermoelectromotive force (*e.m.f.*) and ΔT between the two ends of each sample. As shown in Fig.1, the assemblies of 9 (p-n) couples from these pellets were placed between two alumina plates with the corresponding dimensions of 50 mm × 25 mm, which served as hot and cold ends for the relevant TE pellets. By using Ag paste and Cu plates, the Ag paste–Cu plates–Ag paste electrodes were made on the inner surface of the alumina substrates. The device was then dried at room temperature for one day to metalize the electrodes on the devices.

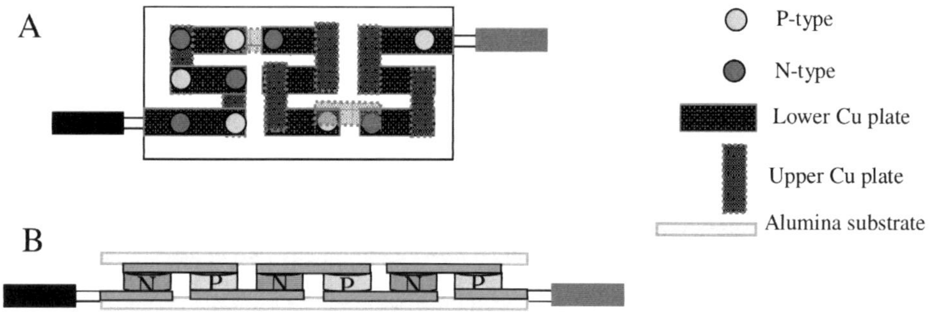

Figure 1. (a) connection of p- and n-type samples by Cu plates, and (b) schematic diagram of the thermoelectric generation device.

To evaluate device performance, the top alumina plate was heated up to 523 K by one brass block as a heater for the device, and the bottom plate was cooled by another brass block with circulated cooling water. ΔT between the hot and the cold sides was measured by two digital K-type E© Sun (ECS820C) thermocouples near the inner surface of the alumina substrates. The current-voltage (*I-V*) lines and the current-power (*I-P*) curves of power generation were performed in air by sweeping the load resistance (R_L) using the variable resistance box. The open circuit voltage (V_{oc}) and many other voltages at the condition of power generation were measured by a voltage meter (Keithley 197).

3. Results and discussion

Typical SEM images of the surface morphology of the samples (ingots and pellets) $Bi_{0.4}Sb_{1.6}Te_3$ and $Bi_2Se_{0.6}Te_{2.4}$ are shown in Figure 2(a) to 2(d). In Fig.2 (a), most of the grains were uniformly arranged and formed by the assembly of micro-sheet grains ($Bi_{0.4}Sb_{1.6}Te_3$ as an ingot). In Fig.2 (b), the FESEM observations revealed the appearance of a typical layered and well-packed structure ($Bi_2Se_{0.6}Te_{2.4}$ as an ingot). In Fig.2 (c) and (d), small pores notice at the grains boundaries in the samples, explaining that the compact texture was obtained. Therefore, the surfaces show uniform grains and high density ($Bi_{0.4}Sb_{1.6}Te_3$ and $Bi_2Se_{0.6}Te_{2.4}$ as pellets).

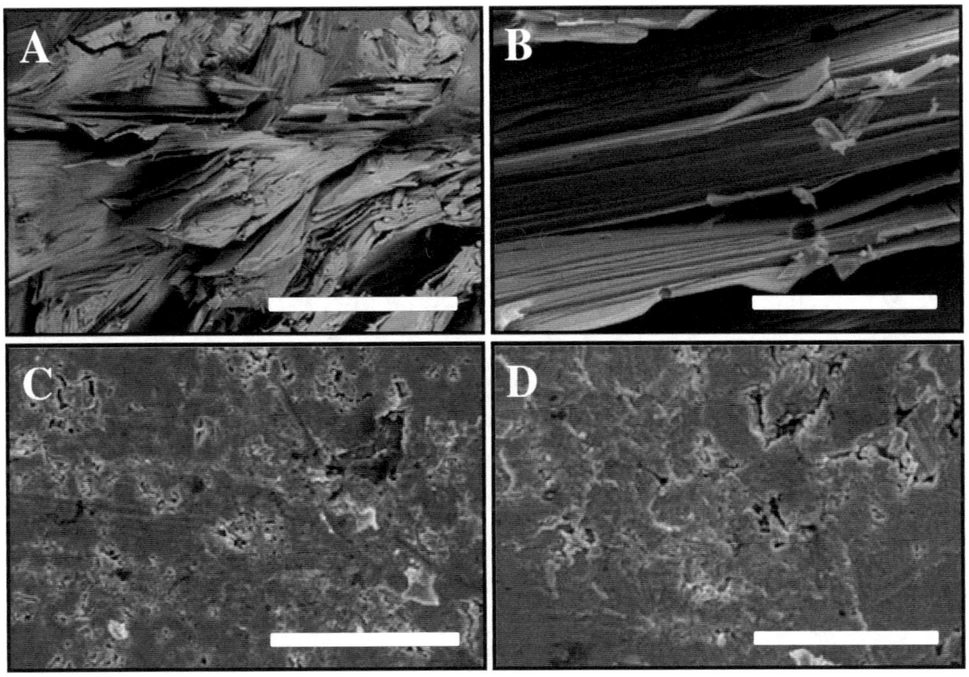

Figure 2. (a) and (b) are the FESEM images of $Bi_{0.4}Sb_{1.6}Te_3$ and $Bi_2Se_{0.6}Te_{2.4}$ ingots, respectively, whereas (c) and (d) are the FESEM images of $Bi_{0.4}Sb_{1.6}Te_3$ and $Bi_2Se_{0.6}Te_{2.4}$ pellets, respectively.

XRD experiments were carried out to determine the structure of the powder samples, and the results are shown in Figs. 3 (a) and (b). XRD spectra for $Bi_{0.4}Sb_{1.6}Te_3$ and $Bi_2Se_{0.6}Te_{2.4}$ powders indicate that the powders are polycrystalline and characterized by a rhombohedral structure, with a dominant peak representing the plane (0 1 5). All of the diffraction peak positions and (hkl) values match very well with the standard diffraction data of pure Bi_2Te_3 and Sb_2Te_3 (JCPDS 15-0863 and 15-0874, respectively) for $Bi_{0.4}Sb_{1.6}Te_3$ (Fig.3 (a)), whereas Bi_2Te_3 and Bi_2Se_3 (JCPDS15-0863 and 33-0214, respectively) for $Bi_2Se_{0.6}Te_{2.4}$ (Fig.3 (b)).

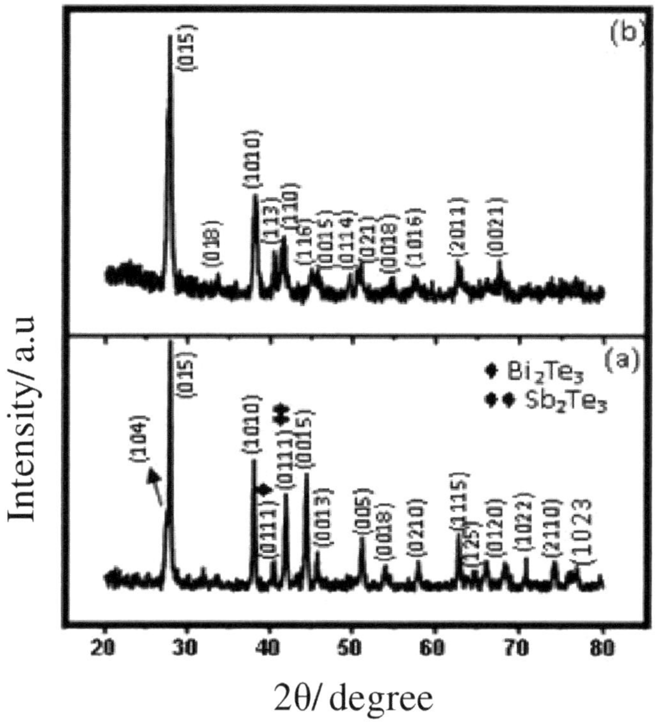

Figure 3. (a) and (b) are the XRD patterns of $Bi_{0.4}Sb_{1.6}Te_3$ and $Bi_2Se_{0.6}Te_{2.4}$ powders, respectively.

The transport properties of p- and n-type samples in terms of σ, S, and P_{factor} were investigated from 300 to 523 K (Figs. 4 to 6). Both sample types had nearly the same behavior for σ, which gradually decreased as the experimental temperature increased, that is, a degenerate semiconductor (Figure 4). σ for the p-type $Bi_{0.4}Sb_{1.6}Te_3$ sample (σ_p) varied from 9.6×10^5 S/m at 300 K to 4.8×10^5 S/m at 523 K, whereas σ for the n-type $Bi_2Se_{0.6}Te_3$ sample (σ_n) was from 1.99×10^4 S/m at 300 K to 1.74×10^4 S/m at 523 K.

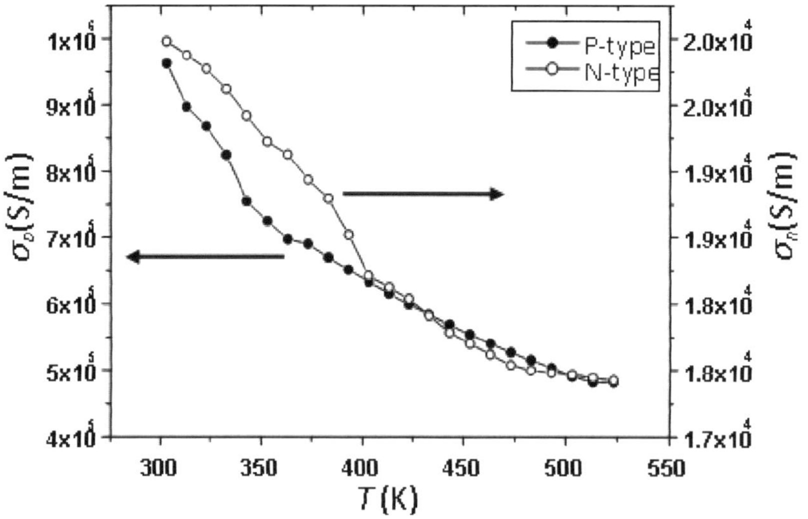

Figure 4. Temperature dependence values of σ of p-type and n-type samples.

As shown in Figure 5, S for the p-type sample (S_p) was 90 at 443 K and for n-type (S_n) was -330.6 μV/K at 423 K. The value of S increases almost rapidly with temperature from 300 to 443K for p-type (300 to 383K for n-type), which is consistent with the Mott formula [4]. Due to the thermal excitation of extrinsic charge carriers at higher temperatures a decrease in S (with rising temperatures) is observed for both types of samples.

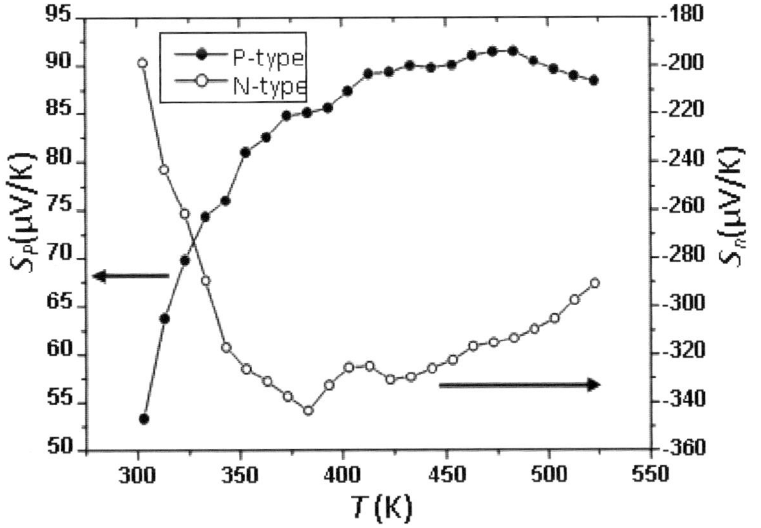

Figure 5. Temperature dependence values of S of p-type and n-type samples.

As evident in Figure 6, the temperature behaviors of P_{factor} for the p-type and n-type samples were similar. $P_{factor-p}$ values obtained for the p-type samples were larger than those for the n-type samples ($P_{factor-n}$) within the entire temperature range. The maximum $P_{factor-p}$ measured was 4.96 mW/mK^2 at 373 K, which is larger than that reported by Li et al. [10], whilst the maximum $P_{factor-n}$ was 2.22 mW/mK^2 at 383 K, which is comparable to that reported by Wang et al. [4]. These two maximum values for p- and n-type contribute to a maximum out put power for our devices.

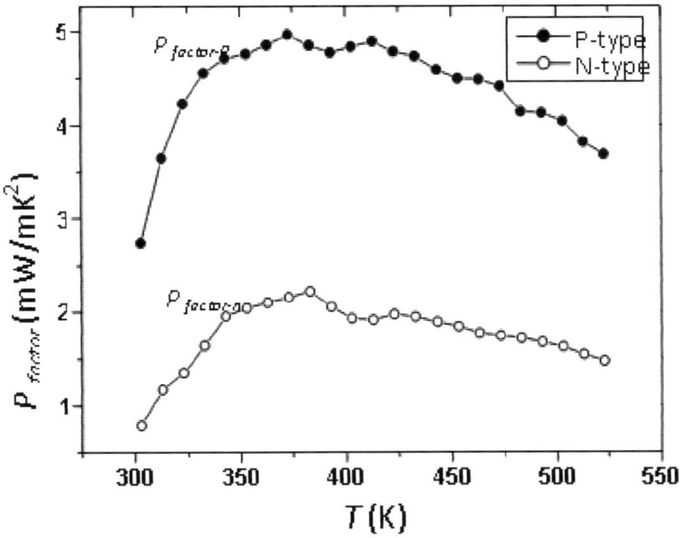

Figure 6. Temperature dependence values of P_{factor} of p-type and n-type samples.

The output voltage and the output power of the fabricated 9 couples versus the current were measured by sweeping $R_L$at several temperature conditions, as shown in Figure 7. The V_{oc} that is equal to the intercept of the *I-V* line reached 355 mV at ΔT of 184 K and T_H of 523 K, which are in agreement with the expression $V=V_{oc}-R_L I$. It is lower than that calculated *S–T* curves (Figure 5) of both p-and n-type legs ($V_{calculated} = (S_p-S_n) \times \Delta T \times N$, where *N* is the number of couples). This voltage loss could have originated from many factors including low thermal conductivity of alumina substrate [11] and unfavorable junctions between the TE legs and the electrodes. *I-P* curves illustrated in Figure 7 exhibit the parabolic curves of the output power (P_{out}); an analysis to plots of I-*V* lines allows the observation of an increasing in the P_{out} with the ΔT. The explanation for this observation results from the rise of the ΔT, whose consequence is an increase in the output voltage (V_{oc}). The higher the value of this V_{oc}, the higher will be the output current (I_{out}) for a given R_L, and therefore will be the dissipated power in the external load ($P_{out} = R_L I_{out}^2$). The maximum output power (P_{max}) values were 39.4 mW at the thermal condition of 523 K T_H and ΔT = 184 K, which means these results could be comparable with the results of Wang et al. [4]. It was investigated that the powers of the devices improved by increasing the temperature. The internal resistance (R_{in}) of each device which corresponds to the slope of the *I-V* lines was directly obtained by the measured system. The ideal internal resistance (R_{id}) was calculated by the sum of the resistance values of p-type and n-type samples. With R_{in} and R_{id}, contact resistance (R_c) can be obtained by $R_c=R_{in}-R_{id}$ [2]. The resistance values of the device (R_{in}=0.8Ω and R_c=0.4 Ω). This result demonstrates that, with the relationship between R_c and P_{max}, R_c should be minimized for each device because it plays a key role in TEG device performance. Based on the data obtained, the good TEG properties originally came from the relatively high electrical properties of p-type and n-type samples that were prepared by solid-state microwave synthesis. Two methods were adopted to optimize the device performance. First, the surface of the alumina plates was treated with NaOH solution to increase roughness and to enhance both mechanical strength and electrical contact between the alumina plates and the Cu electrodes. Second, the ends of p-type and n-type samples were grooved to increase the surface area, also improving the mechanical and the electrical properties of the contacts [2]. Based on these results, the device using p-type $Bi_{0.4}Sb_{1.6}Te_3$and n-type $Bi_2Se_{0.6}Te_{2.4}$ worked successfully and it was stable with satisfactory TE performances.

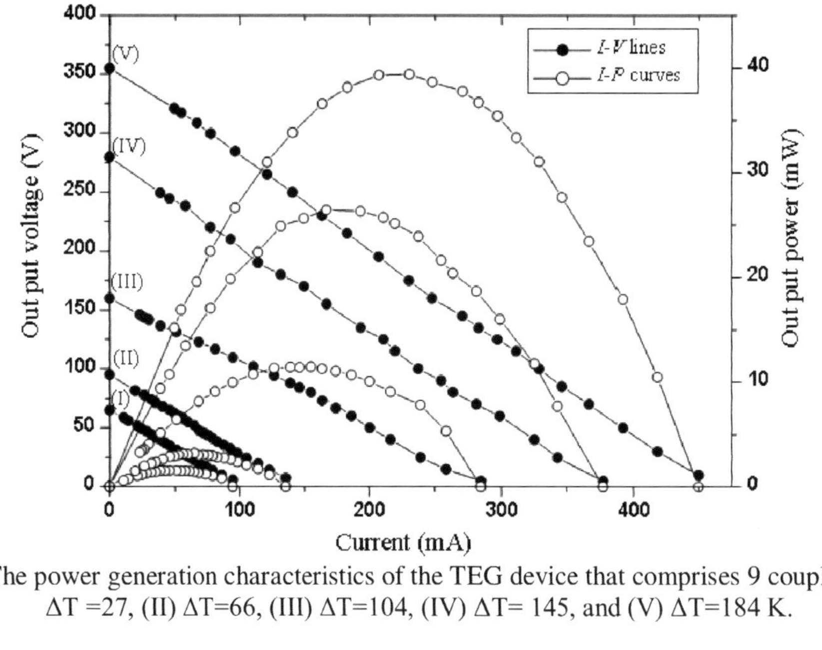

Figure 7. The power generation characteristics of the TEG device that comprises 9 couples, where (I) ΔT =27, (II) ΔT=66, (III) ΔT=104, (IV) ΔT= 145, and (V) ΔT=184 K.

4. Summary

TE materials p-type $Bi_{0.4}Sb_{1.6}Te_3$ and n-type $Bi_2Se_{0.6}Te_{2.4}$ were prepared via solid-state microwave synthesis. TEG devices were fabricated and characterized in terms of high open circuit voltage (V_{oc}) and maximum output power (P_{max}). A maximum V_{oc} and P_{out} (P_{max}) of 355 mV and 39.4 mW was achieved in the device with 9 p-n couples with T_H of 523 K and ΔT of 184 K. V_{oc} and P_{max} systematically increased with ΔT. As a result, the device using p-type $Bi_{0.4}Sb_{1.6}Te_3$ and n-type $Bi_2Se_{0.6}Te_{2.4}$ worked successfully and it was stable with satisfactory TE performances. The successful demonstration of good thermoelectric performances of the as-prepared device suggests the great potential of these low temperature doped thermoelectric materials towards future applications.

Acknowledgments

We are extremely grateful to the School of Physics at the University Sains Malaysia for the general funding provided for this project.

References

[1] L. Gaowei, Z. Jiemin, H. Xuezhang, Analytical model of parallel thermoelectric generator, Appl. Energy 88 (2011) 5193–5199.

[2] H. Ling , J. Yang, L. Shanying, S. Huangming, L. Xinzheng, Q Kaixuan. et al., High temperature thermoelectric properties and energy transfer devices of $Ca_3Co_{4-x}Ag_xO_9$ and $Ca_{1-y}Sm_yMn_{O3}$, J. Alloys Compd. 509 (2011) 8970-8977.

[3] A. Kadhim, A. Hmood, H. Abu Hassan, Novel hexagonal rods and characterization of $Bi_{0.4}Sb_{1.6}Se_{3x}Te_{3(1-x)}$ using solid-state microwave synthesis, Mater. Lett. 81 (2012) 31–33.

[4] W. Shanyu, X. Wenjie, L. Han, T. Xinfeng, Enhanced performances of melt spun $Bi_2(Te,Se)_3$ for n-type thermoelectric legs, Intermetallics 19 (2011) 1024–31.

[5] A. Kadhim, A. Hmood, H. Abu Hassan, Characterizations of solid-state microwave-synthesize Sb_2Te_3-based alloys with various compositions of bismuth in $Bi_{2x}Sb_{2(1-x)}Te_3$, Mater. Sci. Semicond. Process.15 (2012) 549–554.

[6] K. Joohoon, N. Jin-Seo, L. Wooyoung, Simple two-step fabrication method of Bi_2Te_3 nanowires, Nanoscale Res. Lett. 6 (2011) 277.

[7] B. Y. Yoo, C. –K. Huang, J. R. Lim, J. Herman, M. A. Ryan, J.-P. Fleurial et al., Electrochemically deposited thermoelectric n-type Bi_2Te_3 thin films, Electrochim. Act. 50 (2005) 4371–4377.

[8] A. Kadhim, A. Hmood, H. Abu Hassan, Significant influences of selenium on the electrical properties of Bi_2Te_3 compounds synthesized using solid-state microwave irradiation, Advanced Materials Research 501 (2012) 126-128.

[9] D. M. Rowe, Thermoelectrics handbook: macro to nano, CRC Press Taylor & Francis Group, USA, 2006.

[10] D. Li, R.R. Sun, X.Y. Qin, Thermoelectric properties of p-type $(Bi_2Te_3)_x(Sb_2Te_3)_{1-x}$ prepared by spark plasma sintering, Intermetallics 19 (2011) 2002-2005.

[11] Z. Degang, T. Changwen, T. Shouqiu, L. Yunteng, J. Likun and C. Lidong, Fabrication of a CoSb3-based thermoelectric module, Mater. Sci. Semicond. Process. 13 (2010) 221-224.

Electrochemical Synthesis and Characterization of Palladium Nanostructures

S H Ilias[1], K Y Kok, I K Ng and N U Saidin
Malaysian Nuclear Agency, Bangi, 43000 Kajang, Malaysia.

E-mail: suhaila@cc.nuclearmalaysia.gov.my

Abstract. Palladium (Pd) is a good candidate material for hydrogen gas sensing at room temperature due to its high affinity to hydrogen. Pd in the form of nanostructures has demonstrated better performance in its gas sensing property as compared to that in the form of bulk metal because of its large surface area to volume ratio. In this work, various Pd nanostructures were synthesized via electrochemical route. Pd in the form of nanowires can be produced directly via template-assisted electrodeposition while Pd nanotubes were synthesized by galvanic displacement with morphology and composition governed by electrolyte temperature and reaction time. Using similar electrodeposition conditions, Pd in the form of nanoparticles were electrochemically decorated on the surfaces of ZnO rods and Si nanowires arrays. Growth study on various Pd nanostructures produced is presented here, using microscopy, diffraction and probe-based techniques for microstructural, morphological and chemical characterizations.

1. Introduction

Nanostructures such as nanowires and nanotubes have been used as critical components in the latest sensor devices due to their capabilities in improving the sensor efficiency and charge transport property [1, 2]. In addition, due to its low dimensional nanostructures, high-density arrays of sensor elements can also be assembled for better detection accuracies and signal-to-noise ratios as compared to the conventional sensor instrumentations [3]. Palladium nanostructures for examples have been used in many of the hydrogen sensors. Palladium has been a favorite candidate for hydrogen gas sensing because it selectively absorbs hydrogen gas to form chemical palladium hydride.

This work focuses on the preparation of various palladium nanostructures using electrochemical deposition techniques. These techniques offers great advantages in preparing nanostructures, including precisely controlled operation at near-room temperature, low energy requirement, rapid deposition rates, low cost, scalable with easily maintained equipment and flexibility in geometry design. Template-assisted electrodeposition, for example, allows the deposition of high aspect ratio nanostructures such as single elemental or heterostructure nanowires [4, 5] while electrodeless deposition by galvanic displacement of the materials allows the transformation of materials to nanostructures such as nanotubes or coaxial nanowires [6, 7] to achieve the desired chemistry, morphology and microstructures.

In this work, Pd nanostructures were fabricated directly via template-assisted electrodeposition technique and by galvanic displacement of Ni nanowires. Porous anodic alumina (PAA) was used as the templates for confining the electrodeposited materials within their nanopores in order to produce nanowires with controllable dimensions. Using similar electrodeposition conditions, Pd in the form of nanoparticles can also be electrochemically decorated on the surfaces of other nanowires such as ZnO nanorods or silicon nanowire arrays. The microstructures and morphology of the nanostructures were examined using a scanning electron microscope (FEI QUANTA 400) equipped with an X-ray Energy Dispersive Spectrometer (EDS) and operating at an accelerating voltage of 30 kV. The crystallographic information was obtained using X-ray diffractions performed on an X-ray diffractometer (Panalytical X-Pert PRO MPD) operating at 40 kV and 20 mA using Cu Kα radiation.

2. Materials and Methods

Using commercially available anodized alumina (Whatman Anodisc 13) as templates, Pd nanowires with approximately 200 nm diameters were synthesized chronoamperometrically using a three-electrode system with platinum as the counter electrode, Ag/AgCl as reference electrode and alumina template as the working electrode from electrolyte consisting of 0.047 M $Pd(NH_3)_2Cl_2$ + 0.1 M NH_4Cl at -0.8 V and ambient temperature. Prior to electrodeposition, the templates were sputter-coated with a layer of Au on the back. After electrodeposition, the gold-sputtered layer was removed using gold etching solution constituting 2.5 g KI, 10 ml I_2 and 90 ml H_2O. The template was subsequently dissolved in a 5 M NaOH solution. The nanowires were concentrated by centrifuging and rinsing in nanopure water and the process was repeated several times to remove the NaOH solution before finally suspended in isopropyl alcohol.

Similarly, Ni nanowires were prepared potentiostatically at fixed current densities of 10 mA cm^{-2} from Ni electrolyte consisting of 1 M $NiSO_4.6H_2O$ + 0.5 M H_3BO_3. A 2 μL of the Ni nanowire suspension was dispensed on a silicon wafer to allow the solvent to evaporate followed by a dispension of Pd electrolyte onto the nanowires, to galvanically displace the nanowires with Pd. Temperature and time are varied to yield fully displaced Pd nanotubes. Using the same electrodeposition condition, Pd nanoparticles were formed on superhydrophobic surfaces containing ZnO nanorods and silicon nanowires array as working electrode.

3. Results and Discussion

Figure 1 shows the typical scanning electron micrograph of Pd nanowires fabricated by template-assisted electrodeposition and the corresponding X-ray diffractogram shows the presence of Pd with face-centered cubic (FCC) structure. Pd of sub-100 nm diameter can be similarly obtained using template of equivalent pore size fabricated by anodisation process. Figure 2 shows the formation of Pd nanotubes with ~ 20 nm thickness after galvanically displacing Ni nanowires. The inner diameter of the Pd nanotubes was controlled by the diameter of the Ni nanowires. Increasing reaction time or temperature yields rougher and grainier morphologies due to aggregation of more Pd particles along the surface of the nanowires. The driving force behind the galvanic displacement reaction comes from the difference in the reduction potential gaps between Ni^{2+}/Ni and Pd^{2+}/Pd according to the following reaction equation:

$$Ni^0\,(s) + 2\,Pd^{+}\,(aq) \rightarrow 2\,Pd^0\,(s) + Ni^{2+}\,(aq)$$

The exchanged reaction is initiated from the external surfaces of the nickel nanowires with Pd ions accepting electrons from Ni to form Pd nucleus which chained up to form a coating. As the reaction proceeds, Pd evolves into a very thin nm thick shell at the edges of Ni nanowires while nickel nanowires dissolve to form a uniform channel along the nanowires axis. Longer reaction time or higher temperature tends to thicken the Pd surface with aggregation of Pd particles to form rougher Pd nanotubes.

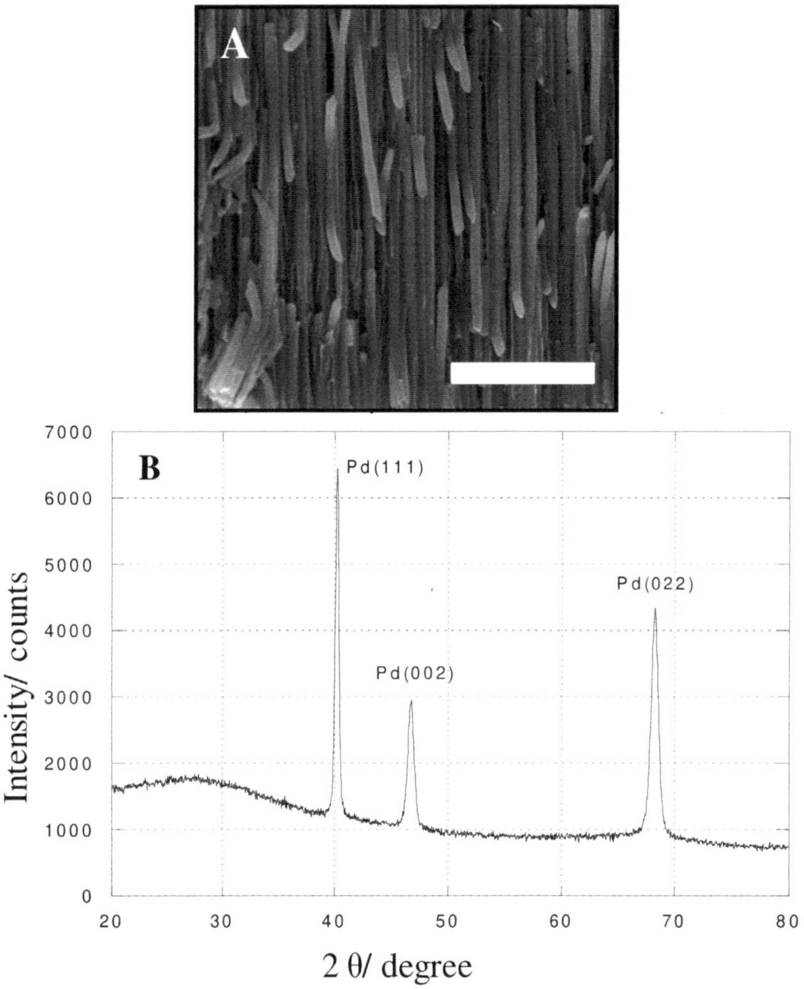

Figure 1. Pd nanowires fabricated by template assisted electrodeposition (a) cross-sectional SEM view of Pd nanowires arrays embedded in the template (b) X-ray diffractogram of the nanowires. Scale bar is 5 μm.

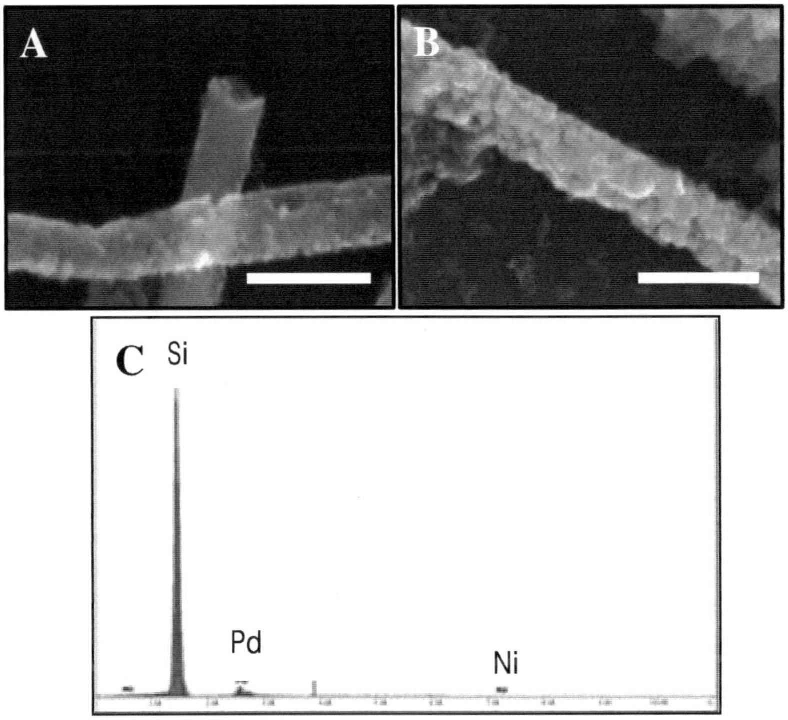

Figure 2. (a) SEM image of Pd nanotubes obtained after fully displacing nickel nanowires for 30 min at room temperature (b) SEM image of Pd nanotubes obtained after fully displacing nickel from nanowires for 30 min at 60°C. Scale bars are 500 nm. Both Pd nanostructures shows absence of Ni in the EDX spectrum as exemplified in (c).

Pd can also be electrochemically deposited on surfaces of semiconductor nanowire arrays such as silicon nanowires or zinc oxide rods. Figure 3 shows the SEM images of quasi-oriented ZnO rods which were electrochemically grown on indium-tin-oxide glass substrate using a three-electrode configuration with Ag/AgCl as reference electrode, platinum as counter electrode and ITO-coated glass substrate as the working electrode respectively using electrolytes consisting of ZnO^{2+} ions [8], and ZnO rods which has been decorated with Pd. The thickness of Pd coating increases with deposition time at a rate of 2.8 nm/min, resulting in the increase in diameter of rod to micon size. The Pd-ZnO rods not only exhibits larger diameter compared to the bare ZnO rods, the tips of the nanorods were also become more rounded and circular as compared to the original hexagonal tips of the bare ZnO rods. It was found that these Pd modified ZnO rods exhibits improved physical properties such as better UV photoresistivity and faster response time as compared to bare ZnO rods [9]. Pd modified ZnO rods are also good candidate materials for hydrogen gas sensing.

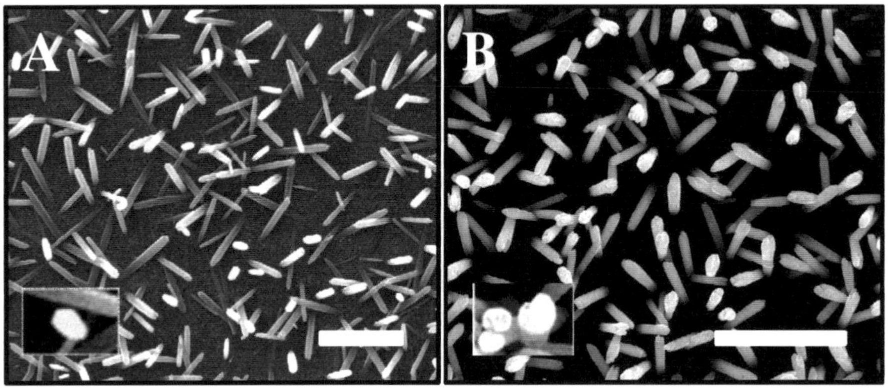

Figure 3. SEM images of the (a) ZnO rods and (b) Pd-ZnO rods. Insets show the tips of the rods. Scale bars are (A) 2 μm and (B) 5 μm.

Unlike ZnO rods where Pd was attached on its walls, Pd was suspended on the tips of the vertically-oriented Si nanowires as nanoparticles without affecting the diameter of the Si nanowires. Silicon nanowires were grown by electroless etching [10]. The SEM images in Figure 4 compare the silicon nanowires before and after Pd deposition. As the free-ends of the Si nanowires of average diameter 70 nm tend to bundle into clusters, the Pd nanoparticles of average size 92 nm were also agglomerated to form micron sized clusters. The addition of the Pd nanoparticles was found to improve the effectiveness in catalytic dissociation of hydrogen molecules into atomic hydrogen, thereby increasing the sensitivity of the sensor device [11]. The size and density of the nanoparticles, which affects the gas sensing capability, can be tuned by the deposition time. Longer deposition time yields higher density of particle and bigger particle size.

Figure 4. SEM images of (a) silicon nanowires (b) deposition of Pd on top of the Si nanowires. Inset shows the Pd nanoparticles decorated at the free-end of the Si nanowires. Scale bars are 2 μm.

4. Conclusion

Various one dimensional Pd nanostructures ranging from nanowires to nanotubes had been synthesized using a combination of template-assisted electrodeposition and galvanic displacement techniques. We have also demonstrated that Pd in the form of nanoparticles was electrochemically decorated on the surfaces of ZnO and Si nanowires arrays. These Pd-modified surfaces should enhance the gas sensing efficiency and charge transport property of the materials.

Acknowledgements
The authors would like to acknowledge the Malaysian Government for funding this work (Science Fund Grant No: 03-03-01-SF083).

References

[1] Hernández-Vélez M 2006 Nanowires and 1 D Arrays Fabrication: An overview, *Thin Solid Films* **495**, pp 51 – 63.
[2] Asefa T, Duncan C T and Sharma K K 2009 Recent Advances in Nanostructured Chemosensors and Biosensors, *Analyst*, **134**, pp 1980 – 1990.
[3] Cui Y, Duan X, Huang Y, Lieber C M 2003 *Nanowires and nanobelts materials, properties and devices, Vol. 1: Metals and semiconductor nanowires*, Wang Z.L. (ed.) (Kluwer Academic Publishers, Boston MA) pp 3 - 68.
[4] Fert A and Piraux L 1999 Magnetic Nanowires *J. of Mag and Mag Mat.*, **200**, pp 338 – 358.
[5] Laocharoensuk R, Sattayasamitsathit S, Burdick J, Kanatharana P, Thavarungkul P and Wang J 2007 Shape-Taoilored Porous Gold Nanowires: From Nano Barbells to Nano Step-Cones, *ACS Nano*, **1** (5), pp 403 – 408.
[6] Xiao F, Yoo B, Lee K H and Myung N V 2007 Synthesis of Bi_2Te_3 Nanotubes by Galvanic displacement *J. Am. Chem. Soc.*, **129**, pp 10068 – 10069.
[7] Lv S, Suo H, Zhoud T, Wang C, Jing S, Fu Q, Xu Y and Zhao C 2009 Effect of Synthesis Route on the Morphologies of Silver Nanostructues by Galvanic Displacement Reaction, *Solid State Communications*, **149**, pp 227 – 230.
[8] Kok K Y, Ng I K, Saidin N U and Ahmad Bustamam F K 2012 Fabrication of ZnO Nanostructures with Self-cleaning Functionality *Advanced Materials Research* **364**, pp 100 – 104.
[9] Bera A and Basak D 2011 Pd-nanoparticle-decorated ZnO Nanowires: Ultraviolet Photosensitivity and Photoluminescence Properties *Nanotechnology* **22**: 2665501 (7pp).
[10] Ng I K, Kok K Y, Zainal Abidin S S, Saidin N U, Choo T F and Myung N V 2010 Synthesis of Silicon Nanowire Arrays and Core-shell Heterostructures via Electrochemical Galvanic Displacement *Journal of Industrial Technology*, **19** (1), pp 209 – 217.
[11] Kim K T, Sim J and Cho S M 2006 Hydrogen gas sensor using Pd nanowires electro-deposited into anodized alumina template) *Sensors Journal IEEE*, **6**(3), pp 509 – 513.

Morphology and Optical Properties of Zinc Oxide Films Grown on Metal Coated Glass Substrates by Aqueous Chemical Growth

M A Bakar[1,a], M A A Hamid[*,1,b], A Jalar[2,c], and R Shamsudin[1,d]

[1]School of Applied Physics, Faculty of Sciences and Technology, Universiti Kebangsaan Malaysia (UKM), 43600 Bangi, Selangor, Malaysia.
[2]Institute of Microengineering and Nanoelectronic (IMEN), Universiti Kebangsaan Malaysia (UKM), 43600 Bangi, Selangor, Malaysia.
[*,b]azmi@ukm.my

Abstract. Zinc oxide films were deposited on three different metal coated substrates (gold, nickel and platinum) by aqueous chemical growth method. This paper discusses the effect of metal coated substrates on the morphology and optical properties of grown ZnO films. X-Ray Diffraction (XRD), Field Emission Scanning Electron Microscopy (FE-SEM) and UV-visible spectroscopy (UV-vis) were employed to characterize the samples. All the as-deposited ZnO films exhibit crystalline hexagonal wurzite structure. The crystallite size of the ZnO films were in the range of 29 to 32 nm. FESEM micrographs revealed hexagonal rod, oval-like and flower-like ZnO structures formed on all metal coated substrates. The Pt coated film contains higher density hexagonal rod as compared to others metal coated substrate. Most probably the Pt lattice parameter is the nearest to ZnO compared to nickel and gold. The optical band gap energy, E_g of ZnO films were estimated to be 3.30 eV which is near to bulk E_g, 3.37 eV. This indicates that the ZnO grown by aqueous chemical growth is able to produce similar quality properties to other conventional method either films or bulk size.

1. Introduction

Zinc oxide has been recognized as a promising semiconductor for various applications in electronics because of its wide band gap (3.37 eV at room temperature) and large exciton binding energy of 60 meV [1]. Recent research developments have revealed that the morphology of the films plays a key role in the application of ZnO in specific fields [2]. Via the chemical solution route, tube-, wire-, and flower-like ZnO nanostructures can be obtained [3]. The aqueous chemical growth has been widely proved for the large scale production of nano/microscale materials because it is less expensive, simple, fast, and requires a low growth temperature [4].

In the present work, we investigate the effect of different metal coated glass substrate on ZnO films growth by aqueous chemical growth method. These three metal coatings; gold (Au), nickel (Ni) and platinum (Pt) are selected since these metals are often used in electronic devices. The morphology of ZnO growth on each substrate and potential growth mechanisms are proposed. It is believed that the type of metal coating on the substrate plays a role in the morphology, shape and size and optical properties of ZnO. The ZnO films grown on the metal coated glass substrates (gold, nickel and platinum) could be used in various applications such as as solar cell, optical coatings, sensors, ultraviolet optical detectors and etc.

2. Materials and Method

ZnO films were grown on metal coated glass substrates by the aqueous chemical growth method. This method is based on heating the aqueous solution which containing the substrate immersed in the solution. All chemicals were used in the as-received form with no further purification. 100 ml aqueous solution with concentration of 0.05 M was prepared using 0.74 g of zinc nitrate hexahydrate ($Zn(NO_3)_2.6H_2O$) and 0.35 g hexamethylenetetramine ($C_6H_{12}N_4$). The chemicals were dissolved in deionized water. The aqueous solution was magnetically stirred until complete dissolution for 1 h.

The glass substrates used for this experiment were ultrasonically cleaned with acetone and ethanol, followed by deionized water, for 15 mins. The cleaned glass substrates were then dried in the air. The three glass substrates were coated with different metals, namely gold (Au), nickel (Ni) and platinum (Pt), using a sputter coater for 2 mins with a current of 30 mA. Each of metal coated glass substrate was placed horizontally at the bottom of a sealed beaker. The samples were then heated at 90 °C for 3 hours in an oven without any stirring. Subsequently, the samples were removed and washed thoroughly with deionized water, to eliminate any residual salts, and were then dried in air.

The morphology and film thickness of the deposited ZnO films were analyzed using field emission scanning electron microscopy (FE-SEM-SUPRA 55VP). Crystal phase and orientation were determined by X-ray diffractometer (XRD, BRUKER D8 ADVANCE) with CuKα radiation source (40 kV, 40 mA, λ = 0.15406 nm). The optical transmittances of the ZnO thin film were measured using UV/VIS/NIR spectrometer (LAMBDA 900).

3. Results and Discussion

X-ray diffraction (XRD) patterns of ZnO films deposited on different metal coated glass substrates at 90 °C for 3 h is shown in Figure 1. All of observed peaks indicated that ZnO were successfully grown on all substrates. The XRD patterns of the ZnO films shows the characteristic peaks corresponding to planes (100), (002), (101) with high intensities and (102), (110) with low intensities planes, which matched well with reference data (JCPDS No. 36-1451). All of characteristic peaks are assigned to hexagonal ZnO wurtzite structure with cell constant of a = 3.24982 Å and c = 5.20661 Å, indicating the product is pure ZnO. Moreover, there are no other visible peaks or impurities from other phases were detected implying the pure ZnO were grown on the substrates. The average crystallite size, (D) is estimated using Scherrer's equation [5], $D = 0.9\lambda / \beta \cos \theta$. Where λ is the X-ray wavelength (0.154056 nm), β is FWHM (in radians) of the diffracted peak 2θ and θ is the Bragg diffraction angle (in radians). The calculated crystallite size is about 29 nm for ZnO film deposited on Au and Ni coated glass substrates and 32 nm is for Pt coated glass substrates.

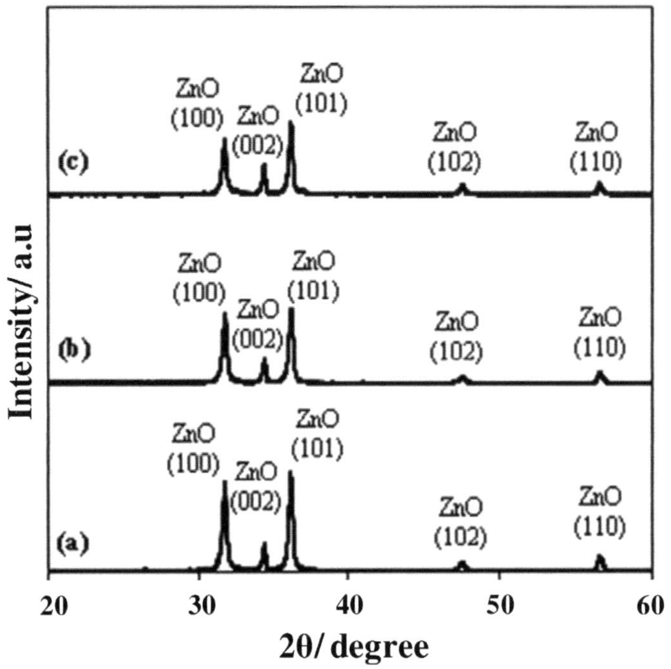

Figure 1. XRD patterns of ZnO films grown on: (a) Au, (b) Ni and (c) Pt coated glass substrates at 90 °C for 3 h.

Figure 2 illustrates the FESEM micrographs of ZnO films grown on different metal coated glass substrates at 90 °C for 3 h. Multiple type of ZnO structures have been observed grown on the different metal coated glass substrates such as hexagonal rods, flower-like, oval-like and special feature of nanorod array. This may due to the metal coated substrate (Au, Ni and Pt) act as catalyst for the growth site of ZnO. Furthermore the lattice difference between ZnO and each metals may results in different ZnO morphology that obtained. Figure 2 (a) shows the morphology of ZnO film grown on Au coated glass substrate. A mix of flower-like and oval-like morphologies were grown on the substrate with specific direction. Similar morphologies were also found in ZnO film grown on Ni coated glass substrate (Figure 2 (b)). The oval-like structure was most likely as a result of coalescence between small hexagonal structures. The flower-like structure are seen to be grown on top of oval-like structure which suggests secondary growth are preferred only on ZnO rather than metal substrate.

Figure 2(c) shows hexagonal rods, oval-like structure and nanorod arrays coexisting on Pt substrate. Higher magnification in Figure 2(d) shows ZnO nanorod arrays structure directly grown on metal substrate which typically are observed grown using ZnO seed layer or substrate. According to Yamada et al. [6], the existence of this nanorod-array is because of a small lattice mismatch that exists between ZnO and Pt-substrate, ~1.4%, which allows this structure to be formed. This type of structure was not observed on both ZnO film grown on Au- and Ni-coated substrate. It is also can be seen that not all ZnO nanorod array region provide secondary growth site as indicated by flower-like structures selectively growth on nanorod region.

Figure 2. FE-SEM micrographs of ZnO films deposited on: (a) Au; (b) Ni; (c) Pt at; (d) nanorod array grown on Pt-coated glass substrate at 90 °C for 3 h.

The average of film thicknesses were measured from the cross-sectional view of ZnO film micrographs using FE-SEM. The average thicknesses of ZnO films were found to be ~8.3 μm, ~8.2 μm and ~8.7 μm for Au, Ni and Pt coated glass substrates, respectively.

The proposed growth mechanism of flower-like structures of ZnO is shown in Scheme 1. Step (a) is the growth process of hexagonal rods that bundle together. Step (b) is the hexagonal rods coalescence become a single oval-like structure; step (c) is the oval-like structure rapidly grow and become brancing out structures and step (d) is the flower-like structure from hexagonal rods begins to grow at the top of the oval-like structure. These is in good agreement with Li et al. [7] who observed that nanorods grow initially as a bundle but in an independent way and then coalesce, thus finally becoming a single nanorod.

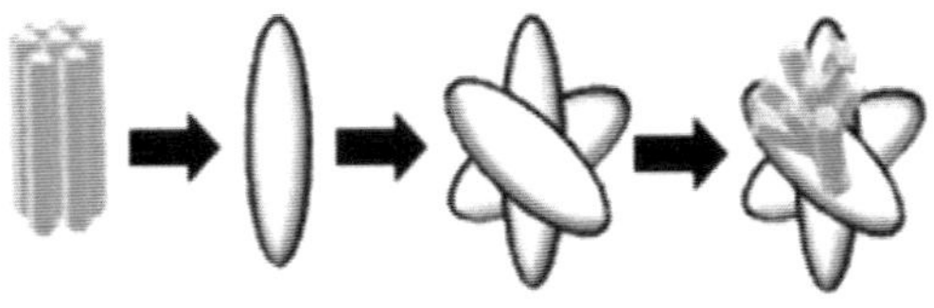

(a) (b) (c) (d)

Scheme 1. Growth mechanism of ZnO flower-like structures: (a) hexagonal rod structures grow as a bundle, (b) oval-like coalescence, (c) oval-like branching out and (d) flower-like structure begins to grow on the top of oval-like structure.

The transmittance spectra are recorded over 300-800 nm. The variations of transmittance with wavelength for ZnO films deposited on different metal coated glass susbtrates are shown in Figure 3. It is clearly seen that the transmittance value are vary with the different metal coated glass substrate. This is corresponding to the substrate colour appearance as the Au coated glass substrate is yellow in appearance; Ni coated glass substrate is transparent while Pt coated glass substrate is dark grey in appearance. Low transmittance value is observed for Pt coated glass substrate, 30-54% compared to Au coated, 31-60% and Ni coated with higher transmittance value, 40-79%. It is found that substrate colour appearances are strongly influenced the transmittance values.

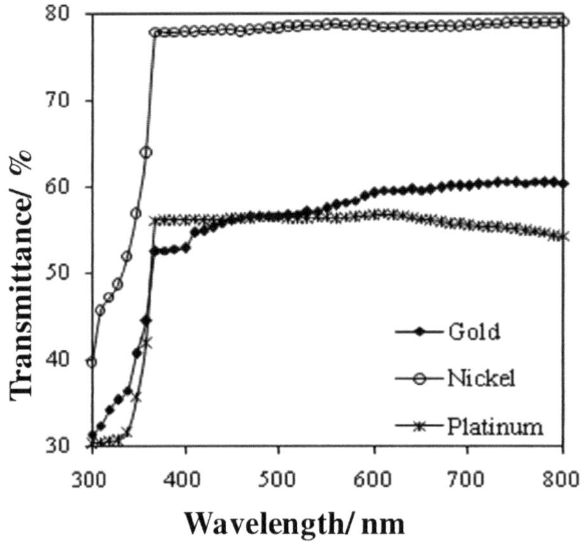

Figure 3. A typical transmittance spectra of ZnO films deposited on different metal coated glass substrates at 90 °C for 3 h.

The optical band gap energy (E_g) can be determined by calculating the absorption coefficient, α. It is related to the band gap energy, E_g and photon energy, $h\upsilon$ as follows, $\alpha = \alpha_o (h\upsilon - E_g)^n$. Where α_o is a constant, and the exponent of n = ½ is for ZnO which have direct allowed transition. The values of band gap energy (E_g) is obtained by extrapolating the linear potion of the curves to $(\alpha h\upsilon)^2$ =0. The E_g values for ZnO film were found to be 3.30 eV for Ni coated and 3.29 eV for Au and Pt coated glass substrate. The differences of E_g values obtained were attributed to the changing in their surface microstructures, morphologies of the grown samples and crystallite size [8].

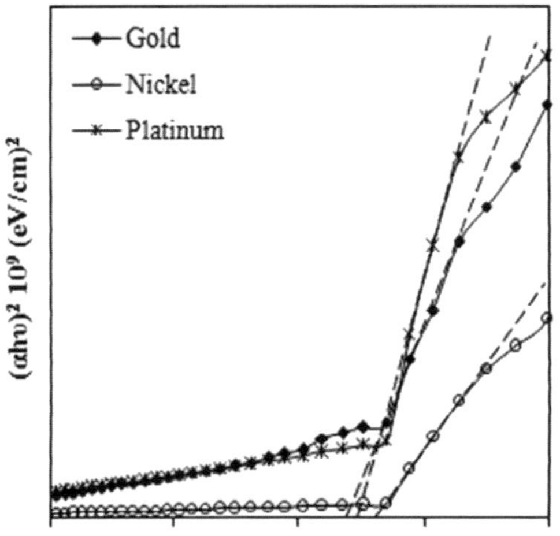

Figure 4. Plot of hυ vs. $(\alpha h\upsilon)^2$ of ZnO films deposited on different metal coated glass substrates at 90 °C for 3 h.

4. Conclusion

In this work, pure zinc oxide films were successfully produced by aqueous chemical growth method on different metal coated glass substrates at relatively low temperature, 90 °C. XRD patterns show well-defined ZnO peaks without any visible impurities for all samples proved the as-synthesized product is pure. The calculated crystallite sizes are found to be in between 29-32 nm. The results presented in this paper revealed that different types of metal coating have a profound effect on the morphology of ZnO growth using this technique. It has been observed that the as-synthesized ZnO present various interesting morphologies of such as hexagonal rods, flower-like, oval-like and nanorod array. Moreover, the ZnO nanorod array is only observed for Pt coated glass substrate which prove the metal coating have influence on the ZnO morphology. The optical band gap, E_g for ZnO films are found to be around 3.29-3.30 eV which is in good agreement with bulk value of ZnO, 3.37 eV.

Acknowledgements

The authors acknowledge the financial support from Universiti Kebangsaan Malaysia (Research fundings: DIP-2012-14 and UKM-RRR1-07-FRGS0257-2010) and Ministry of Science, Technology & Innovation, Malaysia (MOSTI) for this research.

References

[1] S. Sepulveda-Guzman, B. Reeja-Jayan, E. de la Rosa, A. Torres-Castro, V. Gonzalez-Gonzalez, M. Jose-Yacaman, *Mater. Chem. Phys.* **115,** (2009) pp. 172–178.

[2] P. S. Kumar, A. D. Raj, D. Mangalaraj, D. Nataraj, Appl. Surf. Sci. **255,** (2008) pp. 2382–2387.

[3] Q. Cui, Y. Huang, Z. Zhu, *Cur. Appl. Phys.* **9**, (2009) pp. 426–430.

[4] L. Vayssieres, Adv. Mater. **15** (2005), pp. 464.

[5] A. Tanusevski and V. Georgieva. *Appl. Surf. Sci.* **256**, (2010) pp. 5056.

[6] H. Yamada, Y. Ushimi, M. Takeuchi, Y. Yoshino, T. Makino and S. Arai, *Vacuum.* **74**, (2004) pp. 689-692.

[7] Q. Li, V. Kumar, Y. Li, H. Zhang, T.J. Marks, R.P.H. Chang. *Chem. Mater.* **17**,(2005) pp. 1001-1006.

[8] N. Samaele, P. Amornpitoksuk, S. Suwanboon. *Powder Technology* **203** (2010) pp. 243–247.

Journal of Physics: Conference Series **431** (2013) 012005 doi:10.1088/1742-6596/431/1/012005

Submicron Wire Fabrication on Silicon Substrate Based on Atomic Force Microscopy Technique

M T Asmah[1,2,3], S D Hutagalung[1] and O Sidek[2]
[1]School of Materials & Mineral Resources Engineering (SMMRE), Engineering Campus, Universiti Sains Malaysia, Seri Ampangan, 14300 Nibong Tebal, Penang, Malaysia.
[2]Collaborative Microelectronic Design Excellent Centre (CEDEC), Engineering Campus, Universiti Sains Malaysia, Seri Ampangan, 14300 Nibong Tebal, Penang, Malaysia.

E-mail: asmahmattaib@gmail.com

Abstract. This article presents on a simple sub-micron wire fabrication by means of local anodic oxidation atomic force microscopy technique. The silicon-based structure consists of two adjacent terminals serve as probing pads and connected with a wire width less than 200 nm. The pad dimension and wire length are 49 μm^2 and 11 µm each, respectively. The fabrication process is conducted at room temperature 24 – 27 °C with 50 – 60 % relative humidity and operated by commercial atomic force microscopy without any modification done. Furthermore, the local oxidation is performed using gold coated AFM probe tip and assisted by the used of special language supported by the equipment software. In addition, 9.0 V applied voltage with 2 µm / sec writing speed is used to realize the oxidation process. I-V characteristic of bare device shows there is a current flow through the device when a range of voltage is applied. By using this local anodic oxidation technique, the silicon wire-based structure is fabricated and used as gas sensor sensing part. The silicon device demonstrates an increase in resistance as introduced to gas environment. The silicon wire device shows relative sensitivity of 35 % to the oxygen gas.

1. Introduction

Atomic Force Microscopy (AFM) nanolithography is a proven technique to pattern materials at nanometer scale precisely which divided to bias-assisted nanolithography and force-assisted nanolithography [1]. Local oxidation or local anodic oxidation (LAO) which is in bias- assisted class has been used to define structures, modify and develop nano-electronic devices fabrication process with low cost and simple approaches [2]-[3]. In this technique, the bias is applied to the AFM probe tip to generate localized electric field with a distance of nanometer gap between tip and sample [1].

This local oxidation can be used to create oxide patterns on semiconductors and metals such as single electron transistors (SETs), nanomechanical structures, quantum electronic devices and molecular template [2]-[3]. This process is an electrochemical process which sample surface serves as an anode, AFM probe tip acts as a cathode and water molecules serve as electrolytes to the oxidation process [4], [5]-[6]. Under a controlled humidity, water molecules absorbed on substrate surface and AFM probe tip and water layer thickness depends on the relative humidity introduced to the air in the sealed environment [7]. This article will discuss on the device fabrication process by means of atomic force microscopy local anodic oxidation, physical and electrical characterization of fabricated device and device testing in the target sample environment.

2. Experimental

The p-type SOI (silicon-on-insulator) wafer was used as substrate for structure patterning. Prior to the experiment, the sample was cleaned using standard cleaning process. In this study, a device with two adjacent probing pads that connected with a wire will be fabricated by the aforementioned technique. The fabrication process was performed by commercial atomic force microscopy and operated at temperature and relative humidity of 24 – 27 °C and 50 – 60 %, respectively. Furthermore, the lithography process was assisted with the used of special language supported by the AFM software for process recipe, applied voltage (9 V), writing speed (2 µm / sec) and gold-coated AFM probe tip. The created oxide layer during lithography process will serve as a masking layer for etching process. Therefore, unwanted and underlying silicon layer will be removed using tetramethylammonium hydroxide solution. The masking layer was removed by using diluted hydrofluoric acid to obtain final device. Figure 1 exhibits the schematic diagram of the device fabrication process, starting from oxidation process by using AFM, followed by etching process and masking layer removal.

Figure 1. Schematic diagram of device fabrication process; (a) oxidation process by means of AFM, (b) etching process, (c) masking layer removal process.

3. Result and discussion

After fabrication process, the device was physically and electrically characterized. Atomic force microscopy and scanning electron microscopy (SEM) were used for physical characterization. For the electrical testing, semiconductor parameter analyzer, (SPA) was used to characterize the fabricated device. Figure 2 shows the SEM image of the final fabricated device. The inset image shows the measured wire diameter using SEM. From the measurement, the wire diameter of fabricated device was approximately 187 nm. In addition, the pad size and wire length were approximately 7 µm x 7 µm and 11 µm each, respectively.

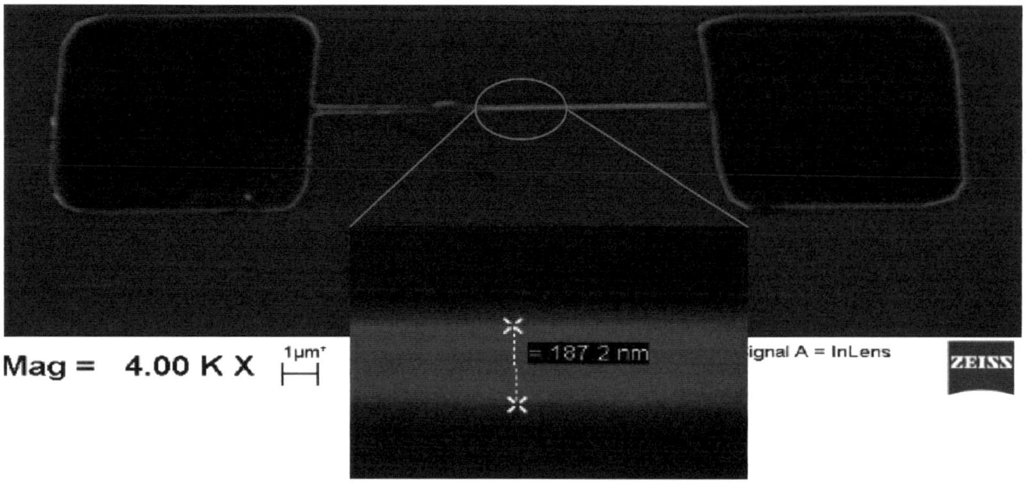

Figure 2. SEM image of fabricated device. The inset image shows the measured wire diameter which around 187 nm.

The fabricated device was electrically characterized by using semiconductor parameter analyzer to measure the current-voltage (I-V) properties. Figure 3 shows the current characteristic of fabricated device (Device_A) as a function of applied voltage which was scanned from -2 V to 2 V. The observed result depicts a current flow through the device from one contact point to another point. However, small current was observed at reverse applied voltage than forward applied voltage. Another set of device with different wire width but has similarity in device structure, namely as Device_B, was fabricated to verify the current-voltage properties. The Device_B wire width has an approximately 40 nm larger than the Device_A. From the graph shown in Figure 4, Device_A conducts more current compared to Device_B, where, Device_A has smaller wire width than Device_B. From the obtained result, therefore, smaller wire width has better conductivity than larger wire width.

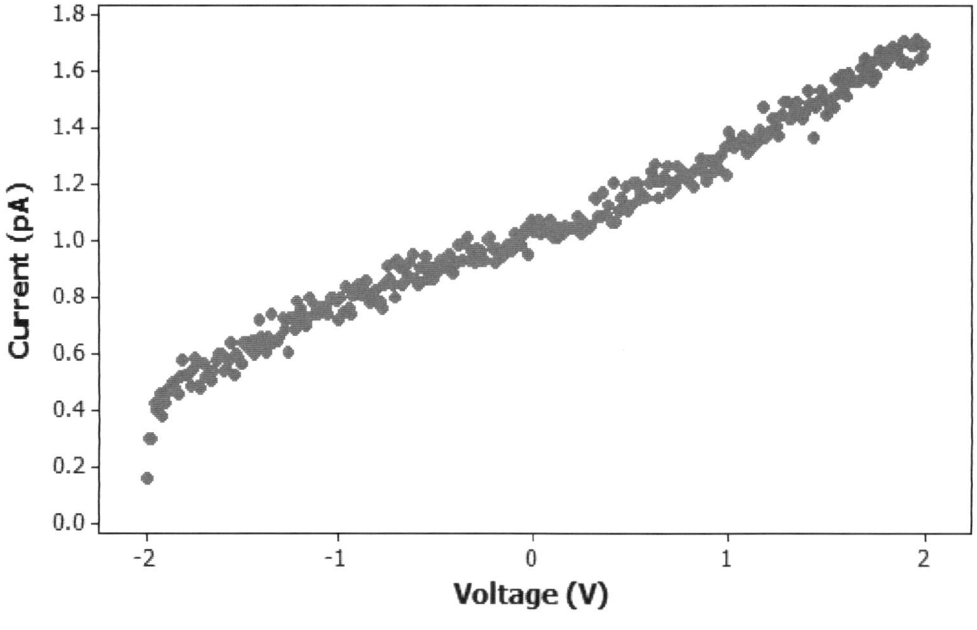

Figure 3. Current-Voltage characteristic of fabricated bare device for Device_A.

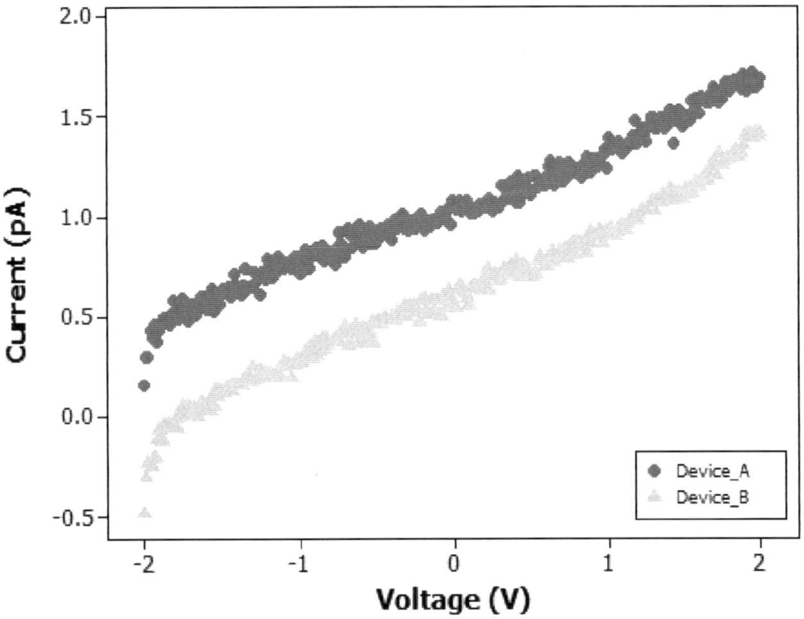

Figure 4. Current-Voltage characteristic of different set of device. Device_A consist of smaller wire width compared to Device_B.

In order to prolong the functionality of fabricated device, the device (Device_A) was tested under gas environment. The testing was conducted at room temperature and oxygen (O_2) gas was selected as target gas. The device was exposed to the gas environment in a closed chamber and current values were recorded. Figure 5 depicts the difference in current distribution for bare device and exposed device to the gas environment. The plotted boxplot shows small current distribution for the device that was exposed to the environment containing target gas. The obtained result shows that the resistance of the device leads to increase after target gas exposure. Therefore, different electrical characteristic was observed between bare and gas tested device. In addition, the fabricated device shows relative sensitivity of 35 % to the oxygen gas at room temperature. The percentage of absolute relative sensitivity, S, is defined as (1), where, I_o and I correspond to current obtained before and after gas exposure, respectively [8]-[9].

$$S\ (\%) = (|\Delta I|/I_o)*100 = [(I-I_o)/I_o]*100 \quad (1)$$

At room temperature, a new type of physiochemical phenomenon such as incomplete covalent bonds and dangling bonds will affect the gas sensing behaviour at room temperature [10]. In p-type semiconductor, the majority carriers are holes. As the electrons from the target gas are injected to the valence band and recombine with the holes, this phenomenon leads to increase the resistance by reducing the number of charge carriers. The Fermi level energy will shift from the valence band if more surface donor states ionized and lead to slow down the process. At last, the Fermi level will stop at new energy value due to the dynamic equilibrium is set up between gas molecules diffused out from the surface to rejoin the gas flow and other gas molecules diffuse into the surface. Therefore, this phenomenon leads to increase the resistance of the tested device [11].

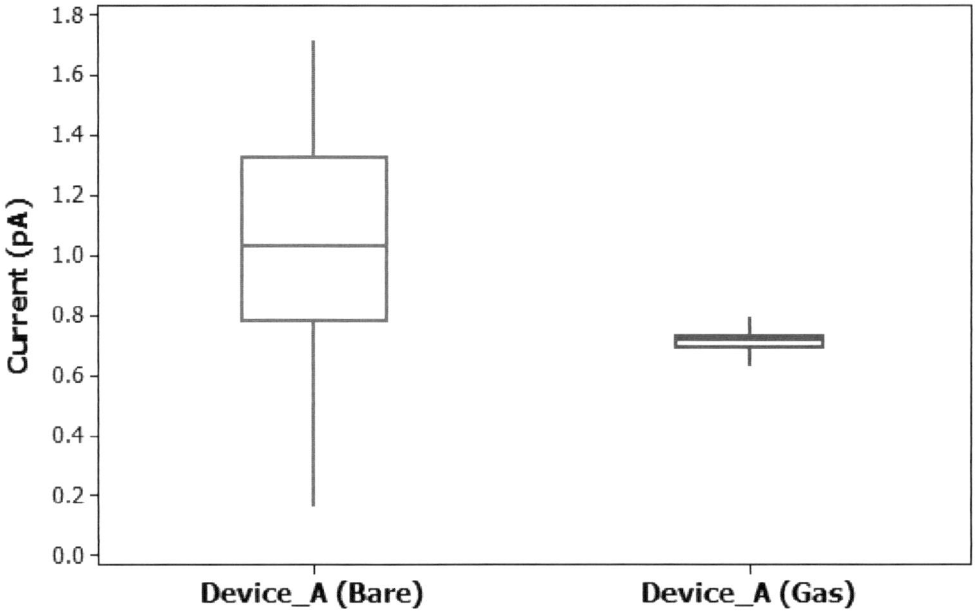

Figure 5. Current distribution of bare device and gas tested device.

4. Conclusion

As conclusion, a device with two probing pads connected with a wire width less than 200 nm was successfully fabricated and tested. Furthermore, the current-voltage characteristic of the fabricated device was verified with different set of wire width device. As the wire width increases, the device conducts less current. In addition, as the device was introduced to the target gas, the current distribution was decreased and higher resistance was observed. Future research will focus on fabrication of various types of device structure and usage of different target gases for device sensing.

Acknowledgement

The author (M T Asmah) would like to thank to Ministry of Science, Technology and Innovation (MOSTI) for the scholarship awarded. The same author would like to thank to Collaborative Microelectronic Design Excellent Centre (CEDEC) USM (311/PCEDEC/440201) and School of Material and Mineral Resources Engineering (SMMRE) USM (1001/PBAHAN/814167) for the financial support and research facilities.

References

[1] Xie X N, Chung H J, Sow C H and Wee A T S 2006 Nanoscale materials patterning and engineering by atomic force micrsocopy nanolithography *Materials Science and Engineering R* **54** 1-48

[2] Fernandez-Cuesta I, Borrisé X and Pérez-Murano F 2005 Atomic force microscopy local oxidation of silicon nitride thin films for mask fabrication *Nanotechnology* **16** 2731-37

[3] Avouris Ph, Martel R, Hertel T and Sandstrom R AFM-tip induced and current-indduced local oxidation of silicon and metals 1998 *Appl. Phys. A* **66** S659-67

[4] Bartošík M, Škoda D, Tomanec O, Kalousek O, Jánský P, Zlámal J, Spousta J and Šikola T 2007 The influence of humidity on the kinetics of local anodic oxidation *Journal of Physics: Conference Series* **61** 75-79

[5] Červenka J, Kalousek R, Bartošík M, Škoda D, Tomanec O and Šikola T 2006 Fabrication of nanostructures on Si (100) and GaAs (100) by local anodic oxidation *Applied Surface Science* **253** 2373-78

[6] Sathiyaraj K, Harshiny M, Banu B N, Rajendran K and Kumaran S 2011 A review on techniques to fabricate silicon oxide arrays for biomolecules patterning *Superlattices and Microstructures* **49** 581-90

[7] Fang T -H and Wu K –T 2006 Local oxidation characteristics on titanium nitride film by electrochemical nanolithography with carbon nanotube tip *Electrochemistry Communications* **8** 173-78

[8] Demami F, Ni L, Rogel R, Salaun A C and Pichon L 2011 Silicon nanowires based resistors as gas sensors *Sensors and Actuators B: Chemical* doi:10.1016/j.snb.2011.04.083

[9] Gabouze N, Belhousse S, Cheraga H, Ghellai N, Ouadah Y, Belkacem Y and Keffous A 2006 CO_2 and H_2 detection with a CH*x*/porous silicon-based sensor *Vacuum* **80** 986-89

[10] Joshi R K and Kumar A 2011 Room temperature gas detection using silicon nanowires *Materials Today* **14** 1-2 52

[11] Mabrook M and Hawkins P 2001 A rapidly-responding sensor for benzene, methane and ethanol vapours based on films of titanium dioxide dispersed in a polymer operating at room temperature *Sensors and Actuators* **75** 197-202

Fabrication of Au/Ni Multilayered Nanowires by Electrochemical Deposition

N U Saidin[1], K Y Kok, I K Ng and S H Ilias
Malaysian Nuclear Agency, Bangi, 43000 Kajang, Selangor, Malaysia

E-mail: ubaidah@nuclearmalaysia.gov.my

Abstract. Electrochemical deposition of Au/Ni multilayered nanowires using template-assisted growth technique from electrolyte containing nickel chloride and gold solution was studied in details. 60 µm-thick anodized aluminum oxide (AAO) with pore diameter of 200 nm was used as the template. Chronopotentiometry experiments were first carried out to determine the deposition conditions and the growth rate of individual Au and Ni layers. Scanning electron microscopy results revealed that the pore channels of AAO were completely filled with Au/Ni multisegmented nanowires. By selectively removing the Ni segments in the multilayered nanowires, high-yield of pure gold nanorods were obtained. Detailed studies on the nanostructures obtained were carried out using various microscopy and probe-based techniques for structural, morphological and chemical characterizations.

1. Introduction

One-dimensional nanostructures such as nanorods, nanowires, nanotubes and nanoshells, have been intensively studied because of their ability to exhibit many novel effects which are not observed in the bulk materials [1]. Among the various metals from which the nanostructures are made, gold is technologically important due to its low resistivity (2.21 µΩ cm) [2]. Apart from this, gold also has good electrical and thermal conductivity as well as high ductility and chemical inertness [3]. Therefore, gold is often utilized in medical applications especially for tumor treatment [4].

A variety of techniques have been developed for the fabrication of gold nanostructures such as lithography, chemical vapor deposition, and electrodeposition. Electrochemical deposition is a method which allows us to deposit materials in many different forms from the solutions. Different from other deposition techniques, electrochemically deposited nanowires are grown continuously from the pore bottom [5]. In this study, electrochemical deposited Au/Ni multilayered nanowires is used as the base material configuration for the fabrication of Au nanostructures. The nanowires were fabricated by sequential electrochemical deposition of two materials inside the nanochannels of an anodized aluminum oxide (AAO) template. AAO is an ideal template for the fabrication of nanostructured materials because of the many advantages it possesses, including tunable pore dimensions, good mechanical property and good thermal property. Moreover, it is cheap and capable of producing oriented arrays of metals nanostructures.

The objective of this study is to study the use of AAO to fabricate Au/Ni multilayered nanowires by electrochemical deposition.

2. Materials and Methods

Electrochemical deposition was carried out in a typical two-electrode cell with a bath volume of 100 ml. A commercial AAO template (Whatman Anodisc-13, average pore diameter 200 nm, 60 μm thick) was used as the working electrode. The back-side of the template was first sputtered with Au as a seed layer. Multi-segmented Au/Ni nanowires were obtained by electrodepositing Au and Ni sequentially via alternate switching of the electrolyte solutions for Ni and Au until the required number of segments was achieved. The gold segments were deposited from a ready-to-use cyanide-free gold bath (Technic RTU-25) at constant current of -0.34 mA at 40°C, while the Ni segments were deposited from a nickel electrolyte comprising of 1 M $NiSO_4$ + 0.5 M H_3BO_3 using a current of -3.4 mA at ambient condition.

Figure 1 shows a section of the typical current-time curves for the electrodeposition of Au/Ni multilayered nanowires. After deposition, the Au seed layer was removed and the nanowires were released by immersing the template into NaOH solution followed by repeated centrifugation and rinsing until a nanowire suspension was obtained. To yield only gold segments from the nanowires, the Ni segments were removed by selective etching using nitric acid. The Au/Ni multilayered nanowires and Au nanostructures obtained were studied by Scanning Electron Microscopy (SEM) and Transmission Electron Microscopy (TEM) while composition characterization of the materials was carried out using energy dispersive X-ray spectroscopy (EDX).

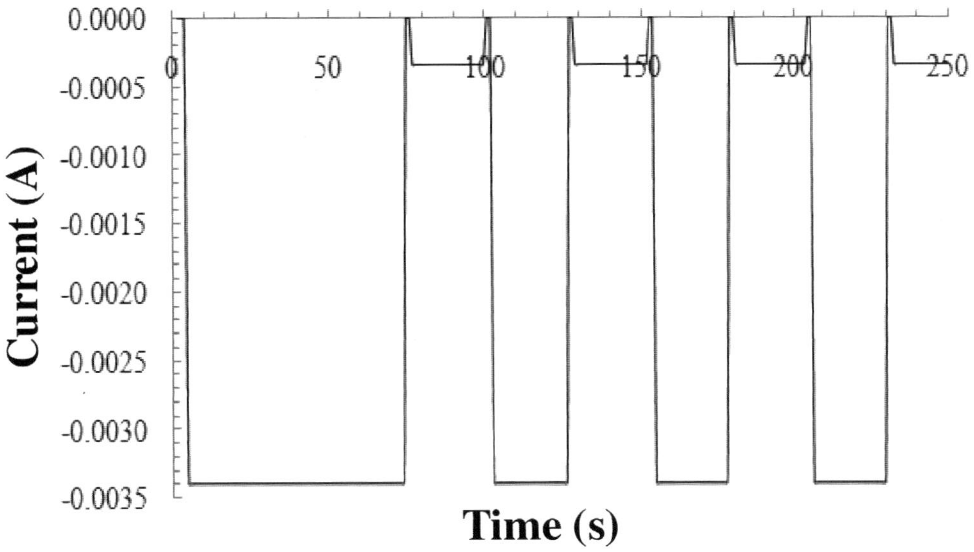

Figure 1. Current-time curve for the deposition of Au/Ni multilayer nanowires.

3. Results and Discussion

Figure 2 shows the typical scanning electron micrograph image of the Au/Ni multilayered nanowires embedded in AAO template. The nanowires filled the AAO nanopores channels continuously and uniformly in an oriented array and exhibit homogenous contours. After removal of the Ni segments, fragmented Au segments were obtained as seen in Figure 2 (b) and (c). As the Au segments were confined in the template, the diameters of the Au segments were determined by the pore size of the template while the lengths were determined by the deposition time. Deposition rate for Au was found to be 88 nm min^{-1} while deposition rate for Ni was found to be 150 nm min^{-1}.

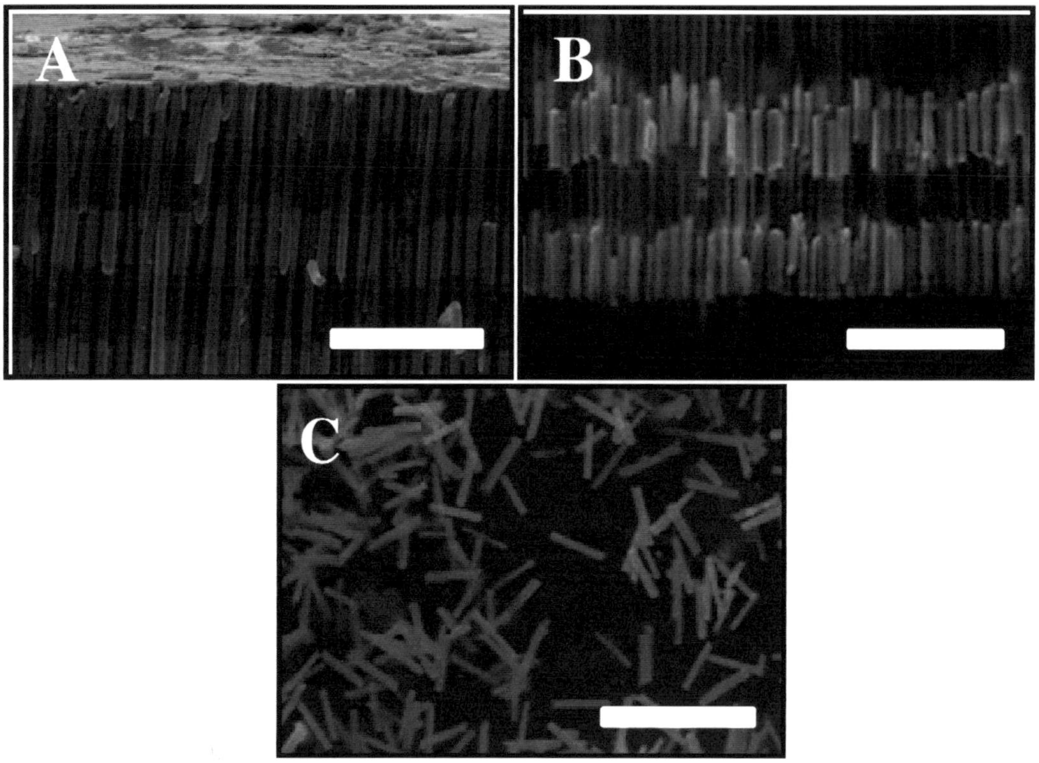

Figure 2. SEM images of (a) Au/Ni multi-segmented nanowires embedded in PAA template (b) after removal of Ni by selective etching of Au/Ni multisegmented nanowires (c) dispersion of the fragmented gold nanorod segments after removal of template. Scale bars are 5 μm.

Figure 3 shows that, instead of nanorods, Au nanodiscs can be obtained from Au/Ni multisegmented nanowires electrodeposited with much shorter segments of Au. Figure 4 provides a closer examination for the Au/Ni interface using TEM. As evident from the Selected Area Diffraction Pattern (SADP) shown in the inset and the EDX spectrum recorded, the Au segment was highly crystalline and consisted only of Au.

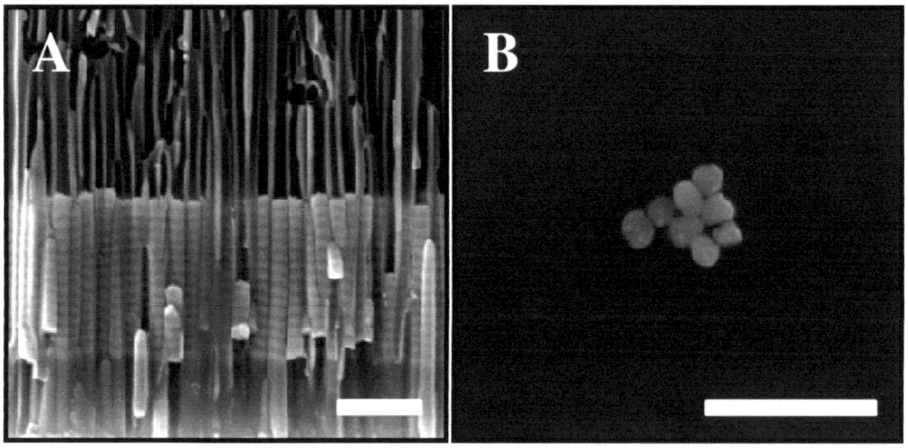

Figure 3. SEM images of (a) Au/Ni multi-segmented nanowires in template and (b) the corresponding dispersed Au nanodiscs after removal of Ni segments and template. Scale bars are 2 μm.

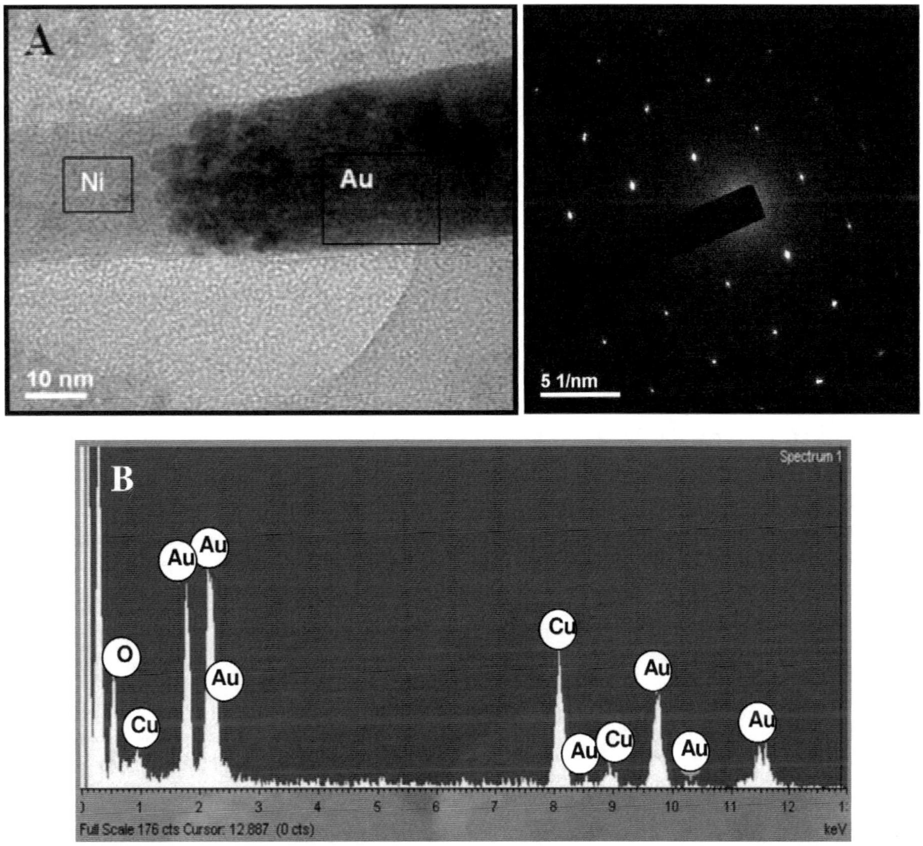

Figure 4. (a) TEM image of the Au/Ni interface from the multisegmented nanowires showing good adhesion between the two materials. The SADP for the Au segment shown in the inset is typical of a single crystalline structure. (b) EDX spectrum indicating high purity of the Au segment (the Cu peak is from TEM sample grid).

Template-assisted deposition also offers additional advantage where various morphologies of the gold nanorods can be obtained by tailoring the corresponding pore geometry of the templates used. Apart from solid gold nanostructures, the technique can be modified to produce nanoporous gold (NPG) structures. This is done by fabricating periodic multisegmented nanowires with Au-Ni composite layers alternating with Ni, followed by selective etching of Ni. Nanoporous gold is good candidate material for high-performance catalyst and sensor due to the large specific surface area it possesses.

4. Conclusion

Au/Ni multilayered nanowire has been successfully fabricated using template assisted electrochemical deposition technique. By using AAO as template, the dimensions of the Au/Ni multilayer can be controlled precisely. Nearly 100% of the pores are filled with Au/Ni multilayer nanowires. We have also successfully demonstrated a novel, high throughput, cost and time effective technique for the synthesis of size-tailored nanomaterials such as gold nanorods. Beside Au/Ni multilayer nanowires, this technique can also be used to fabricate other metals nanostructures and has great potential for commercial exploitation as it involves only simple set up and easily scalable for industrial production.

Acknowledgement
This work was supported by the following funding sources: 03-03-01-SF0083 & NM-R&D-11-28.

5. References

[1] Ng I K, Kok K Y, Zainal Abidin S S , Saidin N U, Choo T F, Goh B T, Chong S K and Abdul Rahman S 2011 Gold Catalysed Growth of Silicon Nanowires and Core-shell Heterostructures via Solid-liquid-solid Process and Galvanic Displacement *Mat. Res. Innovations*, 15(2) S2-55.

[2] Lide D R 1994 *The Handbook of Chemistry and Physics* 74th edn. Boca Raton, FL: Chemical Rubber Company.

[3] Ozturk B, Flanders B N, Grischkowsky D R and Mishima T D 2007 Single-step Growth and Low Resistance Interconnecting of Gold Nanowires *Nanotechnology* 18, , 175707.

[4] Dykman L and Khlebtsov N 2012 Gold Nanoparticles in Biomedical Applications: Recent Advances and Perspectives *Chem. Soc. Rev.* 41 2256.

[5] Nielsch K, Muller F, Li A P and Ilrich G 2000 Uniform Nickel Deposition into Ordered Alumina Pores by Pulsed Electrodeposition *Adv. Mater.* 12, No. 8, 582-586.

Effect of Different Weight Fraction on Dielectric Properties of Barium Ferrite-Barium Titanate Composites

S N Anuar[1], Z Zalita[1], R Shamsudin[1]
[1]School of Applied Physics, Faculty of Science and Technology, Universiti Kebangsaan Malaysia, 43600 UKM Bangi, Selangor, Malaysia.

E-mail: zazai@ukm.my

Abstract. Barium ferrite-barium titanate composite systems, (1-x)$BaFe_{12}O_{19}$ – (x)$BaTiO_3$ with different weight fractions of x = 0.0, 0.1, 0.3 and 0.5 were prepared via the conventional solid state reaction method using nano-sized starting materials. The x-ray diffraction peaks of the composite samples show two phases which can be indexed by the hexagonal barium ferrite and tetragonal barium titanate. Field emission scanning electron microscope (FESEM) image of sample with x = 0.1 shows formation of nano-sized $BaTiO_3$ grains spreading evenly among the micron-sized $BaFe_{12}O_{19}$ grains after sintering at 1000 °C. The dielectric properties studies at room temperature of the composites samples show that sample with $BaTiO_3$ weight fraction of x = 0.1 has the highest dielectric constant, ε' and lowest loss tangent, *tan* δ in the studied frequency range between 1 Hz and 1 MHz. Its highest dielectric constant value is ~ 1.2×10^3 and its lowest loss tangent is 0.06 at 50 Hz.

1. Introduction

Barium ferrite ($BaFe_{12}O_{19}$) is known as a permanent magnetic material which can also be categorize as a ferrimagnetic material that has great properties such as high saturation magnetization, anti-erosion, better chemical stability, great coercivity and low cost material. It has a very wide range of application such as in a high-density recording media and electromagnetic wave absorber [1], [2]. Barium ferrite was also used in microwave communication, microwave dark room and the anti-electromagnetic wave radiation [3]. Barium ferrite has hexagonal structure and can be characterized in magneto-plumbite oxide group and can be classified as hard magnetic materials [4]. Ferrite is one of a ferromagnetic oxide that has both dielectric and magnetic properties when applied with high frequency and is suitable for microwave applications [5]. In barium ferrite compounds the Fe^{3+} ions occupy the sub lattice at different sites that lead to different magnetic properties. Addition of nonmagnetic materials into barium ferrite compound will influence its sub lattice [6] and magnetic properties.

Fine and dense barium titanate ($BaTiO_3$) with nano-sized particles make it an outstanding dielectric material [7]. Barium titanate, is known as a ferroelectric material and has been studied widely because of its potential application such as in multilayer ceramics capacitors (MLCC), positive temperature coefficient of resistance (PCTR) thermistors, piezoelectric sensors, transducers, actuators and ferroelectric random access memories (FRAM) and electro-optic devices [8]. Barium titanate is also an excellent material for field-induced piezoelectric transducer because of its high polarization, high permittivity and high induced strain [9].

Due to the widely usage of barium ferrite and barium titanate respectively, we combined both materials and tried to obtain information whether there are any improvement of the composite system properties compared to the parent materials. However in this paper we only focus on the dielectric propertiesof the composites samples prepared from nano-sized barium ferrite and barium titanate powders.

2. Experimental methods

(1-x)$BaFe_{12}O_{19}$ – (x) $BaTiO_3$ composites have been prepared using the conventional solid state reaction technique. Nano-sized barium ferrite, $BaFe_{12}O_{19}$ and barium titanate, $BaTiO_3$ powders were weighed according to different weight fractions where x = 0.0, 0.1, 0.3 and 0.5. 3 g of the powders mixture was weighed according to the fixed ratios and ground using Fritsch planetary mono mill for 1 hour at a speed of 350 rpm. 1 g of the mixed and ground powder was uni-axially pressed into pellet shape at 290 MPa for 2 minutes. The pellet was then sintered at 1000 °C for 6 hours in air. X-ray diffractometer (XRD) (Bruker D8 Advanced) was used to identify the crystalline structure and phases of the samples with different ratios. The observations were made at angle between 20-80 °. The morphology of the fracture surface of the pellets was observed using field emission scanning electron microscope (SEM) (Gemini Supra Series), which is equipped with energy dispersive X-ray for elemental analysis. Dielectric properties of the samples as a function of frequency were studied at room temperature using High Frequency Response Analyzer (HFRA Solartron 1265) in the frequency range of 1 Hz to 1 MHz. Silver paint was painted on both surfaces of the pellet as electrodes for better electrical contact.

3. Results and discussion

Figure 1 (a) shows the XRD spectrum for all samples. Sample with x = 0.0 shows peaks of a single phase $BaFe_{12}O_{19}$ that can be indexed with the standard JCPDS file number 39-1433 with hexagonal structure. The (hkl) of the prominent peaks are shown in the figure. The peaks of tetragonal structured $BaTiO_3$ (5-0626) emerges when it is added to the $BaFe_{12}O_{19}$. Intensity of the $BaTiO_3$ peaks increase with increasing of its composition in the composites.

Figure 1(b) illustrates the XRD patterns of the samples between 2θ of 31.0 ° - 35.0 ° showing the strongest $BaFe_{12}O_{19}$ peaks of (1 0 7) and (1 1 4); and $BaTiO_3$ peak of (1 1 0). The $BaTiO_3$ peak intensity increases with x but the peak shifted slightly to the right showing a decrement of the lattice spacing due to the increasing $BaTiO_3$ content. On the other hand the $BaFe_{12}O_{19}$ peaks shift slightly to the left when x = 0.1 and shift to right with further x increment. This could be due to the modification of the original crystal structure arrangement when smaller size $BaTiO_3$ substitutes the larger $BaFe_{12}O_{19}$ in the lattice.

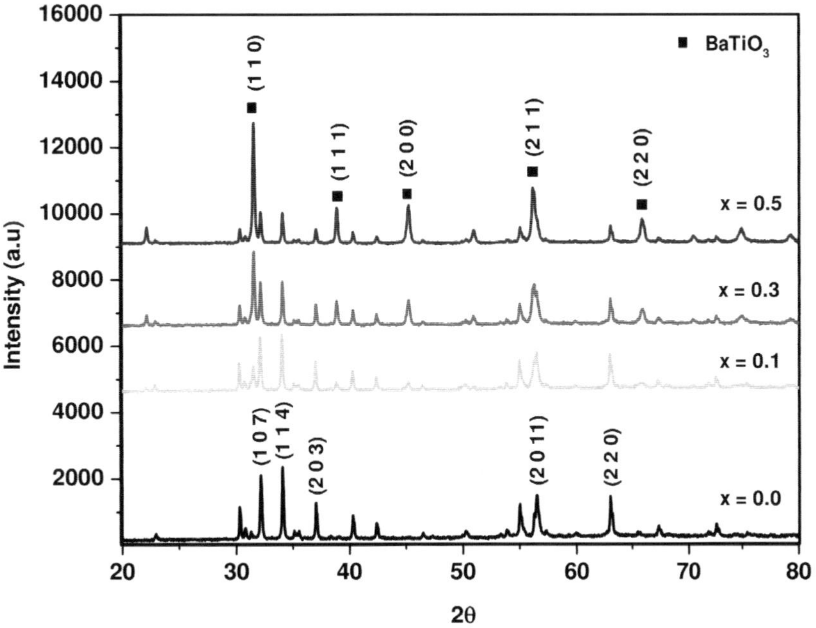

(a)

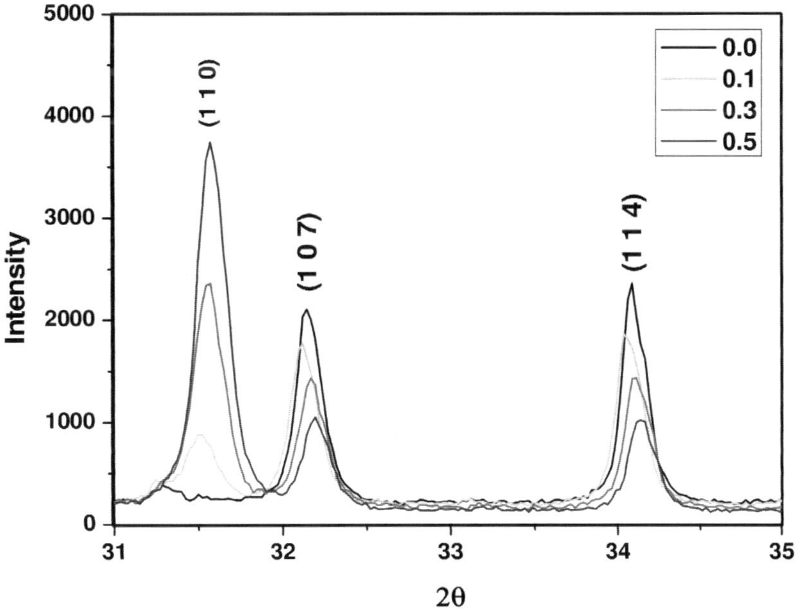

(b)

Figure 1. (a) XRD patterns for (1-x)$BaFe_{12}O_{19}$ – x$BaTiO_3$ samples with x = 0.0, 0.1, 0.3 and 0.5; **(b)** in range of 31.0 – 35.0 °.

The morphology from the fracture surface of composite sample with x = 0.1 can be observed in Figure 2. From the SEM micrograph and EDX mapping analysis it is shown that the spherical nano-sized grains (< 50 nm) are $BaTiO_3$and the micron-sized grains are $BaFe_{12}O_{19}$. The nano-sized $BaFe_{12}O_{19}$ particles change to micron-sized after the sintering process. This is due to the occurrence of diffusion process among the compacted $BaFe_{12}O_{19}$ powders when subjected to a very high temperature. Mallick et al. stated that the grains will combine to form bigger grain because of the thermodynamically driven mass diffusion mechanisms. Bonding of the particles forms larger $BaFe_{12}O_{19}$ with average grain size of 0.2 – 0.3 μm. Ferrite materials can be dense rapidly when subjected to high temperature [10]. The melting point of $BaFe_{12}O_{19}$ is about 1300 °C [11] and the sintering temperature used was 1000 °C while melting point for barium titanate 1620 °C [12].

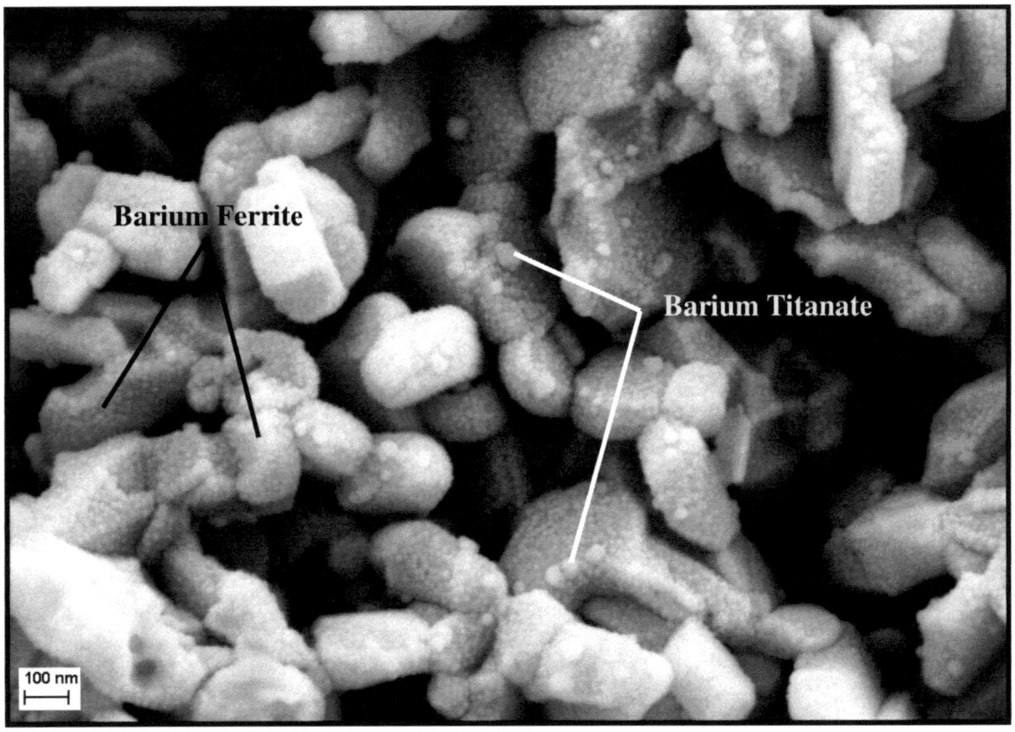

Figure 2. SEM micrograph of (1-x)$BaFe_{12}O_{19}$ – x$BaTiO_3$ sample with x = 0.1

Figure 3 shows the variation of dielectric constant, ε' with frequency for all samples at room temperature. The dielectric constant, ε' represents the ability of a material to polarize and store electric energy. $BaFe_{12}O_{19}$ shows dielectric behaviour by itself [5]. The dielectric constant of all the (1-x)$BaFe_{12}O_{19}$ – (x)$BaTiO_3$ composites decreases with frequency, which is a common behavior for dielectric materials. The dielectric constant becomes nearly constant at higher frequency with the largest constant range (10^4 - 10^6 Hz) and lower dielectric constant value for sample without $BaTiO_3$ content. The high dielectric constant at lower frequency up to around 100 Hz can be interpreted on the basis of the Maxwell-Wagner interfacial space charge polarization which arises due to the heterogeneity of the sample caused by pores, grain boundaries and layered structures. At low temperature the large well conducting grains that are separated by the thin poorly conducting or insulating intermediate grain boundaries promotes polarization in the samples. Charge carriers may also accumulate at the sample-electrode surface.

Graph of the dielectric constant versus weight fraction, x for (1-x)$BaFe_{12}O_{19}$ – (x)$BaTiO_3$ at different frequencies is shown in Figure 4. It is found that $BaTiO_3$ addition improves the dielectric constant of the composite matrix ceramic. At 50 Hz the dielectric constant for x = 0.0 is in the power of three and increases two order of magnitude to the power of five for x = 0.1. The composite mixture with inhomogeneous microstructure creates dipole because of the changes of valence state of cations and the space charge polarization [13]. Sample with x = 0.1 has the highest dielectric constant which is 1.2×10^3 at 50 Hz at room temperature. The dielectric constant is nearly constant at very high frequencies for samples with $BaTiO_3$. When external field is applied to the sample, the domains of the samples were reoriented and became polarize that will lead to straining in the dielectric of composites [14]. Ferrite is known as a material that can make the values of dielectric constant of some materials decrease because of the exchange interaction. Electrons in ferrite begin to polarize due to electron hopping that can make the displacement of charges that will affect polarization in the direction of external applied field [15]. A study by Lokare et al. (2008) showed that dielectric constant increases when barium titanate ratios increases because of the ferrimagnetic region has surrounded the ferroelectric regions.

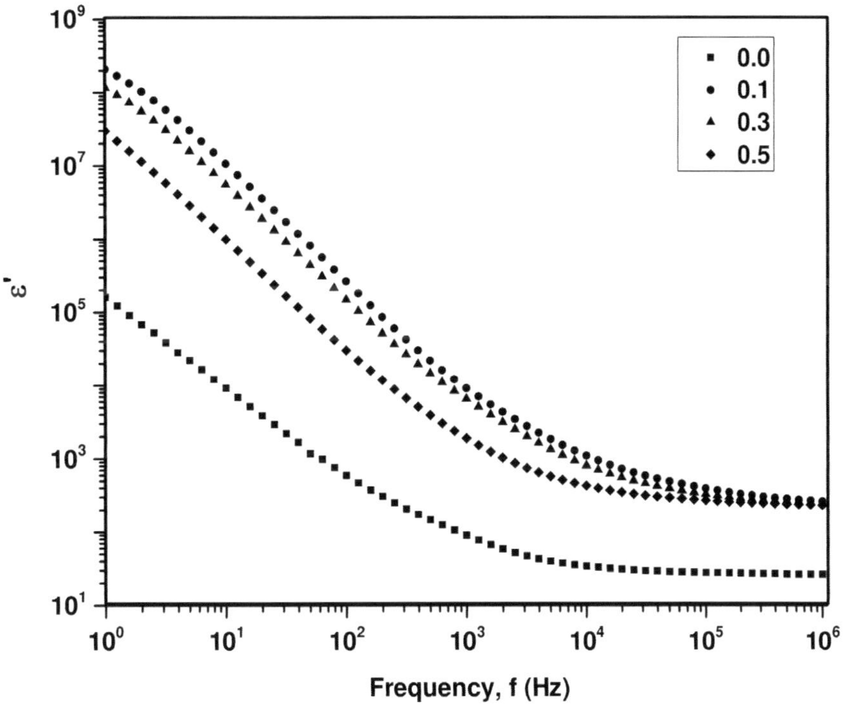

Figure 3. Frequency dependence of the dielectric constant for (1-x)$BaFe_{12}O_{19}$ – x$BaTiO_3$ samples with different weight fraction (x = 0.0, 0.1, 0.3 and 0.5)

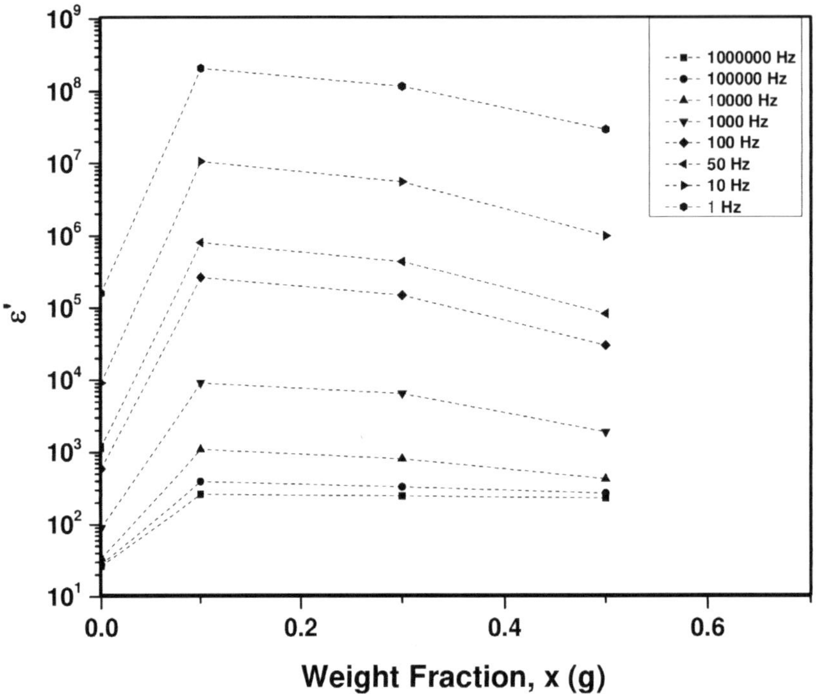

Figure 4. Weight fraction, x dependence of the dielectric constant at different frequencies for $(1-x)BaFe_{12}O_{19} - xBaTiO_3$

Figure 5 shows the variation of loss tangent, *tan* δ with frequency for all samples. Loss tangent is the loss of electrical energy in the form of heat due to the oscillation of the charge carriers. For all samples the loss tangent increases with frequency and forms a peak at a very high frequency, except for sample with x = 0.1. The peak represents the relaxation process where the charges cannot keep up with the frequency of the applied ac electric field. The lowest loss tangent value at 50 Hz is for sample x = 0.1 which is 0.06 at room temperature.

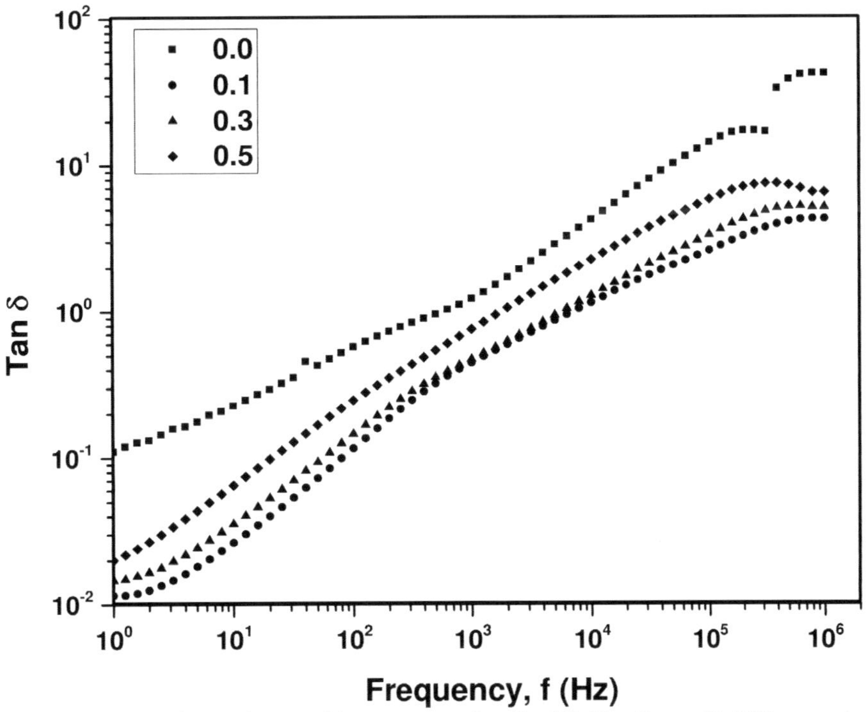

Figure 5. Frequency dependence of loss tangent for (1-x)$BaFe_{12}O_{19}$– x$BaTiO_3$ samples with difference weight fraction (x = 0.0, 0.1, 0.3, and 0.5)

4. Conclusions

The effect of BaTiO3 addition on the dielectric properties of (1-x)$BaFe_{12}O_{19}$ – (x)$BaTiO_3$ composites system with different weight fraction have been studied. XRD results of the sintered composites show the formation of BF and BT phases without other impurity phase. The FESEM micrograph shows nano-sized BT particles are spread among the micron-sized BF. The frequency variation of dielectric constant shows the common dielectric behaviour, which is the dielectric constant decreases when frequency increases. At low frequency the interfacial polarization contributes to the very high dielectric constant. Dielectric constant of the composites increases from the power of three for sample without $BaTiO_3$ at 50 Hz, up to the power of five for sample with x = 0.1. However, higher $BaTiO_3$ fraction (x = 0.3 and 0.5) results in a slight decrement of the dielectric constant. The loss tangent of all samples is very low and increases with frequency. Further study will be done on sample with x = 0.1 due to its potential as a high dielectric constant and low loss material.

Acknowledgment.
The authors would like to thank Universiti Kebangsaan Malaysia for the UKM-GUP-2011-163 grant. Ministry of Science, Technology and Innovation is also acknowledged for the FRGS/1/2011/SG/UKM/02/20 grant and MyMaster scholarship. Thanks to School of Applied Physic, Faculty of Science and Technology, UKM for providing the research facilities.

References

[1] Mu G, Pan X, Chen N, He C and Gu M 2008. Synthesis and characterization of hard magnetic composites-Hollow microsphere/titania/barium ferrite. *Appl. Surf. Sci.* **254**: 2483-2486.
[2] Wang C, Han X, Xu P, Wang X, Li X and Zhao H 2009. Magnetic and dielectric properties of barium titanate-coated barium ferrite. *J. Alloy Compd.* **476** 560-565.

[3] Li Y, Wang Q and Yang H 2009. Synthesis, characterization and magnetic properties on nanocrystalline $BaFe_{12}O_{19}$. *Curr. Appl. Phys.* **9** 1375-1380.

[4] Mishra D, Anand S, Panda R K and Das R P 2004. Studied on characterization, microstructures and magnetic properties nano-size barium hexa-ferrite prepared through a hydrothermal precipitation-calcination route. *Mater. Chem. Phys.* **86** 132-136.

[5] Mallick K K, Shepherd P and Green R J 2007. Dielectric properties of M-type barium hexaferrite prepared by co-precipitation. *J. Eur. Ceram. Soc.* **27** 2045-2052.

[6] Qui J, Lan L, Zhang H and Gu M 2008. Effect of titanium dioxide on microwave absorption properties of barium ferrite. *J. Alloy Compd.* **453** 261-264.

[7] Marković S, Miljković M, Jovalekić Č, Mentus S andUskoković D 2009. Densification, microstructure, and electrical properties of BaTiO3 (BT) ceramics prepared from ultrasonically de-agglomerated BT powders. *Mater. Manuf. Process* **24** 1114 – 1123.

[8] Kumar P, Singh S, Juneja J K, Prakash C andRaina K K 2009. Ferroelectric properties of substituted barium titanate ceramics. *Physica B* **404** 1752-1756.

[9] Sonia Patel R K, Kumar P, Prakash C and Agrawal D K 2012. Low temperature synthesis and dielectric, ferroelectric and piezoelectric study of microwave sintered $BaTiO_3$ ceramics. *Ceram. Int.* **38** 1585-1589.

[10]Roy P K andBera J 2012. Study on electromagnetic properties of MgCuZn ferrite/$BaTiO_3$ composites. *Mater. Chem. Phys.* **132** 354-357.

[11] Stabik J, Dybowska A, Pluszyński J, Szczepanik M and Suchoń L 2010. Magnetic induction of polymer composites filled with ferrite powders. *Arch. Mate. Sci. Eng.* **41** 13 – 20.

[12] Sarkar S K and Sharma M L 1989. Dielectric strength and dielectric constant of BaTiO3 - NaNbO3 composites at room temperature. *J. Mater. Sci. Lett.* **8** 1365- 1367.

[13] Hrib L M and Caltun O F 2011. Effect of the chemical composition of the magnetostrictive phase on the dielectric and magnetoelectric properties of cobalt ferrite-barium titanate composites. *J. Alloy Compd.* **509** 6644-6648.

[14] Kim Y J, Hyun J W, Kim H S, Lee J H, Yun M Y, Noh S J and Ahn Y H 2009. Microstructural characterization and dielectric properties of barium titanate solid state with donor dopants.*Bull. Korean Chem. Soc.* **30** 1267-1273.

[15] Lokare S A, Patil D R, Devan R S, Chougule S S, Kolekar Y D and Chougule B K 2008. Electrical conduction, dielectric behaviour and magnetoelectric effect in (x)$BaTiO_3$ +(1-x)$Ni_{0.94}Co_{0.01}Mn_{0.05}Fe_2O_4$ ME composites. *Mater. Res. Bull.* **43** 326-332.

Influence of Europium Ion on Structural, Mechanical and Luminescence Behavior of Tellurite Nanoglass

E S Sazali, M R Sahar and S K Ghoshal
Advanced Optical Material Research Group, Department of Physics, Faculty of Science, Universiti Teknologi Malaysia, 81310, Skudai, Johor Bahru, Johor, Malaysia.

E-mail: mrahim057@gmail.com

Abstract. Understanding the mechanism of enhanced luminescence of rare earth doped glasses in the presence of nanocrystallites and growth kineics is fundamentally important for optical devices. Tellurite nanoglasses of composition (80-x) TeO_2 - 5 Na_2O – 15 MgO – (x) Eu_2O_3, over the concentration region of 0 to 2.5 mol% are prepared using conventional melt-quenching technique. The nanocrystalline particles are obtained by heating the as-cast glass at temperature 15–20°C above the glass crystallization temperature (T_c). The sizes of nanocrystallites are estimated from the X-Ray Diffraction (XRD) pattern using the Scherrer equation having average diameter ~68.7 nm. SEM studies revealed the nanocrystal glass morphology associated with the existence of crystalline phase. The glass density is determined by Precisa Densitometer and the hardness by the Vickers micro-hardness method. The density of tellurite nano-glass is found to be in the range of 5.2413 to 5.4933 g cm^{-3} while the Vickers microhardness varies from 2.77 to 2.93 GPa depending on the dopant concentration. The photoluminescence (PL) spectra exhibits five peaks around 568 nm, 600 nm, 628 nm, 664 nm and 712 nm assigned to $^5D_0 \rightarrow {}^7F_0$, $^5D_0 \rightarrow {}^7F_1$, $^5D_0 \rightarrow {}^7F_2$, $^5D_0 \rightarrow {}^7F_3$ and $^5D_0 \rightarrow {}^7F_4$ transitions respectively. Interestingly, the FWHM and the inverse quality factor of the heat-treated glass are found to decrease with increasing concentration of Eu^{3+} dopants. Our observation may contribute towards the development of solid state lasers.

1. Introduction

The uses of tellurite glasses as a host material have incredibly increased due to their potential applications in laser and fiber [1]. Besides, the understanding of their microscopic mechanism of structural and optical behavior gave much thrust and fundamental interests for both academia and the industries [2]. Interest in tellurite glass containing rare earth element are expected for nonlinear optical devices as for their large third-order nonlinear optical susceptibility [3]. Moreover, the assimilation of rare earth ions can stabilize the metastable crystalline phase which leads to a development of optical devices [4, 5]. Among all glasses, the Eu^{3+} doped tellurite glass has attracted great attention as they can perform persistent spectral hole-burning in the $^7F_0 \rightarrow {}^5D_0$ transition [6]. The integration of rare-earth ions into some kinds of glasses elucidated the significance of rare earth ion-glass host interaction for engineering waveguide devices [7]. Rare earth ions play important role in modern technology as optically active elements in solid state luminescence materials due to the energy levels possessed by these ions when incorporated into a solid state matrix.

The synthesis of nanocrystalline Eu^{3+} doped TeO_2 based glass has not been extensively reported and the growth dynamics of these glass system is far from being understood [8]. Therefore, their synthesis and understanding the growth behaviors is of considerable technological importance [5]. Moreover, properties of Eu^{3+} doped TeO_2 glasses which undergo heat treatment process are studied only recently. We prepare the tellurite glass doped with europium (Eu^{3+}) via melt quenching technique. The nano-crystalline glass is prepared by the heat-treatment process with nanocrystallites in the host matrix. The effect of dopant concentration on the structural mechanical and optical properties is investigated to study the properties in terms of their compositions.

2. Experimental

The ability of glasses for crystallization is first measured by the values of the glass crystallization temperature (T_c) to obtain transparent glass ceramics. A portion of the glass sample are heat treated for 30 minutes at temperature 15–20°C above T_c in an electrical furnace for all samples. The crystallization of the heat treated samples is identified using XRD (Siemens Diffractometer D5000). The powdered form of the samples is used for diffraction studies. Date are collected in the 2θ range from 15 ° to 75°, in steps of 0.05° and 1s counting time per step using Cu Kα as a radiation source of wavelength λ=1.54056Å. The particle size corresponding to the XRD peak of (111) plane with a cubic structure is estimated from the full width at the half maximum (FWHM), β using Scherrer's equation given by,

$$d = 0.89 \lambda / \beta \cos \theta \qquad (1)$$

where d is the crystallite size, λ is the wavelength and θ is the diffraction angle.

To investigate the kind of the formed crystallites was examined by SEM. The crystallized portion of glass was polished and then successively diamond paste. The specimen is then was coated with a thin layer of carbon by an evaporation technique. An electron accelerating voltage of 8 kV was used and the micrographs were obtained using back-scattered imaging. For the mechanical analysis, the density measurement of glass is made using the Archimedes principle. The density is calculated in gcm^{-3} using,

$$\rho = \frac{W_a}{W_a - W_l}(\rho_l - \rho_a) + \rho_a \qquad (2)$$

where ρ is sample density, W_a is weight of sample in air, W_l is weight of sample in liquid, ρ_a is air density (0.001 g cm^{-3}) and ρ_l is liquid density (toluene = 0.8669 g cm^{-3}). By using the value of density, the molar volume, V_m is calculated from,

$$V_m = (\frac{M}{\rho}) \qquad (3)$$

where M is the molecular weight of glass which is calculated from the glass composition. Moreover, the hardness is performed by using the Shimadzu Microhardness Tester HMV-2 with an applied load of the indentor ~9.807N. The Vickers microhardness, H_v is determined from the load applied and the width of the impressed area.

The room temperature luminescence measurement is performed on Nanosecond Luminescence Spectroscopy System, Ekspla Model NT340/1 tunabled Nd: YAG laser system. Each sample in the powder form is placed in the spectrometer and scanned for radiation spectral wavenumber in the range of 200–900 cm^{-1}. The Xenon lamp (300 < λ < 1300nm) was used as a pumping source.

3. Results and Discussion

A series of unheat-treated and heat-treated TeO_2 - Na_2O – MgO glass system doped with Eu^{3+} are successfully prepared by the melt-quenching technique with excellent transparencies.

3.1. X-ray Spectra

Figure 1 shows the XRD pattern for TeO_2 – Na_2O – MgO glass with 1.5 mol% Eu_2O_3 dopant concentrations in which the expected reflections from crystalline particles after heat-treatment are clearly evidenced. The values of diffraction angle (2θ) of 28.8°, 35.4°, 40.1°, 44.6°, 55.2° and 61.0° confirm the presence of the α-TeO_2 phase (paratellurite) and γ-TeO_2 phase [9]. The particle size in the crystalline phase is estimated from the (FWHM) as shown in the inset of Fig.1 is ~68.7 nm. Indeed the crystallites are nanosized particles. The results indicate that the presence of TeO_2 might play a major role in the phase formation.

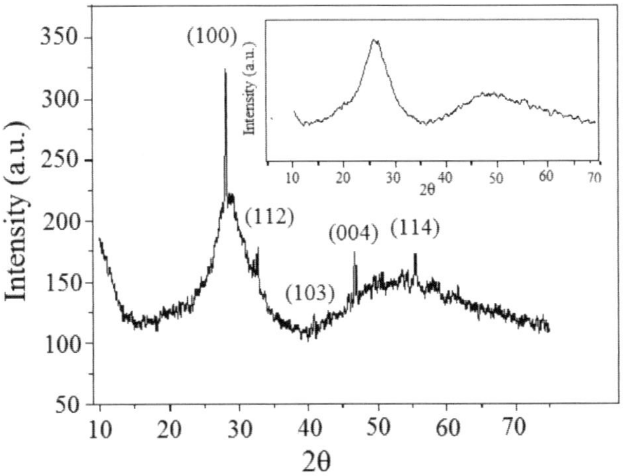

Figure 1. The room temperature XRD spectra for 78.5 TeO_2 – 5 Na_2O – 15 MgO – 1.5 Eu_2O_3 glass after heat-treatment at 442.7°C. The inset shows the XRD patterns for the same nanocrystalline glass in the as-cast condition.

3.2. Structural Properties

The existence of crystalline phase was verified by SEM analysis. Figure 2(a) - (d) shows the SEM micrographs of some TeO_2 – Na_2O – MgO glass systems in varying Eu_2O_3 dopant concentration. Some of them are spherical in shape, some have star-like structure and some of them are irregular in shape. All the micrographs confirm that the crystalline region is composed of TeO_2 phase is dominant.

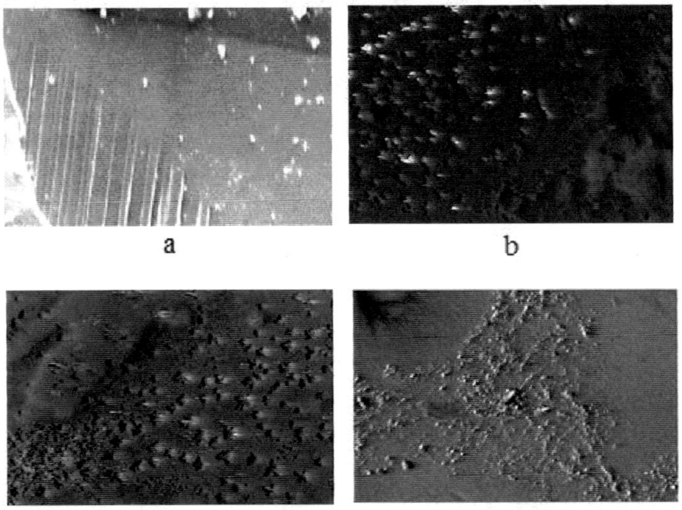

a b

c d

Figure 2. SEM pictures obtained for heat-treated Eu_2O_3 (x) doped glass system above T_c with (a) x = 0, (b) x = 0.5, (c) x = 1.0 and (d) x = 2.0.

Throughout the heat-treated of samples, although the nucleation and growth of nano-crystalline are developed in the same heat-treat operation, but the distribution of nucleates and growth of crystal cannot attain the same extent as the dopant concentration is different. This shows the incorporation of Eu^{3+} dopants with the synthesis of nano-crystalline TeO_2 – Na_2O – MgO glass system. Crystal immersed in the amorphous material, with no observable clustering and a size distribution of are observed.

3.3. Mechanical Properties

Table 1 shows the strong dependence of the glass density on the Eu_2O_3 concentration. The glass density is found to be in the range of 5.2413 g cm^{-3} to 5.4933 g $cm^{-3.}$ As the molecular masses of Eu_2O_3 (351.92) is higher than TeO_2 (159.61), this contributes to higher packing density. The increase in the number of oxygen atoms increases the cation radius. The introduction of Eu^{3+} ions having higher charges and coordination number tend to develop tightly packed glass structure. This result on increasing density is in agreement with others [10]. However, the molar volume in the range of 26.11-25.79 cm^3mol^{-1} shows a decrement as a function of dopant concentration. This might be due to the reduction of total volume size that contributes to the compactness of the glass which is reduced. As the glass undergo the heat-treatment process, the space distance will be decreased leading to a decrease in molar volume.

Meanwhile, it is evident that the Vickers hardness linearly increases from 2.77–2.93 GPa as the concentration of Eu_2O_3 is increased. This may be due to the increasing number of strong covalent bonds from rare-earth dopants causes an increase in the network cross linking in the glasses [11]. The incorporation of Eu^{3+} into the glass system changes the tellurite structure from TeO_4 tbp to TeO_3 bp as more bridging oxygen (BO's) is created that gives more rigidity to the structure and leads as increase in Vickers microhardness [12].

Table 1. Density of Eu^{3+} doped TeO_2 – Na_2O – MgO heat-treated glass system.

Eu_2O_3 (mol %)	Density, ρ (g/cm^3)	Molar Volume, V_m (cm^3/mol)	Hardness, H_v (GPa) ±0.1
0	5.2413 ± 0.016	26.11	2.77
0.5	5.2487 ± 0.019	26.25	2.81
1	5.2827 ± 0.027	26.27	2.86
1.5	5.3011 ± 0.021	26.36	2.88
2	5.4406 ± 0.016	25.89	2.91
2.5	5.4933 ± 0.003	25.79	2.93

3.3. Luminescence Spectra

Figure 3 shows the luminescence spectra of Eu^{3+} doped TeO_2 – MgO – Na_2O for heat-treated glass system as the concentration of Eu_2O_3 increased.

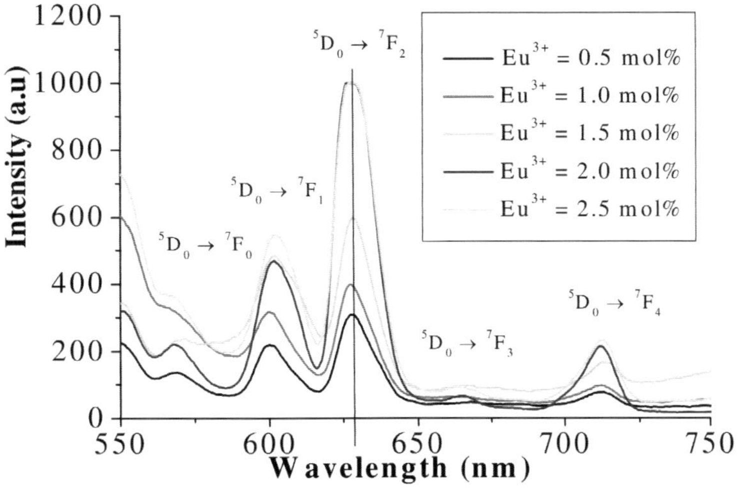

Figure 3. Luminescence spectra of Eu^{3+} doped TeO_2 - Na_2O - MgO heat-treated glass system excited at 400 nm.

Five emission bands corresponding to $^5D_0 \rightarrow {}^7F_0$, $^5D_0 \rightarrow {}^7F_1$, $^5D_0 \rightarrow {}^7F_2$, $^5D_0 \rightarrow {}^7F_3$ and $^5D_0 \rightarrow {}^7F_4$ are evidenced at an excitation wavelength λ_{exci} = 400 nm. Previous report identified the excitation bands at around 355 nm to 395 nm (6, 13, 14). The observed emission peaks of Eu^{3+} doped tellurite glass system at the corresponding transitions cofirmed the presence of europium trivalent state in this glass system which is responsible for the large luminescene efficiency [15].

Figure 4 shows the luminescence spectra of the glass with and without heat treatment at 1 mol% concentration of Eu_2O_3. The detected emission spectra for heat-treated glass sample with the same Eu_2O_3 concentration is similar to the as-cast glass sample which contributes to the same transition. It is clear that the shape of the emission band does not change but there is a slight shift of the peaks position indicating a strong interaction between Eu^{3+} of ions [16].

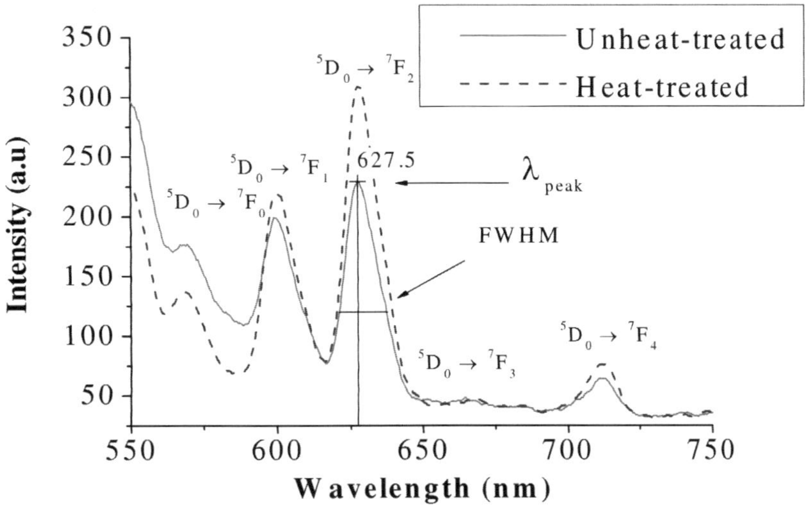

Figure 4. Emission Spectra of glass at 1 mol% of Eu_2O_3 with and without heat-treatment.

The luminescence spectra are further interpreted using the FWHM corresponding to the peak wavelength. This method can be used to indicate the band width ($\Delta\lambda$). Figure 5 shows the average value of FWHM for different transitions for untreated and heat-treated glass samples. The heat-treated glass samples shows more linear changes of FWHM values compared to the un-treated one and is decreased as the Eu_2O_3 dopant concentration is increased except for the $^5D_0 \rightarrow {}^7F_0$ transition. This change is due to the variation of the local structure around lanthanide ion. The increase in the Eu^{3+} ion concentration leads to the changes in the structural unit of the host glass from BO to NBO that contributes to the maximization of Q factor gaving a favorable transition condition for photon to be emitted. Most of the transition that occurs around 600 nm makes the samples best suited for red laser.

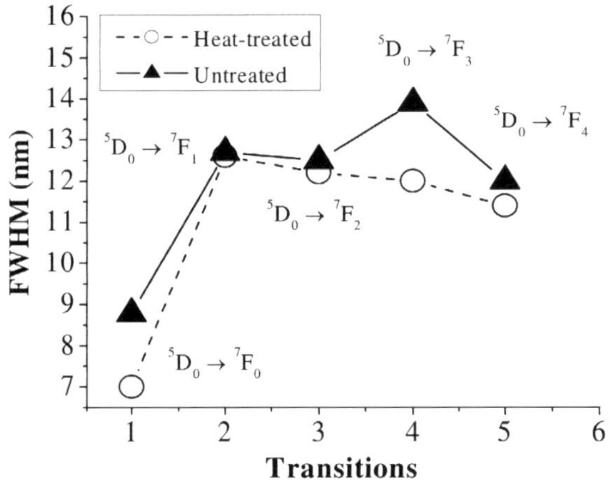

Figure 5. FWHM versus different transitions.

Furthermore, the inverse quality factor Q^{-1} method can be used for emission as described in the Judd Ofelt analysis [17]. A plot of inverse quality factor versus transition for the heat-treated glass samples (Figure 6) shows the decreasing value for the emission transitions $^5D_0 \rightarrow {}^7F_1$, $^5D_0 \rightarrow {}^7F_2$, $^5D_0 \rightarrow {}^7F_3$ and $^5D_0 \rightarrow {}^7F_4$ compared to the un-treated glass system. This result asserts that the Eu^{3+} in the heat-treated tellurite glass system is much promising for laser evelopment. This contributes to higher probability for laser emission of any pumping process.

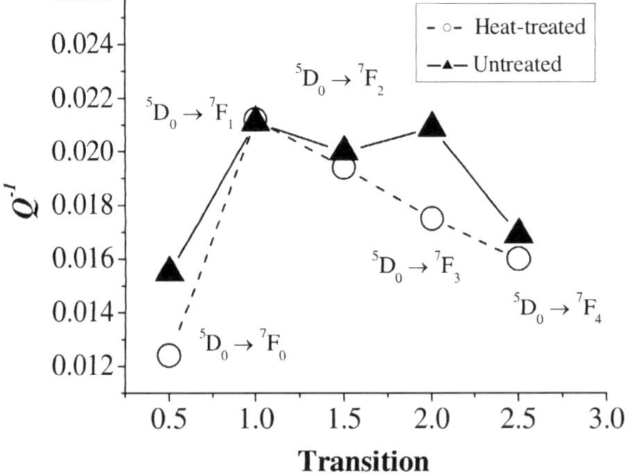

Figure 6. Q^{-1} versus transition for both un-treated and heat-treated samples.

4. Conclusion

A series of nanoglass system based on (80-x) TeO_2 - $5Na_2O$ - 15MgO - (x) Eu_2O_3 for both unheat-treated and heat-treated over the concentration region from 0 to 2.5 mol% are successfully prepared using conventional melt quenching technique. They are mechanically (density and molar volume), structurally and optically (luminescence) characterized. These glasses being largely transparent are of good quality. For the crystallization investigation, XRD spectra confirms the presence of crystalline phase of nano glass which the diameter is estimated around 68.7 nm and SEM studies revealed the nano-crystal glass morphology which is associated to the existence of crystallized phase. Crystal immersed in the amorphous material, with no observable clustering and a size distribution of are observed.The densities are found to be in the range of 5.2413 to 5.4933 g cm^{-3} and increases with increasing the Eu_2O_3 concentration. However, the molar volume varies in the range of 26.11 to 25.79 cm^3mol^{-1}shows a decrement as a function of dopant concentration. In addition, the Vickers hardness is found to increase linearly from 2.77 – 2.93 GPa as the concentration of Eu_2O_3 increased. Luminescence spectra consists of five emission peaks for $^5D_0 \rightarrow {}^7F_0$, $^5D_0 \rightarrow {}^7F_1$, $^5D_0 \rightarrow {}^7F_2$, $^5D_0 \rightarrow {}^7F_3$ and $^5D_0 \rightarrow {}^7F_4$ transitions respectively. The enhancement of Eu^{3+} luminescence especially for the highest emission intensity, $^5D_0 \rightarrow {}^7F_2$ transition lies in the red region as compared to the precursor glass. The heat-treated glass samples shows more linear changes in FWHM and inverse quality factor and is found to decrease with the increase of Eu_2O_3 dopant concentration in comparison to the un-treated glass samples. The role of europium ion of mechanical and spectroscopic properties is understood. Our results may provide valuable information for making nanoglass useful for lasers.

Acknowledgements.
The authors gratefully acknowledge the financial support from Ministry of Higher Education through grant Universiti Teknologi Malaysia under Vot. 06J39, Vot. 06J33 (GUP/MOHE) and MyMaster.

References

[1] Tanabe S *et al* 1990 *J. Non-Cryst. Solids* **122** 79
[2] El-Mallawany R A H *et al* 1995 ESR and electrical conductivity studies of $TeO_2$095CeO2005 semiconducting glasses Materials Chemistry and Physics **41** 87–9
[3] Wang J S *et al* 1994 13μ emission of neodymium and praseodymium in tellurite-based glasses *J. Non-Cryst. Solids* **178** 109-113
[4] Matthew J Dejneka and Alexander Streltsov 2002 Science and Technology Division Corning Incorporated Corning NY 14831
[5] K Hirano *et al* 2001 Rare earth doping into optical nonlinear nanocrystalline phase in transparent TeO2-based glass-ceramics *J. Phys. Chem. of Solids* **62** 2075-2082
[6] Giridhar P *et al* 2011 Spectroscopic studies of RE^{3+} RE =Eu Tb Sm & Dy: lithium lead boro tellurite glasses *ferroelectrics letters* **38** 1–10
[7] Animesh Jha *et al* 2007 Rare-Earth Doped Glass Waveguides for Visible Near-IR and Mid-IR Lasers and Amplifiers *J. Mater. Sci.: Mater Electron* **18** 315–320
[8] Ravi Kumar V *et al* 2001 Structural and Optical Studies of Pr^{3+} Nd $^{3+}$ and Eu^{3+} Ions in Tellurite Based Oxyfluride TeO2-Cif Glasses
[9] Burcu Oz 2006 Thermal Microstructural and Optical Characterization of TeO_2-K_2O Glasses
[10] Yang Li *et al* Hu 2005 *J. Mat. Sci. Tech.* **21** No3 391-394
[11] Kassim A 2010 Physical and Optical Properties of Neodymium and Neodymium Co-Doped Tellurite Glass System PhD Thesis Universiti Teknologi Malaysia
[12] Sahar M R and Sazali E S 2011 Physical and Thermal Properties of TeO_2-Na_2O-MgO Doped Eu_2O_3 Glass UMTAS
[13] Babu P and Jayasankar C K 2000 Optical Spectroscopy of Eu^{3+} Ions in Lithium Borate and Lithium Fluoroborate Glasses *Physica B* **279** 262-281
[14] Luciana R P *et al* 2009 Enhanced luminescence Of Tb^{3+}/Eu^{3+} doped tellurium oxide glass containing silver nanostructures *J. Appl. Phys.* **105** 103505

[15] Ricardo de Almeida *et al* 2008 Eu^{3+} Luminescence in Tellurite Glasses with Gold Nanostructures *Optics Communications* **281** 108–112
[16] Budi Astuti 2005 Study of Optical Properties of P_2O_5 – Sm_2O_3 – MnO_2 Glass System Universiti Teknologi Malaysia
[17] Shaik Abdul Saleem *et al* 2011 Erbium-Doped Fluoroborate Glasses for Near Infrared Broadband Amplifiers International *Journal of Applied Glass Science* **2** 215–221

Journal of Physics: Conference Series **431** (2013) 012009 doi:10.1088/1742-6596/431/1/012009

Optimizing rf Power for Preferential C≡N Bond Formation in a-CN_x Thin Films Prepared by rf-PECVD Technique

N F H Aziz[1a], R Ritikos [2], S A A Kamal[2] and R Awang[1b]
[1]School of Applied Physics, Faculty of Science and Technology, Universiti Kebangsaan Malaysia, 43600 UKM Bangi, Selangor. Malaysia
[2] Low Dimensional Materials Research Centre, Department of Physics, Faculty of Science, University of Malaya, 50603 Kuala Lumpur, Malaysia

E-mail : [1a]farhanahusnaaziz@yahoo.com, [1b]rozida@ukm.my

Abstract. Effects of rf power on the chemical bonding in carbon nitride films deposited using radio-frequency (rf) plasma enhanced chemical vapor deposition in pure methane and nitrogen gas mixtures were investigated. The rf power was varied from 60 to 100 W. The deposition rate of the films increased constantly with increasing rf power up to 80W, before saturating with further increase in rf power. Fourier transform infra-red spectroscopy (FTIR) studies showed a systematic change in the spectra and revealed three main peaks namely the G-peak, D-peak and C≡N triple bond. This work showed that rf power has significant effects on the chemical bonding of the a-CN_x films and the optimum rf power for the high C≡N absorption intensity is 80 W.

1. Introduction

Research on carbon nitride (CN_x) thin films has intensified with the prediction by Liu and Cohen that the properties of crystalline carbon nitride (β-C_3N_4) might be similar or superior to diamond [1]. However, in many cases a-CN_x thin films deposited are mostly obtained in amorphous phase. Nevertheless even these films shows many attracting properties such as extreme hardness, chemical inertness, low friction coefficient, and variable optical band gap [2]. One of the main reason why carbon nitride tends to forms in amorphous phase is because of the presence of hydrogen commonly found in these carbon nitride films. These hydrogen atoms can easily break C≡N and C=N bonds in the film to form C-H and N-H bonds [3]. However, if hydrogen defects were formed intentionally and bonded weakly, a-CN_x thin films can be a good candidate as a humidity sensor [3]. To achieve good a-CN_x thin films for humidity sensor, the films must have high C=N and C≡N bonds content.

There are various techniques capable of producing a-CN_x. These includes rf magnetron sputtering, [4], hollow cathode discharge [5], electron cyclotron resonance microwave plasma chemical vapor deposition [6], graphite hollow cathode system [7] and radio frequency plasma enhanced chemical vapor deposition (rf-PECVD). Among these rf-PECVD is favoured since this technique has the advantage of uniform large area deposition, easy handling, low pinhole density, good adhesion and good step coverage [8]. Furthermore the film characteristice could be tuned simply by controlling the deposition parameters such as rf power, gas flow rate ratios, and substrate temperature. In this owrk we are employing rf-PECVD to obtain a-CN_x thin films. The main focus is to investigate the effect of rf power on the chemical bonding of a-CN_x thin films and to determine the optimum rf power for highest C=N and C≡N bonds.

2. Experimental details

a-CN_xthin films were deposited by rf-PECVD using a mixture of pure (99.999 %) methane, CH_4 and nitrogen, N_2 gas. The type of substrates used in this study was p-type (111) silicon chemically cleaned using acetone and deionized water in ultrasonic baths prior to deposition. The substrate was heated at 100 °C during deposition. Prior to the film deposition, the substrates were treated to hydrogen plasma for 10min with the H_2 flow rate and rf power fixed at 50 sccm and 50 W respectively. This was done to remove impurities on the substrate surface and improve the adhesion of CN layers onto the substrate [9]. The distance between the rf powered top electrode and substrate holder was kept constant at 1 cm while pressure and deposition duration was 0.8 mbar and 90 min. A set of film was prepared at rf power of 60, 70, 80, 90 and 100 W. The films thickness was determined using a surface profilometer. Chemical bonding were characterized using Perkin-Elmer System 2000 FTIR Spectrometer and Horiba Jobin Yvon 800 UV Micro Raman Spectrometer at wavenumber range 1300 – 2500 cm^{-1} and 100 – 3000 cm^{-1}, respectively.

3. Results and discussions

3.1. Deposition rate of a-CN_x thin films

The thickness of the films and corresponding deposition rate for the samples were calculated and presented in Table 1. The deposition rate increased with the increase in rf power up to 80W and begins to saturate and even decrease slightly with further increase in rf power, as shown in Figure 1. The initial increase in deposition rate may be due to the corresponding increase indissociation of reactive gases [9]. The deposition of these radicals formed from this dissociation influences the etching and H atom passivation of the radical sites on the deposition surface. The increase in rf power also increase the concentration of reactive H atoms reaching the deposition surface creating significant increase in the number of nucleation sites [10]. Apparently, the growth of the film through the subplantation of the impinging radicals and neutral species, surpasses the ething of the film resulting in the increase in the deposition rate.

When the rf power exceeds 80W, CH_4, main precursor gas is completed dissociation resulting in the saturated of available radicals and consequently the deposition rate [8]. Normally the radicals formed would react with unreacted molecules in secondary reactions in the plasma. With the complete dissociation of the CH_4 gas in the primary reaction at rf power above 80 W, this supressesthe secondary reaction and thus decrease the radicals in the plasma which otherwise will contribute to the growth of the film.The complete dissociation if the precursor cas also significantly decrease the main species CH_3 which largely contributes to the expecsion of the film. This could be explaining the saturation in deposition rate at rf power above 80 W. Furthermore, due to the decrease in subplantation of these radicals, particularly of CH_3, the ething effects on the film surface becomes significant and this explains the slight decrease in deposition rate above this rf power.

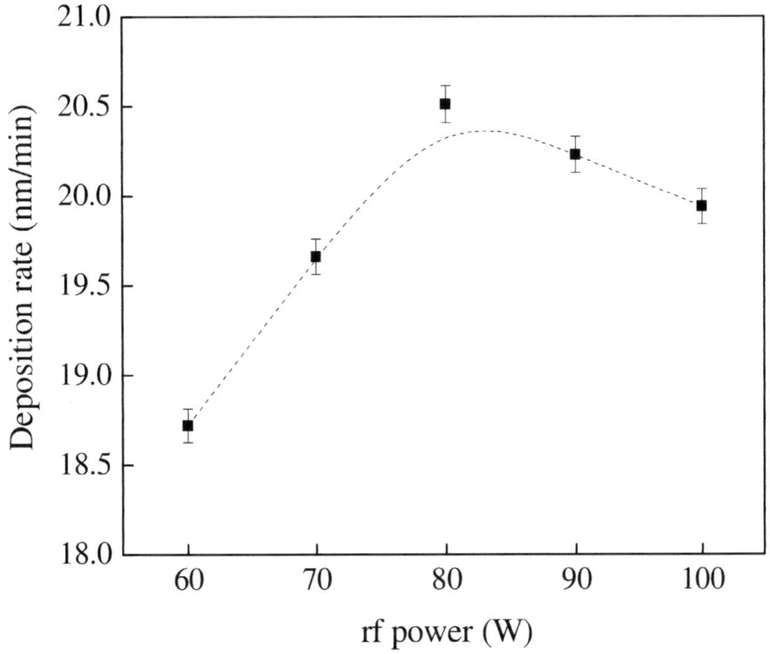

Figure 1. Deposition rate of a-CN_x at different rf power (Line is drawn as guide to the eye)

Table 1. Thickness and deposition rate of the films at different rf power

rf power, P (W)	Thickness, t (nm)	Deposition rate (nm/min)
60	1685	18.72
70	1769	19.66
80	1853	20.51
90	1821	20.23
100	1795	19.94

3.2 FTIR analysis of a-CN_x thin films

The FTIR spectra of the a-CN_x thin films deposited as a function of rf power is shown in Figure 2. The broad FTIR transmission band ranging from 1300-1800 cm^{-1} is due to the presence of sp^2 and sp^3 carbon [11]. The transmission peaks in this range represents the Raman activated peaks and corresponds to the disorder D-band and graphite (G-band). The D band arises from the breathing modes of sp^3 atoms in rings and its intensity is strictly correlated with the presence of the six fold aromatic ring while the G band is associated with the bond stretching of all pairs of sp^2 atoms in both rings and chains [12]. The weak intensity of the D-band at 1350-1450 cm^{-1}in all these films was associated to thge presence of sp^3 C-N bond in the film [10] and clearly indicated content of C-N bonds are low. The broad G-band at 1500–1800 $cm^{-1}$was associated to the presence of C-C and sp^2 graphitic domain. These bands are discussed further in the Raman analysis of these thin films.

The transmission band lying between 2100 and 2200 cm^{-1} suggests the presence of nitrile (-C≡N), isonitrile (-N≡C) and carbodiimide (–N=C=N-) groups [13]. The peak centers at ~2150cm^{-1}which corresponds to the preferential formation of C≡N triple bond. These bonds are network terminators groups. The intensity of this bands increase as the increase in rf power up to 80 W, becoming prominent at rf power of 70 and 80 W. With further increase in rf power their intensities decreases significantly and almost vanishes at rf power of 100W. This indicate that very few C≡N bonds,if any, were formed in this film. It is possible that the strong etching effects on the film surface at rf power above 80W as discussed earlier effectively breaks the C≡N bond in the CN_x films [13]. These results also suggest that the formation of the C≡N bond in the CN_x films can becontrolled by varying the applied rf power.

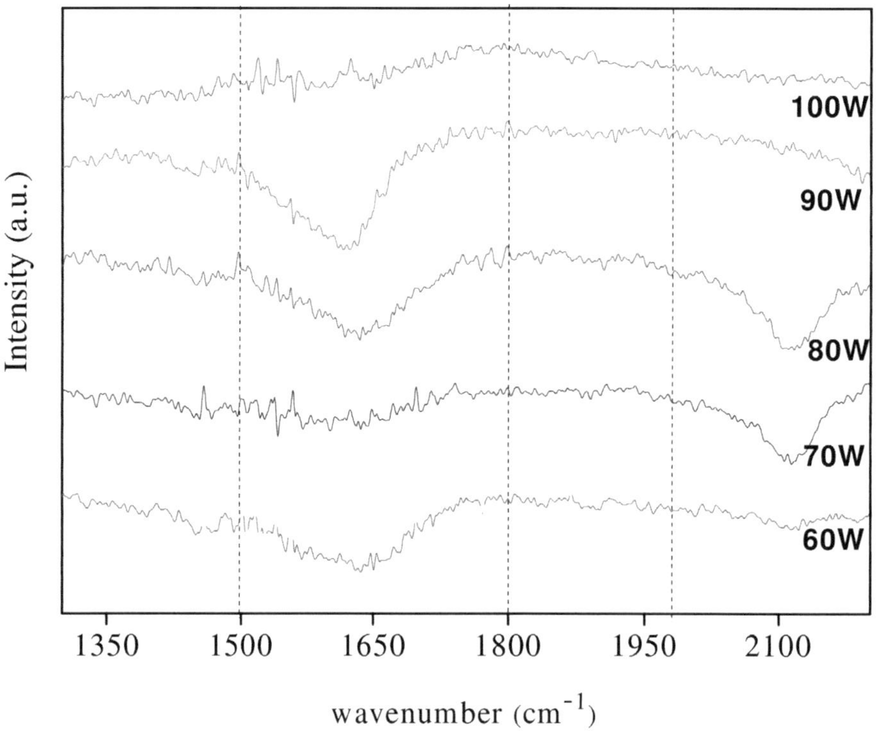

Figure 2. Variation of FTIR spectra as function rf power

3.3 Raman analysis of a-CN_x thin films

In a typical a-CN_xfilms, two broad partly-overlapping bands, centered at approximately 1360 and 1580 cm^{-1} lies in the broad band region between 1300 and 1700 cm^{-1} of the Raman spectra. The Raman spectra can be deconvoluted using Gaussian and Lorentz fitting [15] in order to calculate the Dand G –bands. Similar to FTIR the D band centered at 1360cm^{-1}is due to the presence of disorder and variation in the size of graphitic domains, while the G-band centered at 1580 cm^{-1}is due to carbon-carbon vibration in the aromatic rings. The deconvolution is shown in the example in Figure 3.The G-band is associated to C=N bond originated from the replacement of N atoms at carbon sites in C=C sp^2 bond [14].There is an increase in intensity of the G-band and thus the corresponding C=N bond with the increase in rf power up to 80 W. With further increase in rf power above 80 W,the intensity of G band begins to decrease as shown in Table 2 which could be the result of the hydrogen etching as discussed earlier.

Table 2. The variation of intensity D band (I_D) and intensity G band (I_G)

rf power (W)	I_D (a.u.)	I_G (a.u.)
60	130.42317	273.06389
70	161.02497	341.67020
80	276.23464	617.15321
90	148.43525	308.60477
100	67.59846	155.26008

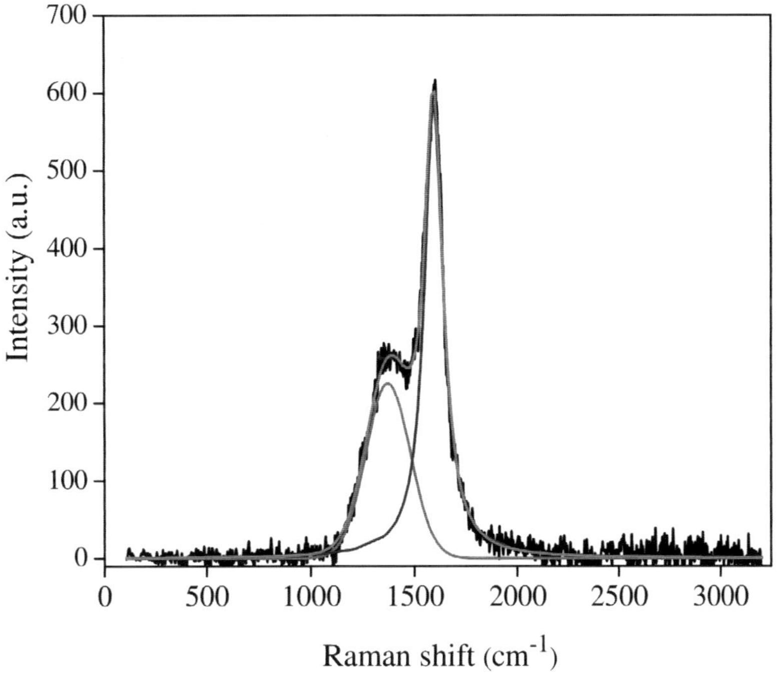

Figure 3. Raman spectra deconvoluted Gaussian and Lorentz fitting

4. Conclusion

a-CN_xfilms were succesfully deposited using a home-built rf-PECVD from the discharge of CH_4 and N_2 gas mixture. The variation of chemical bondingof these a-CN_x films as a function of rf powerwas studied. The deposition rate of the films was influenced by the applied rf power which was associated to the competing effects of dissociation process and surface etching. FTIR spectrum reveals three main peaks at ~1360, ~1570, and ~2150 cm^{-1} correspondingto the D-peak, G-peak and C≡N triple bond, respectively. There are a little changes in the position of the peak when the rf power was changed The results suggest that variation in rf power give effect to the structural properties of the a-CN_x thin films owing to the change in the content of C=N and C≡N bonds in these films.Hence, this will determine the potential of a-CN_x films as a humidity sensor. The optimum rf power that gave high C≡N content is 80 W based on FTIR analysis. In the near future, test will be carried out to confirm if the a-CN_x thin films prepared at 80 W could be used as humidity sensor.

Acknowledgement
The authors gratefully acknowledge the support in the form of a Fundamental Research Grant Scheme FRGS/1/2011/SG/UKM/02/20 by Ministry of Higher Education, Malaysia.

References
[1] A.Y. Liu, M.L. Cohen. 1989. Prediction of new low-compressibility materials. Science **245**: 841.
[2] Y. Aoi, K. Ono, E. Kamijo. *Journal of Applied Physics*. **86**, 2318 (1999)
[3] Ji Gong Lee, Sung Pil Lee. *Sensors and Actuators B*. **108**, 450-454 (2005)
[4] I. Banerjee, Neelam Kumari, Ashis K. Singh, Mukesh Kumar, Pinaki Laha, A.B. Panda, A. K. Pabi, P. K. Barhai, S. K. Mahapatra. *Thin Solid Films*.**518**, 7240-7244 (2010)
[5] M. Balaceanu, E. Grigore, F.. Truica-Marasescu, D. Pantelica, F. Negoita, G. Pavelescu, F. Ionescu. *Nuclear Instruments and Methods in Physics Research B* **161-163**, 1002-1006 (2000)
[6] Wai-Chung Chan, Man-Keung Fung, Kai-Ho Lai, Igor Bello, Shuit-Tong Lee, Chun-Sing Lee. *Journal of Non-Crystalline Solids*. **254**, 180-185 (1999).
[7] Stephen Muhl, Adriana Gaona-Couto, Juan Manuel Mendez, Sandra Rodil, Gonzalo Gonzalez, Alexander Merkulov, Rene Asomoza. *Thin Solid Films*. **308-309**, 228-232 (1997).
[8] Maisara Othman, Richard Ritikos, Noor Hamizah Khanis, Nur Maisarah Abdul Rashid, Saadah Abdul Rahman, Siti Meriam Ab Gani, Muhamad Rasat Muhamad. *Thin Solid Films*.**519**, 4981-4986 (2011).
[9] L. Valentini, J. M. Kenny, Y. Gerbig, A. Savan, H. Haefke, L. Lozzi, S. Santucci. *Thin Solid Films*. **398-399,** 124 (2001).
[10] C. Y. Hsu, F. C. N. Hong. *Diamond and Related Materials*.**8**, 1315 (1999).
[11] A. Crunteanu, M. Charbonnier, M. Romand, F. Vasiliu, D. Panelica, F. Negoita, R. Alexandrescu Surf. Coat. Technol., **125**, 301 (2000).
[12] A. C. Ferrari, S. E. Rodil, J. Robertson. *Physical Review B*. **67**, 155306 (2003).
[13] A. C. Ferrari, S. E. Rodil, J. Robertson. *Physical Review B*. **61**, 14095 (2000).
[14] Sang Hoon Kim, Cheol Min Choi, Kil Mok Lee, Yoon-BongHahn. *Synthetic Metals*. **160**,2442-2446 (2010).
[15] Myung, S. H., Park, S. Y., Hong, B., Han, G. J. dan Shaginyan, R. L. *Thin Solid Films*. **506** 87-91 (2006)

Biodesel Production from *Pseudomonas Fluorescens* Lp1 Lipase Immobilized on Amino-silane Modified Super Paramagnetic Fe_3O_4 Nanoparticles

S Kanimozhi[1,a] and K Perinbam[2]

Department of Biotechnology, Sathyabama University, Jeppiaar Nagar, Old Mamallapuram Road, Chennai - 600 119, Tamil Nadu, (INDIA)
Faculty of of Plant Biology and Biotechnology, Nandanam Arts College (Men), Chennai 600 035 - 600 Tamil Nadu, (INDIA)

Email: skanimo@gmail.com, drperinbam73@gmail.com

Abstract. An extracellular lipase from *Pseudomonas fluorescens* Lp1 isolated from oil contaminated soil was immobilized onto amino silane modified superparamagnetic Fe_3O_4 nanoparticles. The magnetic nanoparticles, magnetite was synthesized chemically by co-precipitation and characterized by Scanning Electron Microscopy (SEM), Fourier Transformed Infrared Spectroscopy (FT-IR) and Powder X-ray diffraction studies (XRD). The structure of the synthesized magnetic nanoparticles was uniform, spherical and the size was determined around 31 nm by powder XRD. The biodiesel production mixture was prepared by addition of waste cooking oil, lipase immobilized magnetite and methanol. The transesterified products were analyzed by Gas Liquid chromatography-Mass spectroscopy (GC-MS). The methyl esters such as Oxiraneundecanoic acid, 3-pentyl-methyl ester, Hexadecanoic acid, methyl ester and 10-Octadecenoic acid, methyl ester were obtained. The study experimentally proved the use of amino silane modified superparamagnetic Fe_3O_4 nanoparticles in biodiesel production from waste cooking oil.

1. Introduction

Biodesel production has received considerable attention in the recent past as a biodegradable and nonpolluting fuel [1]Vegetable oil and animal fat (m) ethyl esters, commonly referred to as ''biodiesel,'' are prominent candidates as alternative Diesel fuels [2]. Biodiesel can be processed from any type of vegetable oils and animal fats: (1) food grade vegetable oils, such as soybean, canola, palm, sunflower and peanut; (2) animal fats, such as lard, tallow, chicken fat and fish oils and (3) used cooking oils from restaurants. Alternative fuels of petroleum solve many of the current social problems and concerns, from air pollution and global warming to other environmental improvements and sustainability issues [3, 4]. The used cooking oils are used as raw material, adaption of continuous transesterification process and recovery of high quality glycerol from biodiesel by-product (glycerol) are primary options to be considered to lower the cost of biodiesel [5].

Lipases (triacylglycerol acyl ester hydrolases; EC 3.1.1.3) are biocatalysts that hydrolyse long chain triglycerides at the water/oil interphase to yield free fatty acids, monoglycerides, diglycerides and glycerols [6].Lipases find promising applications in organic chemical processing, detergent formulations, synthesis of biosurfactants, the oleochemical industry, the dairy industry, the agrochemical industry, paper manufacture, nutrition, cosmetics, and pharmaceutical processing. Development of lipase-based technologies for the synthesis of novel compounds is rapidly expanding the uses of these enzymes [6, 7]. Lipases are widely distributed in nature and found in many species of plants, animals, bacteria, yeast and fungi. The *Pseudomonas* lipases constitute a major group; they have been reported from *P. aeruginosa*; *P. fluorescens, P. glumae* and other *Pseudomonas* sp. [8, 9].

Magnetic nanoparticles offer abundant attractive possibilities in biotechnology. The use of magnetic nanoparticles offers many advantages due to their unique size and physical properties [10]. Besides, Magnetic nanoparticles exhibits the superparamagnetism, not keeping magnetized after the action of magnetic field, provides the highest advantage of reducing risk of particle aggregation [11]. Among various magnetic nanoparticles, Fe_3O_4 nanoparticles have been considered as suitable for biological applications due to super paramagnetic behavior and low toxicity [12]. It is a promising alternative which provide outstanding support materials for the enzymes immobilization because of its striking characteristics, such as large surface area, mobility and high mass transference [13]. Immobilization provides convenient handling of the enzyme, provides for its facile separation from the product, thereby minimizing or eliminating protein contamination of the product. Immobilization also facilitates the efficient recovery and reuse of costly enzymes with enhanced stability under storage and operational conditions [14]. In the present study the lipase from *Pseudomonas fluorescens* Lp1 was immobilized on to the aminosilane modified super paramagnetic Fe_3O_4 nanoparticles and characterized .The biodiesel production was carried out and analyzed by GL-MS.

2. Materials and Method

2.1. Lipase production and purification

Extracellular lipase was obtained from *Pseudomonas fluorescens* Lp1 by submerged production in the medium containing olive oil 1% v/v (emulsified with gum acacia- 0.5%) as the carbon source and inducer of lipase,peptone-5 g/l; yeast extract-5 g/l; NaCl-0.5 g/l; $CaCl_2$-0.05 g/l; pH-8.0. The fermentation was carried out in 100 ml of sterile production broth seeded with 3% inoculum and incubated at 40°C and 150 rpm agitation. Crude lipase was obtained by centrifuging the culture broth 10,000 *g*, 10 min at 4 °C. The lipase purification was carried out by ammonium sulphate precipitation followed by dialysis against 0.05 M Sodium phosphate buffer solution (pH 7.2). Dialyzed lipase was further purified by Sephadex G -100 previously equilibrated with 0.1M Tris- HCl (pH 7.0). Extra cellular lipase activity was estimated by photometric assay [15].

2.2. Synthesis of magnetite nanoparticles, amino functionalization and lipase immobilization

Fe_3O_4 nanoparticles (Magnetite) was prepared through chemical co precipitation method by mixing $FeCl_2$ (0.2 M) and $FeCl_3$ (0.3 M),stearic acid and NaOH (4M) and stirred vigorously. A black precipitate obtained was filtrated, washed and dried. Surface modification of magnetite was done by mixing magnetite, 3-(2-aminoethylamino propyl trimethoxysilane (APTS), methanol and sodium fluoride (NaF) and stirred vigorously for 10 min. To this tetraethyl orthosilicate (TEOS) was added drop wise slowly into the flask and stirred vigorously at room temperature for 24 hrs. The black precipitate of amino functionalized magnetic nanoparticles (AFMNPs) obtained was collected by magnetic decantation, washed thrice with ethanol and water and dried.

The immobilization of lipase was performed using glutaraldehyde as a cross linking agent. To the magnetite, glutaldehyde was added and stirred at room temperature. Then, *Pseudomonas fluorescens* Lp1 lipase in phosphate buffer solution was added and stirred for several minutes at room temperature. The immobilized lipase was separated by centrifugation and washed with phosphate buffer and dried. The immobilization efficiency (η %) was calculated according to the following equation

$$\eta\ (\%) = \frac{Us}{Uo} \times 100$$

where Us is the total enzyme activity recovered in the support and Uo is the enzyme units offered for immobilization.

2.3. Biodesel Production

The biodiesel production mixture was prepared by the addition of waste cooking oil and immobilized magnetite and incubated at 40°C for 48 hours with 150 rpm agitation. For the transesterification process, methanol was added thrice at different interval of time. After 48 hrs, content was mixed vigorously with hexane and centrifuged. The upper layer was subjected to Gas chromatogrphy (GC-MS) analysis for fatty acid methyl esters.

2.4. Characterization of nanoparticles

XRD patterns of magnetic nanoparticles obtained with X-ray Powder Diffractometer model Bruker AXS D8 Advance with Cu (Kαλ=1.5406 Å) radiation, scanning range from 10°–90°.The morphology of the magnetic nanoparticles was revealed by scanning electron microscopy SEM JEOL Model JSM - 6390LV. FT-IR spectra of magnetic nanoparticles, was obtained using Nicolet FT-IR Avatar 370 with the wavenumber ranged from 4,000 cm^{-1} to 400 cm-1 using KBr beam splitter.

3. Materials and Method

Extracellular lipase from *Pseudomonas fluorescens* Lp1 was purified to 18 fold with the specific activity of 1016.78 U/mg by sequential purification methods as described. *Pseudomonas fluorescens* MTCC 2421 lipase was purified 184.37-fold with a specific activity of 424.04 U/mg after anion exchange and gel exclusion chromatography with an apparent molecular mass of 65.3 kDa [16]. Lipase from *Candida rugosa* was purified by ammonium sulphate precipitation, dialysis, ultra-filtration and gel filtration using Sephadex-200 to a 43-fold purification and 64.35 mg/ml specific activity [17]. The magnetite as support was selected because it has some advantages: (i) higher specific surface (ii) lower mass transfer resistance and less fouling, and (iii) the selective separation of immobilized enzymes from a reaction mixture by the application of a magnetic field [18]. Magnetite is one of the famous magnetic materials in common use. As a result of strong magnetic property and low toxicity, its applications in biotechnology and medicine have gained significant attention [19].

In the present study the synthesized magnetite was analyzed by XRD (Fig.1) showed the characteristic spinel structure of Fe_3O_4 confirmed the synthesis of magnetic nanoparticles. The average size of the magnetic nanoparticles was calculated to be about 31nm by Debye–Scherrer equation using the (311) peak, and there are no changes after the consequent lipase adsorption .The results were coinciding the results of [20].

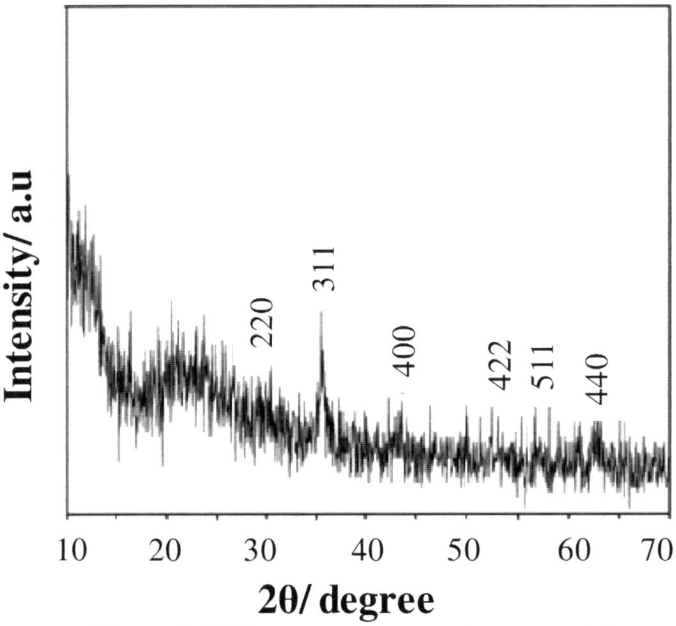

Figure 1. XRD patterns of magnetic nanoparticles

The SEM analysis revealed that magnetite has uniform spherical structures (Fig.2) and was well dispersed. The grain size of the agglomerated spherical magnetite was found to be around 100 nm.

Figure 2. SEM image of magnetic nanoparticle

FT-IR vibrational analysis (Fig. 3) revealed the structure of Fe_3O_4.The vibrational peak at 582 cm^{-1} in the naked nanoparicles corresponds to the Fe–O bond, O-H stretching vibration near 3419 $cm^{-1}$and O–H deformed vibration near 1563 (cm^{-1} [21, 22]. The efficiency of immobilization of *Pseudomonas fluorescens* Lp1 lipase on to magnetic nanoparticles was found to be 58%. The binding efficiency of lipase was 90% and the corresponding activity recovery was 70% [23].

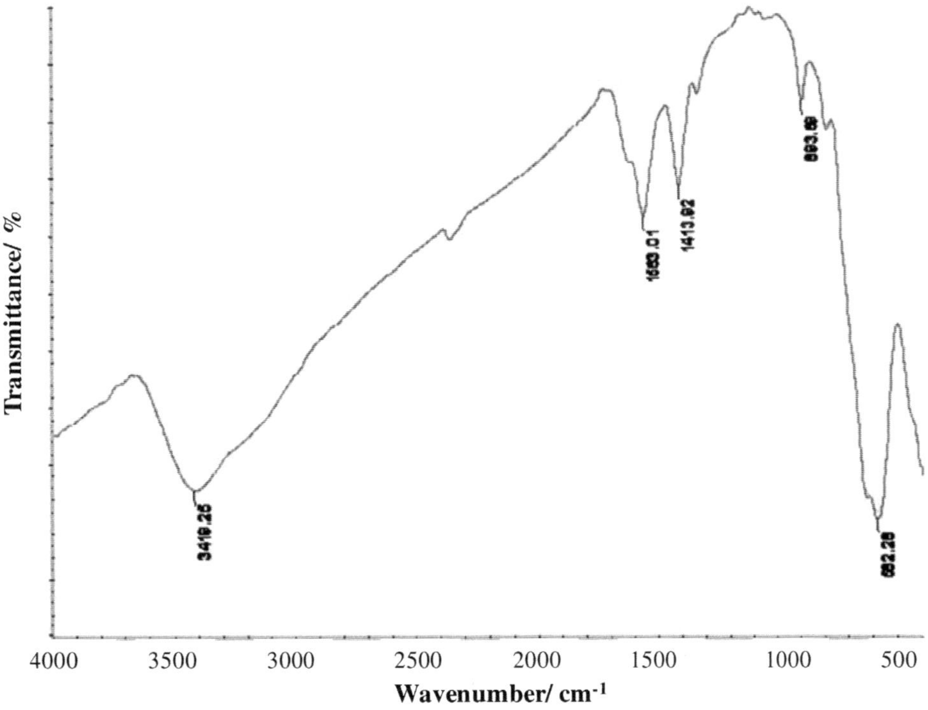

Figure. 3 FT-IR spectra of Magnetite

The transesterification of waste cooking oil into methyl esters was analyzed by GC-MS. The results revealed that the majority of the oil was converted to biodiesel and GC-MS analysis demonstrated the synthesis of Fatty acid methyl esters (FAMEs) such as Oxiraneundecanoic acid, 3-pentyl-methyl ester, Hexadecanoic acid, methyl ester and 10-Octadecenoic acid, methyl ester were obtained (Fig.4). The result was correlated with biodiesel produced by *Photobacterium lipolyticum* lipase [17].FAMEs were produced by the enzymatic transesterification of soyabean oil using immobilized *Pseudomonas cepacia* [24].Waste cooking oil used as a substrate for biodiesel synthesis using immobilized lipase from *Penicillium expansum* [25]. By lipase transesterification process palm oil yielded biodiesel [26].

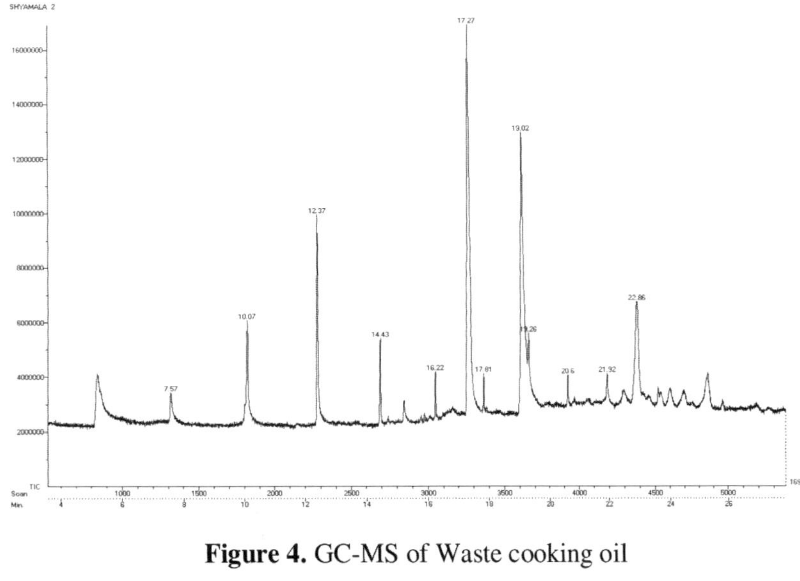

Figure 4. GC-MS of Waste cooking oil

References

[1] S. V. Ranganathan, S.L Narasimhan, K .Muthukumar, An overview of enzymatic production of biodiesel, Biores.Technol.99(2008) 3975–3981.

[2] G. Vicente, M. Martinez, J. Aracil, Integrated biodiesel production: a comparison of different homogeneous catalysts systems, Biores.Technol 92(2004)297–305.

[3] H.L MacLeana, L.B Laveb, Evaluating automobile fuel/propulsion system technologies, Prog. Energy. Combus. Sci. 29 (2003)1–69.

[4] A.S. Ramadhas, S. Jayaraj, C. Muraleedharan, Use of vegetable oils as IC engine fuels-A review, Renew. Energy. 29 (2004)727–42.

[5] F. Ma and M.A. Hanna, Biodiesel production: a review, Biores.Technol.70 (1999)1-15.

[6] R.K. Saxena, P.K. Ghosh, R. Gupta, W.S. Davidson, S. Bradoo, R. Gulati, Microbial lipases: Potential biocatalysts for the future industry, Curr. Sci. 77 (1999)101-115.

[7] A Liese, K Seelbach, C Wandrey, editors. Industrial biotransformations Weinheim: Wiley-VCH, 2000.

[8] E. Kageyama,M. Hirata,T. Nibira, and Y. Yumada, Purificaton, characterizaton and molecular cloning of loctonzing lipase from *Pseudomonas* species, J. Biol. Chem. 266(1991) 18135-18140.

[9] H. Kumura, S. Hirose, H. Sakurai, K. Mikawa, F. Tomita, and K. Shimazaki, Molecular cloning and analysis of a lipase gene from *Pseudomona fluorescen,* BioSci. Biotechnol. Biochem. 62 (1998) 2233-2235.

[10] M.A. El-Sayed, Some Interesting Properties of Metals Confined in Time and Nanometer Space of Different Shapes, Acc. Chem. Res. 34(2001) 257-264.

[11] T.K. Indira and P.K. Lakshmi, Magnetic Nanoparticles – A Review, Int. J. Pharma. Sci. Nanotechnol. 3(2010)1035-1042.

[12] A.K. Gupta and M.Gupta, Synthesis and surface engineering of iron oxide nanoparticles for biomedical applications, Biomater. 26 (2005) 3995–4021

[13] A.H. Lu, W. C. Li, A. Kiefer, W. Schmidt, E. Bill, G. Fink, and F. Schuth, Fabrication of Magnetically Separable Mesostructured Silica with an Open Pore System, J. Am. Chem. Soc. 126(2004) 8616 -8617.

[14] R.A. Sheldona, Enzyme Immobilization: The Quest for Optimum Performance, Adv. Synth. Catal .349(2007) 1289.

[15] U.K. Winkler, M. Stuckmann, Glycogen, hyaluronate and some other polysaccharides greatly enhance the formation of exolipase by *Serratia marcescen,* J. Bacteriol.38 (1979) 663–670.

[16] K.R. Chakraborty, Purification and Biochemical Characterization of an Extracellular Lipase from *Pseudomonas fluorescens* MTCC, J. Agric. Food. Chem. 57 (2009) 3859–3866.

[17] S. Benjamin and A.Pandey, Isolation and Characterization of Three Distinct Forms of Lipases from *Candida rugosa* produced in solid state fermentation, Braz. Arch. Biol. Technol.43(2001)453-460.

[18] S.H Huang, M.H. Liao, Direct Binding and Characterization of Lipase onto Magnetic Nanoparticles, Biotechnol. Prog. 19 (2003)1095-1100.

[19] A.Curtis and C. Wilkinson, Nanotechniques and approaches in biotechnology, Trends Biotechnol. 19 (2001) 97-101.

[20] J .Dussan, O. H. Giraldo and C. A. Cardona, Application of magnetic nanostructures inbiotechnological processes: Biodiesel production using lipaseimmobilized on magnetic carriers, paper presented at the *European Congress of Chemical Engineering (ECCE-6),* Copenhagen, 2007.

[21] B. Hu, J. Pan, H.L. Yu, J.W. Liu, J.H. Xu, Immobilization of *Serratia marcescens* lipase onto amino-functionalized magnetic nanoparticles for repeated use in enzymatic synthesis of Diltiazem intermediate, Process. Biochem. 44(2009)1019–1024.

[22] L.H. Andrade, L.P. Rebeloa, C.G.C.M Nettob, H.E. Toma, Kinetic resolution of a drug precursor by *Burkholderia cepacia* lipase immobilized by different methodologies on

superparamagnetic nanoparticles, J. Mol. Catal B: Enzym. 66(2010) 55–62.

[23] K.H. Mak, C.Y.Yu, I.C.Kuan and S.L. Lee, Immobilization of *Pseudomonas cepacia* lipase onto magnetic nanoparticles for biodiesel production, Sci. Technol. Vision. 5 (2009) 19-23.

[24] H . Noureddini, X.Gao and R.S. Philkana, 2005. Immobilized *Pseudomonas cepacia* lipase for biodiesel fuel production from soybean oil, Biores.Technol.96 (2005)769-777.

[25] N.W. Li, M.H. Zong and H.Wu, Highly efficient transformation of waste oil to biodiesel by immobilized lipase from *Penicillium expansum*, Process. Biochem. 44(2009)685-688.

[26] L. Zhang, S. Sun, Z. Xin, B. Sheng, and Q. Liu, Synthesis and component confirmation of biodiesel from palm oil and dimethyl carbonate catalyzed by immobilized-lipase in solvent-free system, Fuel. 89(2010)3960-3965.

Deposition of Au/TiO_2 Nanocomposite on ITO Surface by Seed-Mediated Liquid Phase Deposition Method

Suratun Nafisah[1], Arfian Oktaviandi[2], Akrajas Ali Umar[1], Muhammad Mat Salleh[1], Aamna Balouch[1,3], and Siti Khatijah Md Saad[1,4]
[1] Institute of Microengineering and Nanoelectronics (IMEN), Universiti Kebangsaan Malaysia, 43600 UKM Bangi, Selangor, Malaysia
[2] Department of Chemistry, Faculty of Mathematics and Natural Science, Padjajaran University, Bandung, Indonesia
[3] National Centre of Excellence in Analytical Chemistry, University of Sindh, Jamshoro, Pakistan
[4] College of Engineering, Universiti Tenaga Nasional, Km 7, Jalan Kajang-Puchong, 43009 Kajang, Selangor, Malaysia

E-mail: akrajas@ukm.my

Abstract. An efficient, simple and new procedure has been performed to synthesis Au/TiO_2 nanocomposite thin film on solid surface of ITO by seed-mediated of liquid phase deposition (LPD) method. The deposition of Au seed was applied by our previously reported seed-mediated growth procedure. The solution was prepared by mixing $HAuCl_4$ and $(NH_4)_2TiF_6$ successfully deposited on to ITO substrate containing Au seed. After one hour well adhered film was obtained by mentioned approach of synthesis and nonocomposite with spherical rod like networks has been successfully grown. The resulting nanocomposites were confirmed by ultraviolet-visible (Uv-Vis) absorption spectroscopy. Further characterization and morphology was checked by field emission scanning electron microscopy (FESEM), energy dispersive X-ray spectroscopy (EDX) and X-ray diffraction (XRD).

1. Introduction

Nanocomposite of metal and metal oxide have been intensively studied due their good optical, photocatalytic, electrical, magnetic and mechanical properties [1]. These metal and metal oxide nanocomposites are widely applied in gas sensor [2], catalyst [3],[4] and anti bacterial [5]. Nanocomposite of metal and metal oxide have been intensively studied due their good optical, phtocatalytic, electrical, magnetic and mechanical properties such as..Many studies have been reported in the literatures on the effect of doping metal like Au, Pd, Ag and Pt on the TiO_2 [5],[6]. TiO_2 is one of the most attractive semiconductor materials due to the lowest cost, non-toxicity, biologically and chemically stable properties [7]. The presence of metal in the TiO_2 facilitates electron transfer from photoexcited semiconductor to the surroundings and decreases recombination rate between the electrons-holes so that good for application in photoelectrochemical and photocatalytic [8]. In recent years, various methods has been used in synthesis of composite TiO_2-metal nanoparticles such as photoreduction [5], magnetron sputtering [1, 9], dip coating [10] and sol gel method [6, 11]. Owing to

[1] akrajas@ukm.my

the simplicity of the process, catalytic and optical properties, the presence method could be potential approach for producing Au/TiO_2 nanocomposite for application in photocatalysis, optical sensing and non-linear optical application.

Liquid phase deposition method (LPD) is widely used in the deposition of TiO_2 on solid surface. In particular, LPD is promising technique for direct preparation of TiO_2 thin film on to solid substrate using boric acid as scavenger. However, metals such as -Al and Fe has also been reported to be used as scavenger for fluoride ion[12]. In this study, we aims to develop a simple and efficient method to synthesis Au/TiO_2 nanocomposite thin film by seed-mediated grow procedure using LPD technique. The obtained composite thin films were characterized by using Uv-Vis absorption spectroscopy, field emission scanning electron microscopy (FESEM), energy dispersive X-ray spectroscopy (EDX) and X-ray diffraction (XRD).

2. Experimental

2.1. Materials

Analytical grade chemicals in this work were used without further purification. $HAuCl_4$ and $(NH_4)_2TiF_6$ were purchased from Sigma Aldrich while trisodium citrate was purchased from Wako Chemical and $NaBH_4$ was from Fluka. The ITO substrate was purchased from CBC Ing. Co. Ltd., Japan. The ITO substrate was cleaned by distilled water and sonicated in acetone followed by 2-propanol for 30 minutes according to the standard procedure. The depositions of Au/TiO_2 nanocomposite on to ITO substrate practically involved two steps, namely seeding process and growth process. The ITO substrate immersed into Au seed solution as reported in our previous work [13].The growth process was carried out by immersing the sample in standard growth solution which containing 5 ml of $HAuCl_4$ 1 mM and 5 ml of $(NH_4)_2TiF_6$ 50 mM. The reaction was performed in bath deposition to keep temperature unchanged during the reaction. The sample was then washed several times by distilled water and dried under nitrogen flow.

Effect of concentration $(NH_4)_2TiF_6$ was performed by preparing four different of variation concentration namely 50, 100, 200 and 250 mM with the fixed concentration of $HAuCl_4$ 1 mM at temperature 90°C for growth time 1 hour. The investigation of formation Au/TiO_2 nanocomposite was then studied by varying the growth time of the synthesized nanocomposite for four different variations namely 15 min, 30 min, 1 hour and 3 hours.

2.2. Au/TiO_2 nanocomposite characterization

The optical absorption of the sample for fixed concentration, variation concentration of $(NH_4)_2TiF_6$, and growth time were studied using Uv-Vis spectrophotometer (Lambda 900 PERKIN ELMER) with the wavelength range 300-900 nm. Morphology of nanocomposite for fixed concentration and variation of growth for synthesized nanocomposite were also studied using FESEM technique (ZEISS SUPRA 55VP). The elemental composition of nanocomposite was studied using energy X-ray dispersive (EDX) method (Oxford instruments INCA) at 30 kV using CuKα radiation (λ = 1.541 Å). X-ray diffraction spectrum was obtained from X-ray diffractometer (BRUKER D8 Advance) with the diffraction angle of 0-60°.

3. Results and Discussions

Au/TiO_2 has been successfully deposited on the surface using the presence approach. As reported by Gutierrez et al. who successly produced TiO_2 using trivalent Al(III) and Fe(III), in this work we tried to use trivalent Au(III) as scavenger ion in the absence of boric acid. The chemical reaction can be expressed as follows:

$$TiF_6^{2-} + 2H_2O \leftrightarrow TiO_2 + 4HF + 2F^- \quad (1)$$

$$Au^{3+} + 4HF \leftrightarrow [AuF_4]^- + 4H^+ \quad (2)$$

In typical process, the Au/TiO_2 nanocomposite formed in standard growth solution at temperature of 90ºC for 1 hour. Figure 1a shows a typical FESEM image of Au/TiO_2 nanocomposite on the ITO surface via seed-mediated LPD method. From the FESEM image, we can see that the shape Au/TiO2 nanocomposites was more towards spherical rod-like with a diameter of 250 nm. These spherical rod-like particles however tends to agglomerate with one another forming a micron size network. In high magnification (Figure 1b), it can be clearly seen that the roughness among the surface of nanoparticles, indicated that the nanoparticles was not only pure Au, but contain dopant (TiO_2). The EDX study was performed on the sample to confirm the existence of TiO_2. The EDX spectrum below (figure 1c) showed Au, Ti and O peak. This confirmed the formation of Au/TiO_2 nanocomposite thin film. However as can be seen from the spectrum, Ti peak was not as high as Au indicates that the composition of nanocomposite was dominated by Au.

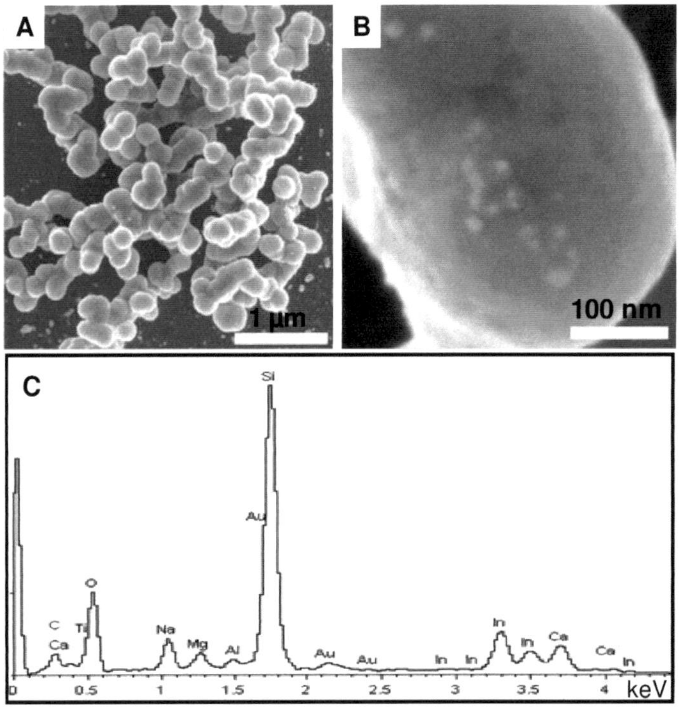

Figure 1. FESEM image of Au/TiO2 nanocomposite deposite on to ITO surface (a) at high magnification (b) and it's EDX spectrum (c)

Figure 2 showed the XRD spectrum of Au/TiO_2 nanocomposite. The peaks observed match to the JCPDS of Au nanoparticles (file number 04-0784) where the peaks related at (111) and (200) corresponding with the diffraction angle 38.185° and 44.393°. However from the XRD spectrum the presence of TiO_2 was not seen. This might due to the fact that the presence of TiO_2 in the Au/TiO_2 nanocomposite was very small. As also can be seen in figure 1, the distribution of investigated nanocomposite was not dense enough.

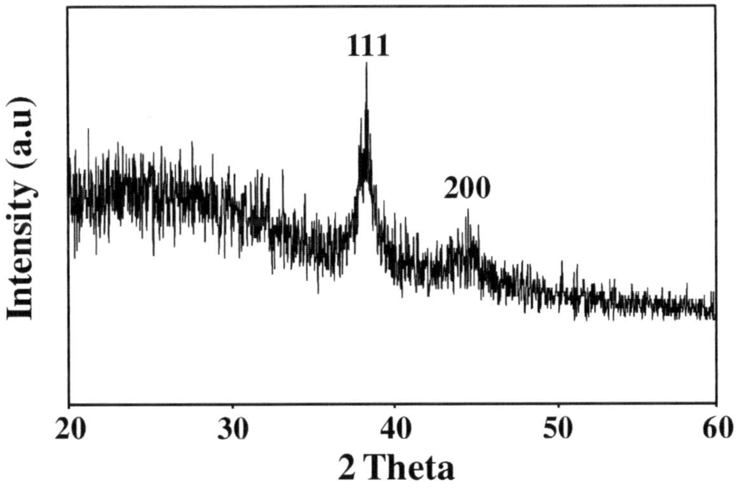

Figure 2. XRD spectrum of Au/TiO_2 nanocomposite deposited on solid surface.

UV-Vis absorption in figure 3 showed that the absorption of Au/TiO_2 nanocomposite centering at wavelength 570 nm. This absorption is slightly difference with the absorption of Au nanoparticles which is normally exhibit plasmonic absorption around wavelength 520 – 530 nm [13]. There are two possible reasons that may be responsible for that bathocromic shift. First, the presence of TiO_2 in the gold nanoparticles (Au/TiO_2 nanocomposite) as it was previously reported that the absorption peak for thin film TiO_2 was found to be at 350 nm[14]. Second, it was caused by the link of spherical nanocomposite that was contributed to the red shifted absorption spectrum.

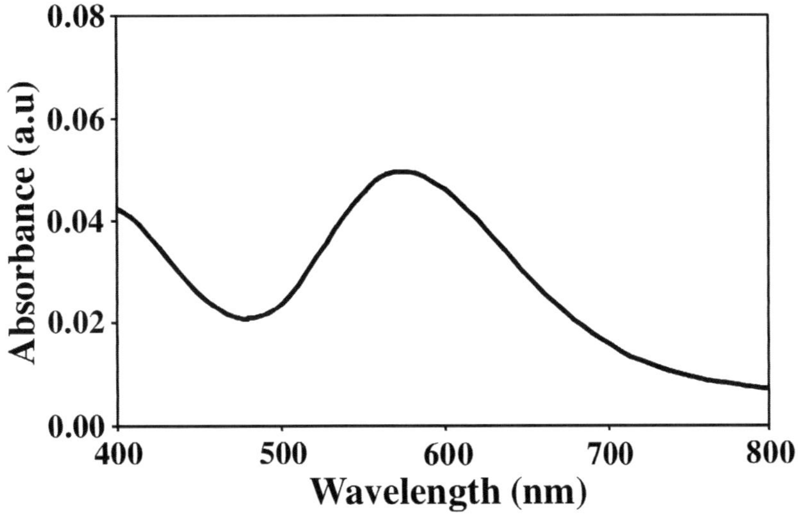

Figure 3. UV-Vis absorption of Au/TiO_2 nanocomposite.

The works was carried out further to optimize the effect of concentration $(NH_4)_2TiF_6$ used on the formation of nanocomposite. Four different concentration of $(NH_4)_2TiF_6$ namely 50, 100, 200 and 250 mM were used in this study. Here, it can be seen when the molarity of $(NH_4)_2TiF_6$ used increased, the optical absorption of the prepared samples seems to decrease. This might due to the formation of TiO_2 complex has hindered the performance of Au in the nanocomposite as it is known better that Au has excellence optical performance contributed by its plasmonic effect. From the spectra profile (figure 4) of UV-Vis absorption, it can be conclude that the optimum $(NH_4)_2TiF_6$ concentration is 50 mM gave the best results as it can be seen that the sample growth within this concentration has the highest optical absorption compared to the other sample.

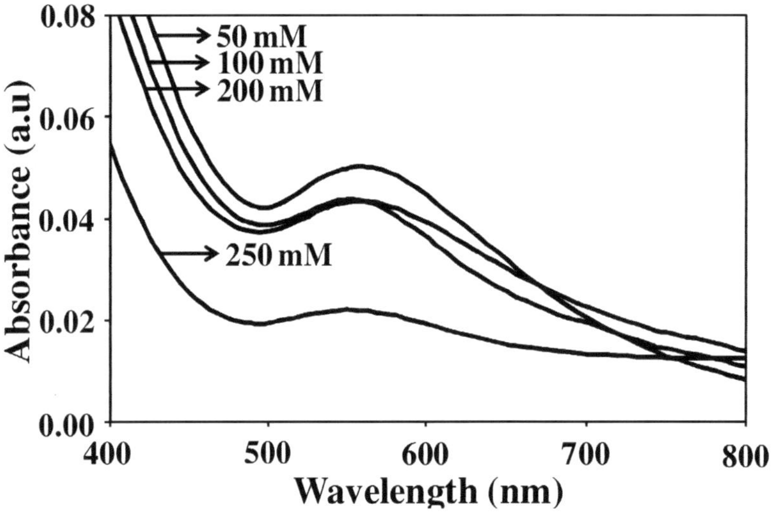

Figure 4. UV-Vis absorption of Au/TiO_2 nanocomposite growth in variation concentrations of $(NH_4)_2TiF_6$.

The optical properties of the Au/TiO_2 nanocomposite synthesis at temperature of 90ºC with variation of growth time were further studied by UV-Vis measurement. Four variation of time growth namely 15 minutes, 30 minutes, 1 hour and 3 hours were chosen. By immersing the attached Au seed substrate in the solution containing $HAuCl_4$ 1 mM and $(NH_4)TiF_6$ 50 mM. The color of the thin film changed from colorless to yellow within 3 hours of the reaction. Figure 5 shows the time-dependent spectral response obtained during the deposition of Au/TiO2 nanocomposite. The intensity of the peak increases monotonically with time, indicating the increase in the amount of the Au/TiO_2 nanocomposite product. The spectra also recorded that the absorption peak broadens and shifts towards longer wavelength from 540 to 700 nm with the increase of growth time. This broadening and red shift are related to particle size as confirmed from FESEM image (figure 6).

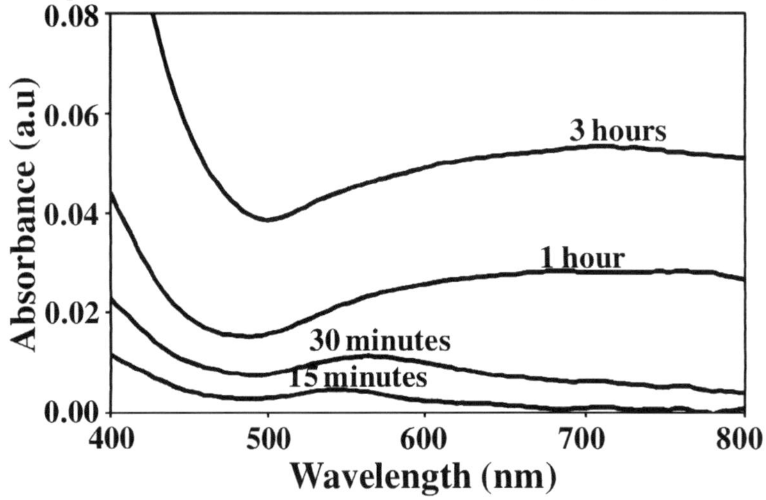

Figure 5. UV-Vis absorption of Au/TiO_2 nanocomposite with variations of growth time.

The morphology of Au/TiO_2 nanocomposite growth at different time intervals shows was presented by the FESEM image shown below. The shape of the nanocomposite formed was more into spherical shape and was bound to agglomerate towards each other. As the growth time increased, the linked agglomeration of the spherical particle of nanocomposite was elongated. This is due to the fact of the increased deposition time give rise to the formation of this nanocomposite. Thus, it is clear why when the growth time increased, the nanocomposite in micron size network increased. From the FESEM images obtained at the magnification of 20 kx, the grain size of the nanocomposite formed were 149, 172, 278 and 371 nm for the growth time 15 minutes, 30 minutes, 1 and 3 hours respectively.

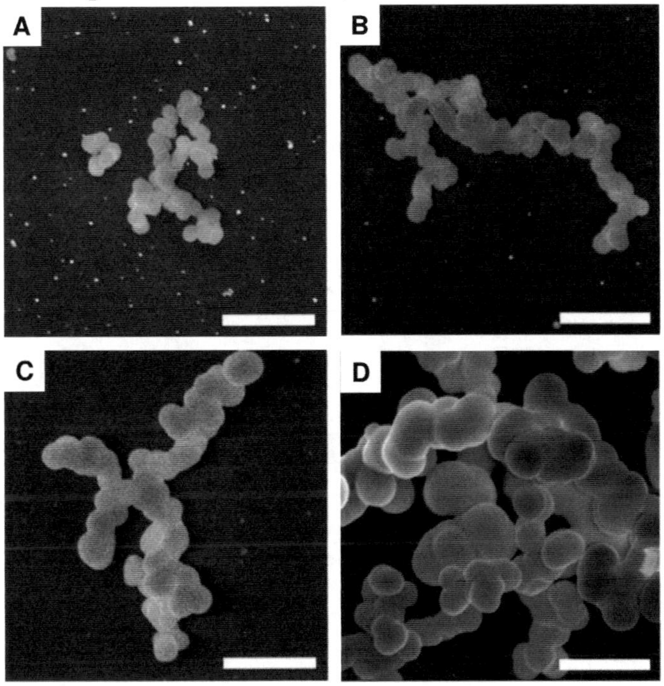

Figure 6. FESEM image of Au/TiO_2 nanocomposite with variation growth time i.e. 15 minutes (A), 30 minutes (B), 1 hour (C) and 3 hour (D) in obtained condition. Scale bars are 1 μm.

4. Conclusion

Au/TiO_2 nanocomposites thin film has been successfully prepared by seed-mediated LPD technique on the surface of ITO substrate. The UV-Vis spectrum shows band of nanocomposite thin films centering at wavelength 540 – 700 nm. FESEM image showed shape of nanocomposites were spherical like nanorods. EDX analysis confirmed the presence of Au, Ti and O atoms in the Au/TiO_2 nanocomposite thin film. From the studies, the optimum growth condition for deposited Au/TiO_2 nanocomposite thin films on the surface by seed-mediated liquid phase deposition for $HAuCl_4$ of 1 mM is $(NH_4)TiF_6$ 50 mM with the temperature of 90°C grown 1 hour.

Acknowledgements

This work was partially supported by the Universiti Kebangsaan Malaysia under research grant UKM-GUP-2011-377 and UKM-GUP-NBT-08-25-086. Author would like to acknowledge Center for Research and Instrumentation Management (CRIM) for the FESEM instrument.

References

[1] Pal, M., T. Sasaki, and N. Koshizaki, 2001. **44**(8–9): p. 1817-1820.
[2] Jia, W., et al. The Journal of Physical Chemistry C, 2009. **113**(37): p. 16402-16407.
[3] Enache, D.I., et al., Science, 2006. **311**(5759): p. 362-365.

[4] Su, L., et al., Electrochemistry Communications, 2009. **11**(11): p. 2199-2202.
[5] Chen, S., et al., Nano Research, 2010. **3**(4): p. 244-255.
[6] Zhao, G., H. Kozuka, and T. Yoko, 1996. **277**(1): p. 147-154.
[7] Hashimoto, K., H. Irie, and A. Fujishima, AAPPS Bulletin, 2007. **17**(6): p. 13.
[8] Subramanian, V., E. Wolf, and P.V. Kamat, The Journal of Physical Chemistry B, 2001. **105**(46): p. 11439-11446.
[9] KIM, S.C., et al., Journal of the Korean Physical Society, 2005. **47**(4): p. 700-704.
[10] Hsiao, Y.C. and Y.H. Tseng, Micro & Nano Letters, IET, 2010. **5**(5): p. 317-320.
[11] Zhao, G., H. Kozuka, and S. Sakka, Journal of Sol-Gel Science and Technology, 1995. **4**(1): p. 37-47.
[12] Gutierrez-Tauste, D., et al., Journal of Materials Chemistry, 2006. **16**(23): p. 2249-2255.
[13] Umar, A.A. and M. Oyama, Applied Surface Science, 2006. **253**(4): p. 2196-2202.
[14] Umar, A.A., et al.,. Applied Surface Science,In Press.

Surface Compositional, Structural and Optical Properties of Nanohybrid TiO_2/ZnO Coated Silicone Surface

M. F. Achoi[1, 2a], M. Z. M. Gambar[1], M. Rusop[2] and S. Abdullah[1, 2b]

[1]Faculty of Applied Sciences, Universiti Teknologi MARA (UiTM), 40450 Shah Alam, Selangor Darul Ehsan, MALAYSIA.
[2]NANO-SciTech Centre, Universiti Teknologi MARA (UiTM), 40450 Shah Alam, Selangor Darul Ehsan, MALAYSIA.

Email: [b]saifolabdullah@yahoo.co.uk, [a]faizlss_choi@yahoo.com

Abstract. This paper reports on preparation and characterizations of nanohybrid TiO_2/ZnO using sol-gel spin-coating techniques. The study involved of modification band-gap energy of Titanium dioxide (TiO_2) at different precursor molecular concentration of Zinc oxide (ZnO). The Field Emission Scanning Electron Microscope (FESEM) for morphological characterization showed the existence of TiO_2 and ZnO particles at 20 nm scale of magnification. The compositional of both particles were then confirmed using Energy Disperse Analyzer X-Ray (EDAX) analysis which indicates the elements of Ti, O and Zn exhibited EDAX spectras at about 0.5 and 1.0 keV while silicone (Si) at 1.75 keV. The X-Ray diffraction (XRD) characterized the structural properties of anatase and rutile for TiO_2, and zincite for ZnO, then crystallite size for both elements. AT optimum molar concentration ratio of nanohybrid TiO_2/ZnO is 1:4 (0.20M), the crystallite size, d, is 23.71 nm with peak of anatase found at diffraction angle of 48.05° and 62.50° while at 33.25° and 55.06° is rutile phase, respectively. We have found the distribution of nanohybrid TiO_2/ZnO particles dispersed homogeneously since the particles were not bonded together with increasing ZnO molar concentration. From Ultra-Violet/Visible-near-infrared spectroscopy (UV-Vis-NiR), we have found the optical band-gap energy of nanohybrid TiO_2/ZnO was increased at 3.38 eV, while the transmittance percentage is 88.67%. This studied was successfully improved the optical band-gap energy of TiO_2 and the results suggests for improvement of optical band-gap energy TiO_2.

1. Introduction.

TiO_2 has shown good optical properties with optical band-gap energy of anatase about 3.1 eV until 3.2 eV while ZnO has optical band-gap energy about 3.37 eV [1, 2]. These two material are suitable to hybrid together not only due to their similar band-gap energy, but they also has a similarities in their properties such as high chemical and thermal stability, high refractive indices and share a high transmittance in the visible region although they transmittance relatively different in UV region [3]. Moreover, both materials have good photocatalytic efficiency. TiO_2 has mostly been used for photocatalyst, photoelectrodes, gas sensors and electrochromic display devices [4] while the attractive application of ZnO used for forming various forms of nanostructures, such as nanorods, nanowires, and nanobelts.

The hybrid of TiO_2/ZnO has been done via many methods such as hydrothermal method [5], magnetron sputtering method [6], and in-situ mixing [7]. In this study, sol-gel spin coating method has been used because this method allow a higher surface activity, low-temperature synthesis and easily control reaction condition besides of safe and low cost method [8]. The purpose in this study was to

Journal of Physics: Conference Series **431** (2013) 012012 doi:10.1088/1742-6596/431/1/012012

improve the optical band-gap and structural quality of TiO_2 with ZnO material grown on Si substrate and further increase the application of this nanohybrid material.

2. Methodology.

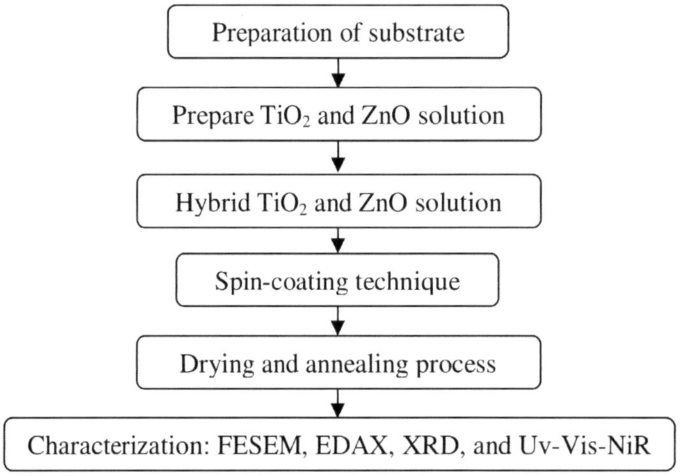

Figure 1. Flow chart process diagram of methodology.

2.1. Material.

The following chemicals were obtained from commercial sources and used as received: Titanium *Tetra* Isobutoxide (TTIB; grade 99.99%, metal basis from Aldrich), Ethanol (EtOH; 99.99% purity, R&M Chemicals), Glacial Acetic Acid (CH_3COOH; J.T. Baker Analyzed; grade 48%), Triton-X 100 (R&M; $C_{34}H_{62}O_{11}$), Zinc Acetate Dihydrate ($Zn(CH_3COO)_2.2H_2O$; 99.5% purity; Merck), Isopropanol ($CH_3CHOHCH_3$; 99.99% purity, Merck), Dietahanolamine (NH $(CH_2CH_2OH)_2$; 99.5% purity; Merck), Distilled Water (DI; H_2O), Acetone (C_3H_6O; ChemAR), Methanol (CH_3OH; 98.99% purity, Merck), Hydrofluoric Acid (48% purity; Merck), Silicon wafer (p-type, thickness 25 μm) and Glass Substrate (Microscope slide borosilicate glass; Sail brand).

2.2. Preparation of the nanohybrid samples.

The Si substrate was cleaned using acetone, methanol and hydrofluoric acid respectively in ultrasonic bath. The glass substrate for deposited TiO_2/ZnO nanohybrid was also used to determine optical band-gap, the cleaning process was involved acetone and methanol solution respectively, and was then cleaned by using distilled water. Lastly, the substrate was dried using nitrogen gas.

The TiO_2 solution was prepared via sol-gel process where TTIB (titanium *tetra* isobutoxide) was used as precursor while ethanol as the solvent. Firstly, TTIB was dissolved in ethanol. Next, glacial acetic acid (GAA), triton-X 100 and distilled water was dropped in the precursor solution. After that, the solution were stirred for 1 hour with applied heating and was then continue stirred without heating with ageing time of 24 hours. The similar method was used to prepare ZnO solution where zinc acetate dehydrate as a precursor was dissolved in isopropanol solution. Next, dietahanolamine (DEA) and distilled water was dropped in the precursor solution. Then, the solution was also stirred for 1 hour with applied heating and was then continues stirred without heating, and leaved 24 hours. After that, the two solutions were directly mixed gradually until finish and was again the solution stirred for 1 hour with heating and were then continue stirred without heat for 24 hours. For spin coating technique, speed of 5000 rpm for 30 second was used to deposit nanohybrid TiO_2/ZnO on Si surface. Next, the sample was dried using microwave oven for 10 minutes at 150°C and was then annealed at 450°C for 90 minutes using furnace.

2.3. Characterization of samples.
After completing the process of producing nanohybrid sample, the characterization techniques were used Field Emission Scanning Electron Microscope (FESEM; Zeiss Supra 40VP) and Energy Disperse Analyzer X-ray (EDAX; 200.0 kV) mapping analysis, applied on the nanohybrid sample. This characterization was done to observe the surface morphology and confirmed the existence element of nanohybrid sample. Then, X-Ray Diffraction (XRD; Siemens D5000) characterization was done to characterize the compositional of crystal structures, peaks and then, determines the crystallite size. The Ultraviolet-Visible Near-infrared spectroscope (UV-Vis-NiR; Varian Cary 5000) characterization was done to determine the changes on optical band-gap energy (E_g) and observe the percentage optical transmittance of nanohybrid.

3. Result and Discussion.

3.1. Structural Properties.
The surface morphology of each sample was characterized using FESEM with an accelerated voltage of 5 kV. Figure 2 shows the FESEM morphologies of pure TiO_2 and pure ZnO, respectively. Figure 2 (a) shows seem smooth surface while Figure 2 (b) shows existence of ZnO particles at 20 nm scale of magnification. Both Figures have small of particle size.

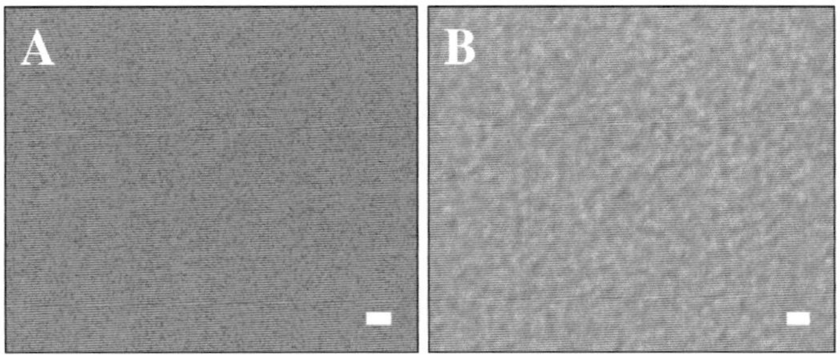

Figure 2. FESEM morphologies of the (a) pure TiO_2 and (b) pure ZnO. Scale bars are 20 nm.

Figure 3 shows the FESEM morphologies of the nanoparticles of TiO_2/ZnO can be seen appeared in each sample. Figure 3 (a) shows low concentration of Zn, the particles formed a small agglomerates and the size of these agglomerated particles increased with an increasing in ZnO sol concentration. As can be observe in Figure 3 (d), the agglomeration of particles was clearly observed. This is due to the particles dispersed homogeneously and exhibited near and close each other of particle size distribution [9]. Since the formation agglomerated particles is due to the electrostatic interaction between the solute particles, hence the increase in ZnO sol concentration, will also increase the amount of solute. While, Figure 3 (e) indicates at higher concentration of sol ZnO, the particles are less likely to form agglomeration, and these particles are rather form only particles or small agglomeration [9].

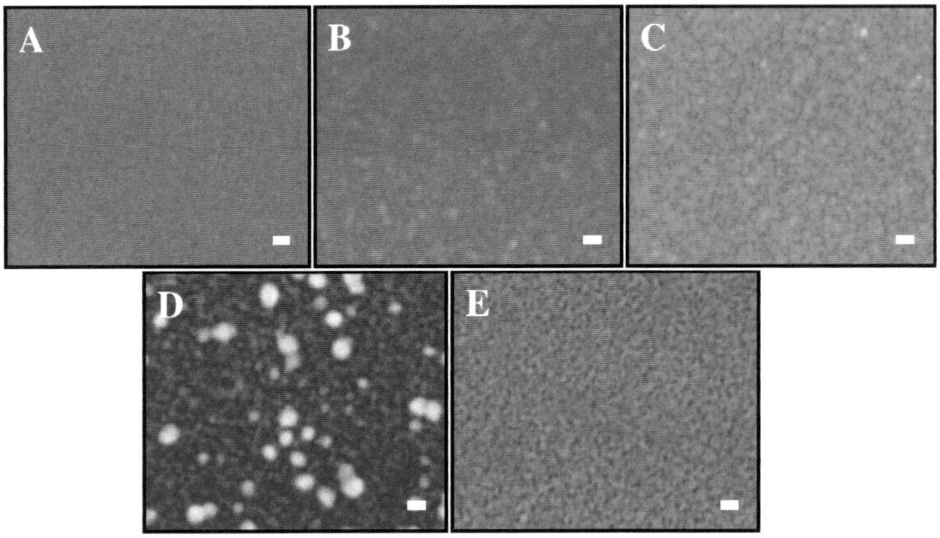

Figure 3. The FESEM morphologies of the nanohybrid TiO_2/ZnO with different precursor molar concentration (a) 1:1 (0.05M); (b) 1:2 (0.10M); (c) 1:3 (0.15M); (d) 1:4 (0.20M) and (e) 1:5 (0.25M). Scale bars are 20 nm.

The EDAX analysis shows the atomic percentage of each element detected. Observe Table 1, the element of Si has highest peak and the energy dispersive of Si element was found in between of 1.6 and 2.0 keV indicates the substrate of Si. It was used because of higher conductivity, it was thus reduced the probability of sample charging. Furthermore, Ti and O elements were found bonded together. Since the energy dispersive between the two elements is relatively close. This result proved the existence both of TiO_2 and ZnO particles that distributed on the Si surface in Figure 3.

Table 1. The analysis data of atomic percentage for each element composition taken from EDAX.

No.	Elements	Atomic percentage for each element composition (%)						
		(a) Pure TiO_2	(b) Pure ZnO	(c) Ratio of 1:1 (0.05M)	(d) Ratio of 1:2 (0.10M)	(e) Ratio of 1:3 (0.15M)	(f) Ratio of 1:4 (0.20M)	(g) Ratio of 1:5 (0.25M)
1.	Ti	12.11	none	0.00	0.00	0.00	0.00	0.00
2.	Zn	none	2.37	0.61	2.38	2.39	9.71	9.71
3.	O	7.36	7.11	6.18	8.90	7.71	19.14	19.14
4.	Si	80.54	90.53	93.21	88.71	89.89	71.15	71.15

Figure 4 indicate the XRD peak of Si at diffraction angle of 62.5°. Observed optimum sample in Figure 4 (d), with ratio of nanohybrid TiO_2/ZnO 1:4 (0.20M), the XRD measurement has detected 4 obvious peaks contains of zincite (002), anatase (200), rutile (211) and silicon (100) at diffraction angle of 33.35°, 48.05°, 55.06° and 62.50°, respectively. These result expressed the particle formation was not bonded together and had less effect to the crystallization of the particles. A narrow peak for anatase and rutile phase was found indicates a smaller crystallite while the broaden peak represent the zincite particle distribution. This statement consistent with the result of surface morphology where ZnO particles found disposed throughout TiO_2 particles. This result in agreement with Tian et. al. [7].

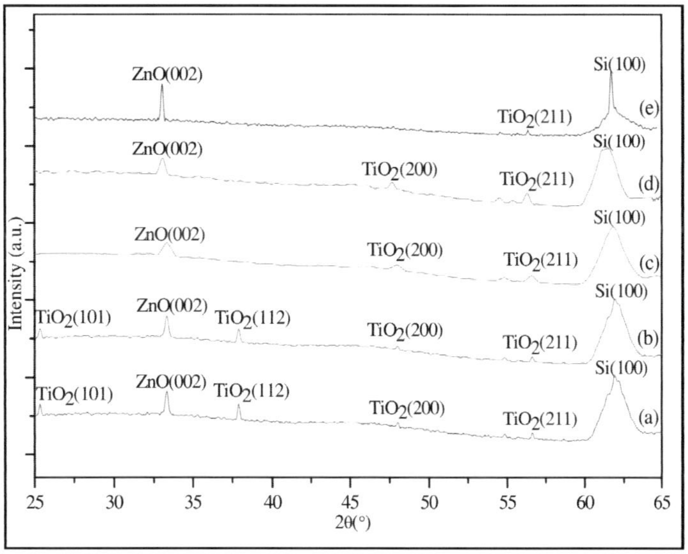

Figure 4. XRD pattern of nanohybrid TiO_2/ZnO with different precursor molar concentration: (a) 1:1 (0.05M); (b) 1:2 (0.10M); (c) 1:3 (0.15M); (d) 1:4 (0.20M) and (e) 1:5 (0.25M).

Figure 5 indicates the effect of increasing ZnO sol concentration on the crystallite size. When the concentration of ZnO increased from 0.05 to 0.20 M, the crystallite size was decreased. The smallest crystallite size was observed for nanohybrid TiO_2/ ZnO 1:4 (0.20M) with an average crystallite size is 23.71 nm and was smaller. It was proved the sample was prepared below 100 nm ranged which indicates a better crystallinity of the surface sample as reported by M Dutta et. al., (2008) [10]. Observed optimum sample of ratio 1:4 (0.20M) was calculated by using the Scherer's formula:

$$d = 0.94\lambda \,/\, \beta \cos \theta \qquad (1)$$

where d: crystallite size, λ: x-ray wavelength, β: full-width at the half-maximum (FWHM) of the peak, and θ: peak of diffraction angle.

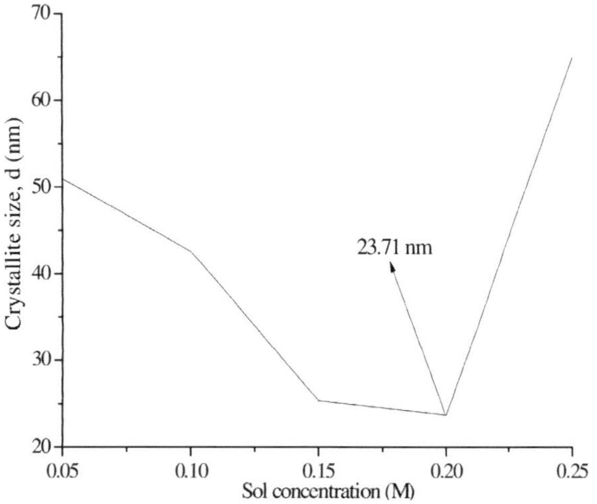

Figure 5. Effect of variation ZnO sol concentration of the nanohybrid TiO_2/ZnO with average crystallite size, d.

3.2. Optical Properties.

Figure 6 shows the transmittance of pure TiO_2 and ZnO and 5 ratios of varied nanohybrid TiO_2/ ZnO. The data roughly emphasize that almost no transmission were detected on the wavelength from range 200 nm to 300 nm. Conversely, in this region, the absorption of UV wavelength occurs with high band-gap energy (E_g). However, obvious increases in transmittance for each nanohybrid ratio as can be seen in the range of ultraviolet (UV) region causes absorbed UV light contain of electron. Then, the maximum and more consistent transmittance were detected in the visible (Vis) region. Figure 6, shows the sample ratio of nanohybrid TiO_2/ZnO 1:4 (0.20M) roughly shows a better transmittance in both regions compares to the other samples. It was proved this sample was mixed well an obviously better than the pure samples.

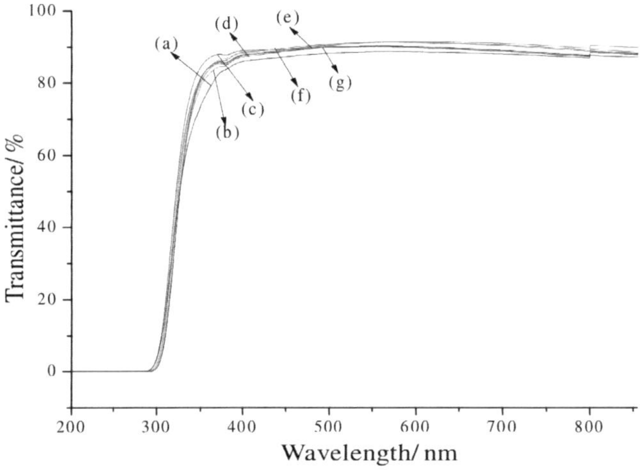

Figure 6. The graph of transmittance (%) versus wavelength (nm) of (a) pure TiO_2; (b) pure ZnO and nanohybrid TiO_2/ZnO with different precursor molar concentration (c) 1:1 (0.05M); (d) 1:2 (0.10M); (e) 1:3 (0.15M); (f) 1:4 (0.20M) and (g) 1:5 (0.25M) in Uv-Vis-NiR region.

Figure 7 shows the transmittance of optimum nanohybrid TiO_2/ZnO (1:4) (0.20M)). Since the optimum result works well in both regions, this sample has remarkably shown the dual abilities and even better in transmittance region compared to pure TiO_2 and ZnO in UV region and visible region, respectively. The shift in UV region of pure TiO_2 and ZnO indicates the differences in the surface sample due to the lower energy needed to be excited that lead to an improvement in irradiation energy efficiency. The optical band-gap determined on Figure 8 by plotting graph of absorption coefficient from transmission data against optical band-gap energy. From this graph, the optical band-gap energy for sample nanohybrid TiO_2/ZnO 1:4 (0.20M) was 3.38 eV which is higher compared with the theoretical band-gap of anatase. This indicates the sample has successfully improved its optical band-gap energy that suitable for coated glass application, where the percentage of transmittance in UV region and visible region increased and subsequently increases the efficiency of photocatalytic activity. The details about the experimental value of optical band-gap for each samples is tabulated in Table 2.

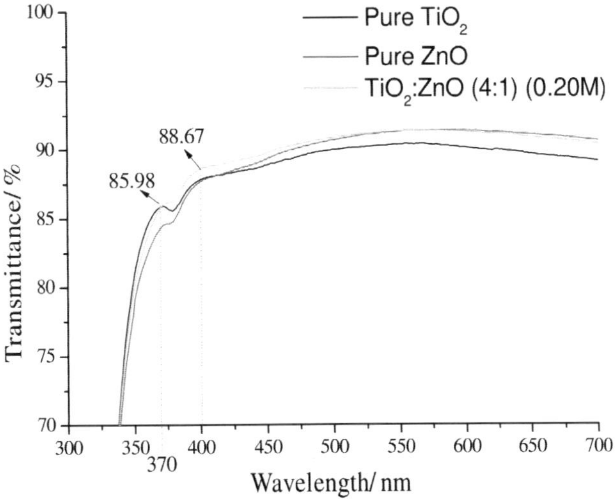

Figure 7. The graph comparison of transmittance (%) against wavelength (nm) compares to pure for optimum sample, ratio of nanohybrid TiO_2/ZnO 1:4 (0.20M).

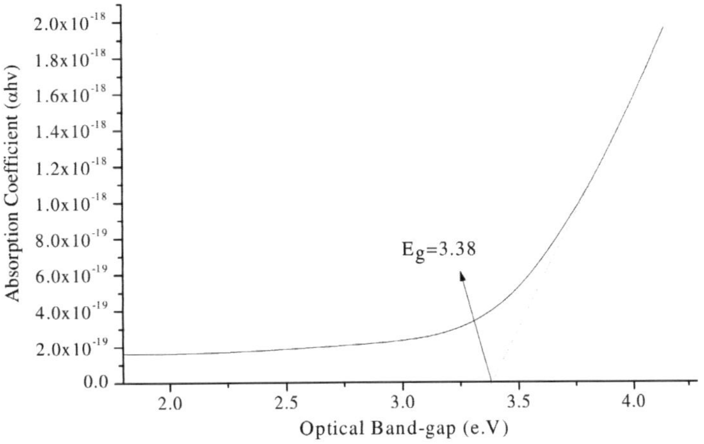

Figure 8. The optical band-gap ratio of nanohybrid TiO_2/ZnO 1:4 (0.20M).

Table 2. Optical band-gap of the nanohybrid TiO_2/ZnO calculated from TαU'c noting and using (2).

Name of samples TiO_2: ZnO.	Ratio (TiO_2:ZnO)	Optical Band-gap, E_g (e.V).
(a)	Pure TiO_2	3.09
(b)	Pure ZnO	3.43
(c)	1 : 1 (0.05M)	3.11
(d)	1 : 2 (0.10M)	3.22
(e)	1 : 3 (0.15M)	3.33
(f)	1 : 4 (0.20M)	3.38
(g)	1 : 5 (0.25M)	3.40

The value of an optical band gap can be obtained by using the formula:

$$E = hv = \frac{hc}{\lambda} \qquad (2)$$

Where hv is $(\alpha \text{ x } hv)^2$, r is considered indirect band gap energy which equal to 2, h is Planck's constant, c is speed of light and λ represent wavelength.

4. Conclusion.

Nanohybrid TiO_2/ZnO has been successfully prepared on top of silicon and glass substrate via hybridization process and spin-coating. For this study, we have been varied the concentration of ZnO and identified the optimum sample was the hybrid ratio of TiO_2/ZnO 1:4 (0.20M). The morphological characterization has confirmed the ZnO particle dispose throughout TiO_2 particle at scale of magnification 20 nm while the EDAX analysis confirmed the atomic percentage of element composition for this sample. From the XRD characterization, we found the peaks of anatase and rutile from TiO_2, zincite from ZnO and Si from Si substrate. This sample also shown a better crystallization and a smaller crystallite size that is much improved compares to TiO_2 itself with value of 23.71 nm. Lastly, the Uv-Vis-NiR characterization has detected this sample has a better transmittance in both region of Uv and Vis region compared pure TiO_2 and ZnO. The optical band-gap energy for this sample was 3.38 eV also increased compared pure TiO_2 that noted the modification of optical band-gap energy has successfully affect the transmittance and efficiency of its photocatalytic activity in Uv and Vis region. Hence, this result suggest for improvement of TiO_2 optical band-gap energy.

Acknowledgement.

The authors would like to thank you to Faculty of Applied Sciences, Universiti Teknologi MARA for financial support and also, NANO-SciTech Centre, Institute of Science, for equipment support.

References

[1] Tian L, Ye L, Deng K and Zan L 2011 *Journal of Solid State Chemistry* **184** 1465-1471
[2] Valenzuela M A, Bosch P, Jiménez-Becerrill J, Quiroz O and Páez A I 2002 *Journal of Photochemistry and Photobiology A: Chemistry* **148** 177–182
[3] Xu L, Shi L, Li X, 2008 *Applied Surface Scienc.* **255** 3230-3234
[4] Huang L, Peng F, Wang H, Yu H and Li Z 2009 *Catalysis Communication.* **10** 1839–1843
[5] Aal A A, Barakat M A and Mohamed R M 2008 *Appl. Surf. Sci.* **254** 4577
[6] Zhang W, Zhu S, Li Y and Wang F 2008 *Vacuum* **82** 328-335
[7] Tian J, Chen L, Dai J, Wang X, Yin Y and Wu P 2009 *Ceramics International* **35** 2261-2270
[8] Zainun A R, Tomoya S, Noor U M, Rusop M and Masaya I 2012 *Materials Letter.* **66** 254–256
[9] Houskova V, Stengl V, Bakardjieva S and Murafa N 2008 *Journal of Physics and Chemistry of Solids* **69** 1623-1631
[10] Dutta M, Mridha S and Basak D 2008 *Applied Surface Science* **254** 2743-2747

Surfactant Assisted Electrodeposition of Nanostructured CoNiCu Alloys

I. Hanafi[1], A. R. Daud[1], S. Radiman[1], M. H. A. Ghani[1] and S. Budi[2]

[1]School of Applied Physics, Faculty of Science and Technology, Universiti Kebangsaan Malaysia, 43600 UKM Bangi, Selangor, Malaysia
[2]Department of Chemistry, Faculty of Mathematics and Natural Sciences, Universitas Negeri Jakarta, Jl. Pemuda No.10, Rawamangun 13220, Jakarta, Indonesia

Email: ismailhanafi89@yahoo.com

Abstract. In order to obtain a desirable deposition thickness, nanostructured ternary CoNiCu alloys have been electrodeposited from salt solutions containing Co^{2+}, Ni^{2+} and Cu^{2+}in the presence of a neutral surfactant alkyl polyglucoside (APG). Electrodeposition of CoNiCu was carried out in a three-electrode cell with indium–tin oxide (ITO) on glass plate, platinum wire and saturated calomel electrode (SCE) as working electrode, counter electrode and reference electrode, respectively. The nanostructured alloys were characterized by means of atomic force microscopy (AFM) and supported by field emission scanning electron microscopy (FESEM). The nanostructured CoNiCu alloys prepared in sulphate solutions in the presence of APG produced a thinner layer of nanoparticles of the alloys compared with the one deposited in the absence of APG. For a fixed deposition time, 160s and at an applied potential of -875mV (SCE) the layer thickness was about 56 nm for electrodepositionin the presence of 3.25 wt.% APG and about 110 nm without APG. Only a small increase in thickness was observed when the potential was increased to more negative values i.e. - 875 to - 950 mV(SCE).

1. Introduction

Electrodeposited magnetic thin films iron- group metals (Fe, Co and Ni) have been developed because of potential applications in electronics, optics, catalyst and magnetic data storage. In recent years, research in magnetic materials has increased and being applied in energy converter devices, sensor, data storage and memory devices[1]. CoNiCu alloys have attractive magnetic properties, magnetoresistive characteristic and sufficient ductility required for forming them into complex shape[2].

For the preparation of the nanoparticles, electrodeposition has been identified as one of the most efficient techniques because of the particles growth parameters can be controlled through potential, current density and temperature [3]. It is considered as a practical technique and widely used for nanoparticles preparation due to its ability to directly attach the nanoparticles on the substrate [4, 5]. Due to its simple and easy process in terms of controlling the experimental parameters[6], economically cheap and wide range industrial applicability[7]put this method as a potential tool for future nanoparticles synthesis.

Controlling size and shape of the electrodeposited nanoparticles is an important issue in the nanomaterials synthesis due to the strong dependence of properties on shape and size. The addition of surfactant into the electrolyte is among a straight forward method in controlling the nanoparticles growth. As the result of the unique roles in modifying the deposition preference such as thinner deposited thickness and properties that in turn enable for obtaining controlled size particles[8, 9], morphology and nanostructure[10].According to Siddig et al. [11], the addition of surfactant into the electrolyte is believed to modify the morphology and structure of materials due to its nonionic characteristic and good psychochemical properties. Alkyl polyglucoside (APG) is a surfactant that possesses environmental friendly features [12-14] such as non-toxic and biodegradable in aqueous media [15, 16].

In this study, we report the results of the synthesis of nanostructured CoNiCu alloys via electrodeposition in the presence of APG surfactant. The effect of APG concentration in the electrolyte on thickness of nanostructured CoNiCu alloys deposited on the substrate was also quantified.

2. Experimental

Technical grade Glucopone 215 CSUP, commercial name of the alkyl polyglucoside (APG), was obtained from Fluka. Analytical grade boric acid (H_3BO_3) and all metal salts (technical grade $CoSO_4.7H_2O$, $NiSO_4.6H_2O$ and $CuSO_4.5H_2O$) were purchased from Sigma-Aldrich. Indium-tin oxide (ITO) coated on glass plate with sheet resistance $10\Omega/cm^2$ was obtained from Praezisions Glas & Optik GmbH.

Electrodeposition of CoNiCu was carried out in a three-electrode cell with ITO on glass plate, platinum wire and saturated calomel electrode (SCE) as working electrode, counter electrode and reference electrode, respectively. The electrolyte containing 0.0018M $CoSO_4$, 0.18M $NiSO_4$, 0.002M $CuSO_4$ and 0.4M H_3BO_3 were prepared using pure water with resistivity of 18.2 MΩ cm. The electrodeposition process was carried out by using electrolyte without APG surfactant and electrolyte with up to3.25 wt. % of APG surfactant under room temperature. Co-electrodeposition potentials werefrom -875 to -975 mV (SCE) and the deposition time was 160 s.

The thickness of deposited CoNiCu alloys was determined using an atomic force microscope (AFM) Veeco Dimension V. A Zeissfield emission scanning electron microscope (FESEM) SUPRA 55VPsupported by an energy dispersive X-ray analyzer (EDX) was used in morphological study and determination of elemental composition of nanostructured CoNiCu alloys deposited on ITO film.

3. Results and discussion

Effect of the presence of APG surfactant on condition of deposited CoNiCu alloy particles was studied at concentration below and above the critical micelle concentration (CMC) value which was $1.3x10^{-2}$ wt.%. The concentrations studied were $2.6x10^{-3}$and 1.3 wt.% respectively. Figure 1 shows typical FESEM images of the CoNiCu alloys nanoparticles prepared in mentioned APG concentrations.

For the concentration of APG below the CMC (figure 1b), it was found that high-density nanoparticles with irregular shape grew on the surface ITO substrate which was almost similar to that formed in electrolyte without APG (figure 1a). The size of the particles was less than 100 nm. However, it can be seen that the nanoparticles seemed to agglomerate and formed a dense surface. This might be due to the limited number of surfactant molecules presence were unable to act as a template in deposition of Co, Ni andCu ions thus the CoNiCu particles were freely formed on the surface of ITO. A significant change was observed when the APG concentration was increased above the CMC value (figure 1c) where spherical shape nanoparticles were formed on the surface of the substrate. The size of the nanoparticles was found to be in the range of 25-70 nm. In this case a sufficient number of APG molecules are available to serve as templates for Co, Ni and Cu deposition on the ITO substrate. The hydrophobic part of APG molecules may also attach to the metal ions consequently control the size and shape of deposited particles. This phenomenon was also observed by Gomes and da Silva Pereira [17] who reported that the addition of surfactant into the electrolyte has affected deposition orientation and morphology.

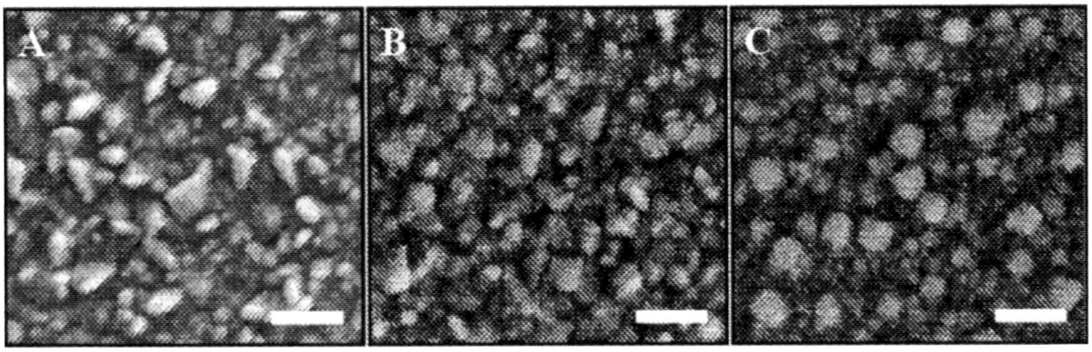

Figure 1. FESEM images of CoNiCu nanoparticles electrodeposited on ITO glass substrate from different concentrations of APG at co-deposition potential -925 mV vs SCE (a) 0 wt.%; (b) 2.6x10^{-3} wt.%; (c) 1.3 wt.%

Figure 2 shows a typical EDX spectrum of the electrodeposited nanoparticles prepared in the presence of 1.3 wt.% of the APG surfactant with deposition time of 160 s. It can be seen that the peaks related to the Co, Ni and Cu were present confirming the CoNiCu alloy nanoparticles were successfully formed on the ITO surface. Other elements such as In, Sn, Si, O and Ca were also observed which were related to the components of ITO substrate coated on glass. Typical elemental composition of the alloy prepared at this condition was 16.01, 31.45 and 52.54 at.% for Co, Ni and Cu, respectively.

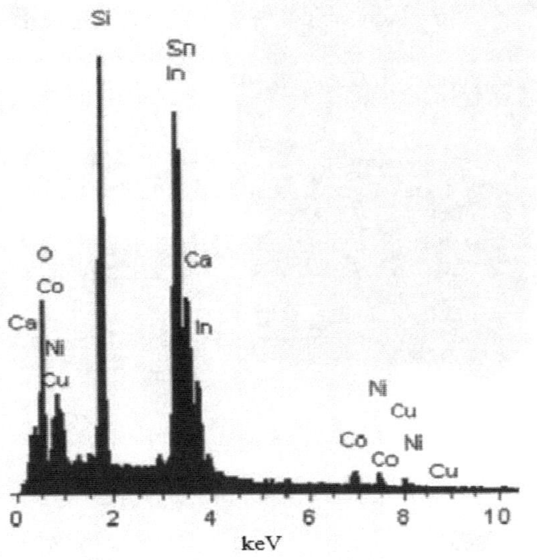

Figure 2. An EDX spectrum of CoNiCu alloy nanoparticles electrodeposited on ITO substrate coated on a glass plate from the electrolyte containing 1.3 wt.% of APG with deposition time of 160 s

Thin layers of nanoparticles of CoNiCu alloy which were electrodeposited on ITO substrate were also observed using FESEM as shown in figure 3 and their thicknesses were determined by AFM. Figure 4 shows the step mode technique based on AFM image used in determination of thickness of an electrodeposited layer of CoNiCu alloy nanoparticles.

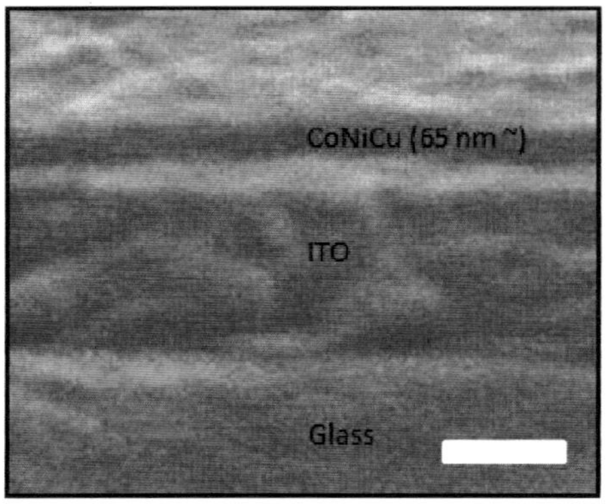

Figure 3. A typical FESEM image of a cross sectional thin layer of CoNiCu nanoparticleselectrodeposited on ITO substrate. Scale bar is 200 nm.

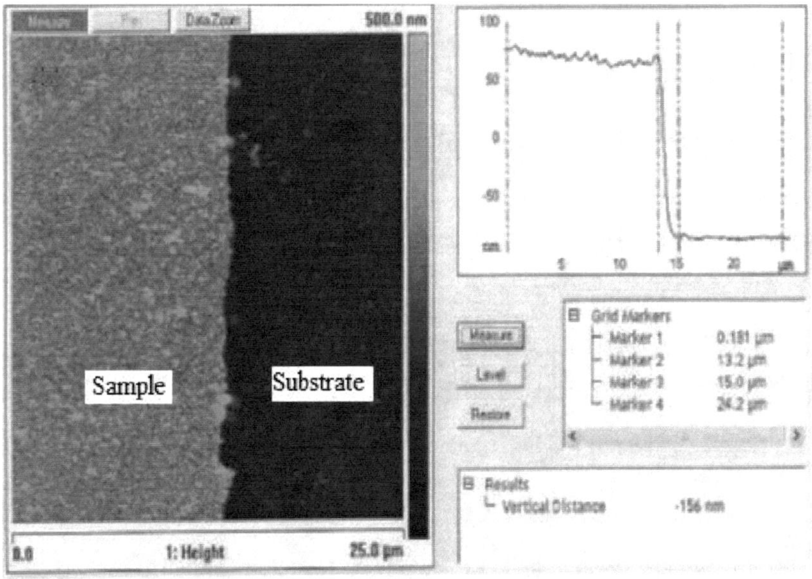

Figure 4. Thickness measurement using AFM for the electrodeposited CoNiCu nanoparticles layer on ITO substrate

The presence of APG in the electrolyte produced a thinner layer of nanoparticles of the alloys compared with the one deposited in the absence of APG. However, for APG concentration of less than the CMC value, insignificant thickness difference was observed compared with the layer of CoNiCu alloy produced in the absence of APG surfactant. According to Budi et al. [18] the presence of limited number of surfactant molecules or in the absence surfactant concentration, Co, Ni and Cu can be freely deposited on the surface of substrate followed by agglomeration of deposited CoNiCu alloy particles. This has resulted in a thicker layer of deposit compared to those deposited with APG concentrations above the CMC value. As mentioned earlier concentration above CMC value has provided sufficient number of surfactant molecules to act as template in electrodeposition effectively. For example, at electrodeposition potential of -925 mV (SCE), the nanoparticles of CoNiCu alloy layers in the absence of surfactant and in the presence of 3.25 wt% surfactant have thickness of 160 and 65 nm, respectively (figure 5).

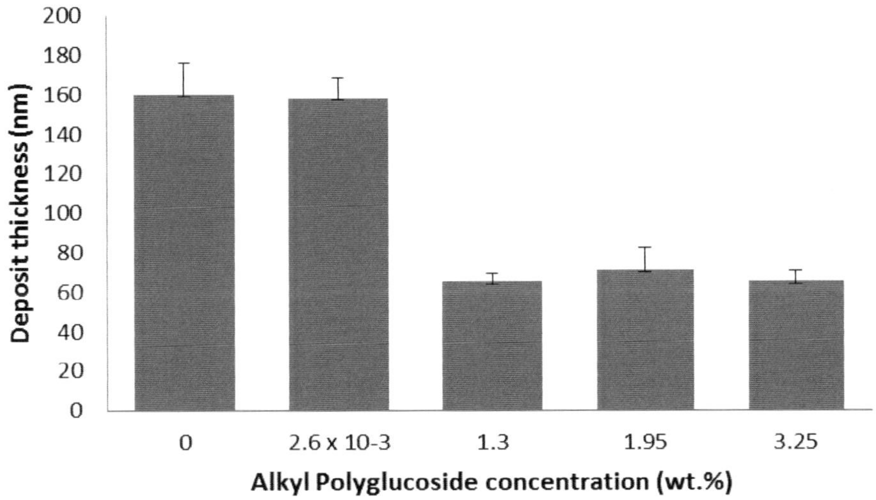

Figure 5. Samples deposition thickness prepared by varyingthe concentration of APG surfactant at deposition potential -925 mV vs SCE

Figure 6 shows thickness variation of layers of the electrodeposited CoNiCu alloys nanoparticles prepared using electrolytes without surfactant and with 3.25 wt.% surfactant and electropotential from -975 to -875 mV(SCE).For a fixed deposition time, 160s and potential of -875mV(SCE) the layer thickness was about 56 nm for electrodeposition with the presence of 3.25 wt.% APG whereas for without APG the thickness obtained was about 110 nm. The thinner deposited layer of nanoparticles obtained in the presence APG was due to lack of electrons transfer for deposition process. This phenomenon was the effect of APG molecules adsorption onto the working electrode surface i.e. ITO substrate whereby hydrophobic part of APG molecules absorbed onto ITO which is hydrophobic. The electrodeposition potentials seem to have little influence on thickness of deposited nanostructured CoNiCu alloys for the electrolyte containing APG especially for potentials from -950 to -875(SCE) except at -975 mV(SCE) whereas in the electrolyte without APG the influence of electrodeposition potentials was significant (figure 6).This is in good agreement with the work of Zhou et al. [19] who report that the electrodeposition current has increased when electrodeposition potential was more negative thus increases reaction rate for deposition consequently increases the thickness of deposit.

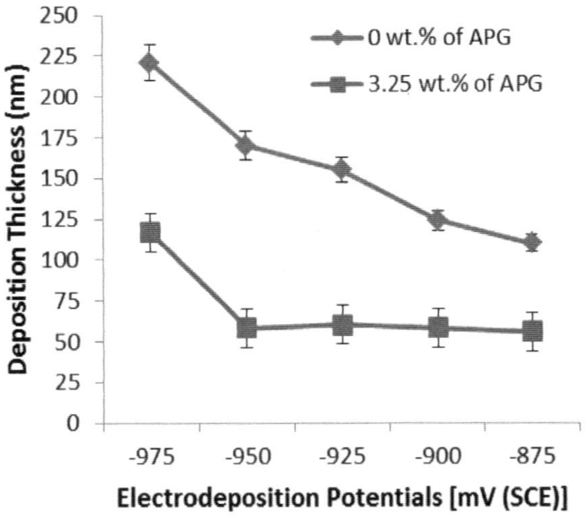

Figure 6. Thickness of deposited nanostructured CoNiCu alloys at different electrodepositionpotentials

4. Conclusion

The effect of alkyl polyglucoside (APG) concentration on the electrodeposition of nanostructured CoNiCu alloys particularly the deposited thickness has been studied. At higher APG concentrations compared with the critical micelle concentration (CMC) value, spherical shape nanoparticles of CoNiCu alloys can be produced. A thin layer of CoNiCu alloy nanoparticles with thickness of about 65 nm can be obtained from the electrolyte containing 3.25 wt.% APG by using electrodeposition potentials from -950 to -875 (SCE). The thinner deposited layer of nanoparticles obtained in the presence of APG is resulted by less electrons transfer during deposition process due to adsorption of APG molecules onto the ITO surface.

Acknowledgements
The authors would like to thank the Ministry of Higher Education (MOHE) and Universiti Kebangsaan Malaysia (UKM) for supporting this project under research grants of UKM-ST-07 FRGS0033-2009 and UKM-OUP-2012-137.

References

[1] D.C. Jiles, Acta Materialia, 51 (2003) 5907-5939.
[2] B.N. Mondal, A. Basumallick, P.P. Chattopadhyay, Journal of Alloys and Compounds, 457 (2008) 10-14.
[3] G. Sasikala, R. Dhanasekaran, C. Subramanian, Thin Solid Films, 302 (1997) 71-76.
[4] A. Bauer, E.L. Gyenge, C.W. Oloman, Electrochimica Acta, 51 (2006) 5356-5364.
[5] M.S. El-Deab, Electrochimica Acta, 54 (2009) 3720-3725.
[6] S. Cheng, G. Chen, Y. Chen, C. Huang, Optical Materials, 29 (2006) 439-444.
[7] I. Gurrappa, L. Binder, Science and Technology of Advanced Materials, 9 (2008) 1-11.
[8] C. Gu, J. Lian, Z. Jiang, Q. Jiang, Scripta Materialia, 54 (2006) 579-584.
[9] E. Sheridan, J. Hjem, R.J. Forster, Journal of Electroanalytical Chemistry, 608 (2007) 1-7.
[10] A.I. Inamdar, S.H. Mujawar, V. Ganesan, P.S. Patil, Nanotechnology, (2008) 1-7.
[11] M.A. Siddig, S. Radiman, L.S. Jan, S.V. Muniandy, Colloids and Surfaces A: Physicochemical and Engineering Aspects, 276 (2006) 15-21.
[12] M. Ferrer, F. Comelles, F.J. Plou, M.A. Cruces, G. Fuentes, J.L. Parra, A. Ballesteros, Langmuir, 18 (2002) 667-673.
[13] K. Hill, W. von Rybinski, G. Stoll, Alkyl Polyglucosides: Technology, Properties and Applications, Wiley, 1997.
[14] K. Fukuda, O. Soederman, K. Shinoda, B. Lindman, Langmuir, 9 (1993) 2921-2925.
[15] J.-Y. Park, H.-H. Lee, S.-J. Kim, Y.-J. Lee, J.-W. Yang, Journal of Hazardous Materials, 140 (2007) 230-236.
[16] I.J.A. Baker, B. Matthews, H. Suares, I. Krodkiewska, D.N. Furlong, F. Grieser, C.J. Drummond, J Surfact Deterg, 3 (2000) 1-11.
[17] A. Gomes, M.I. da Silva Pereira, Electrochimica Acta, 51 (2006) 1342-1350.
[18] S. Budi, A.R. Daud, S. Radiman, A.A. Umar, Applied Surface Science, 257 (2010) 1027-1033.
[19] W.-J. Zhou, D.-D. Zhao, M.-W. Xu, C.-L. Xu, H.-L. Li, Electrochimica Acta, 53 (2008) 7210-7219.

Inactivation of Bacteria using Combined Effects of Magnetic Field, Low Pressure and Ultra Low Frequency Plasma Discharges (ULFP)

A R Galaly[1,2] and H H Zahran[3]

[1] Physics Department, Faculty of Science, Beni - Suef University, Beni- Suef, Egypt
[2] Umm Al-Qura University-Faculty of Community - Engineering Science Department, Azzizia, Makkah, P.O. Box (715), KSA
[3] Department of Botany, Faculty of Science, Beni – Suef University, Beni- Suef, Egypt

ahmed_galaly@yahoo.com

Abstract. Inactivating viable cells at very short application times has been studied using Ultra Low Frequency Plasma (ULFP) at one Kilo Hertz, using an RF source. The targeted fashion is to inactivate *Escherichia coli* (*E. coli*) in the absence and in the presence of magnetic field. Adding oxygen (O_2) to argon (Ar) in the discharge leads to a complete bacterial inactivation, where the inactivation rate increased as the concentration of O_2 increases. Analyses of the experimental data of the initial and final densities of viable cells, using survival curves, showed a dramatic inhibitory effect of plasma discharge to the residual survival of microbial ratio due to the influence of the magnetic field.

1. Introduction

Plasma is currently widely used in different fields due to the wide temperature range enables various applications for plasma technologies [1-4]. These applications include surface etching, biomedical, surface coatings, surface modification, waste destruction, gas treatments, chemical synthesis and machining. Nevertheless, many of these techniques have not been industrialized. Inactivation factors of spore-forming bacteria using low-pressure plasmas and comparative studies of low pressure plasma characteristics between different working gases for sterilization was carried out through different articles [5-6]. So far, depending on the discharge conditions, most of the articles did not take the trend of the influence of the magnetic field on the cold plasma which is one of the oldest problems in plasma physics and remains of great interest in plasma fusion studies. Recently, this has been an important fashion in many plasma discharge used in processing sterilization [7], because the application of a magnetic field results in enhancement of some desirable features of specific plasma sources.

In the present study, it has been suggested to use ultra low frequency plasma (ULFP) radiations for killing some microorganisms adhering to surfaces under the influence of magnetic field. The objective is to investigate the inactivation characteristics of Escherichia coli (E. coli) bacteria, at very short application times, low-pressure and (ULFP), moreover many parameters affecting on the survival curve shapes will be discussed, such as concentration of oxygen to the argon plasma discharge, the presence of the magnetic field, and the direct exposure time.

2. Experimental setup

The CCP plasma system consists of a stainless steel chamber that has few glass ports, connected to vacuum system and evacuated to 10^{-4} torr. The used gases were then admitted to the system via needle valves. As shown in Figure 1, more details about the system, the RF circuit and the plasma cell has been discussed previously in many articles [8-11].

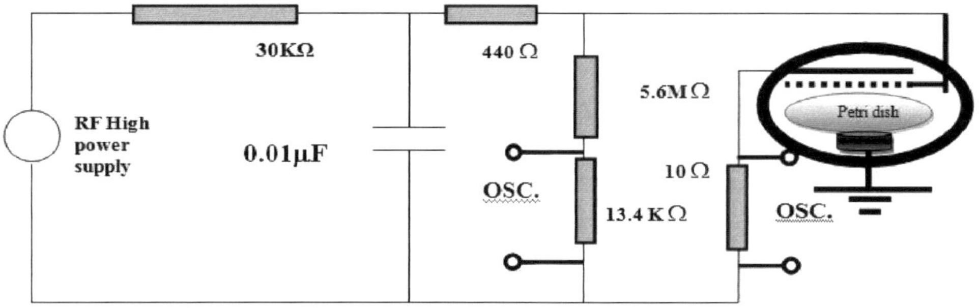

Figure 1. The RF electrical discharge circuit.

Biological material can be exposed to plasma in direct exposure as shown in Figure 2, where the treated Petri dishes (containing *E. coli* bacteria) were in direct contact with the plasma. All plasma generated agents, including charged particles, come in contact with the sample (Petri dishes-containing bacteria). The amount of heat transmitted to the sample is increased, the charged particles play a role since they sputtering the sample, and many of the short-lived neutral reactive species also reach the sample. Thus the main inactivation factors by plasma are heat, charged particles, reactive species, and UV radiation.

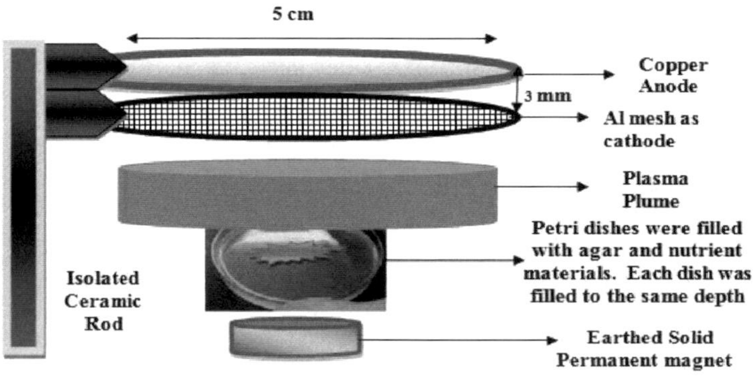

Figure 2. A schematic diagram of the plasma cell.

To show a clear visual effect of the plasma discharge on inactivation process of *E. coli* bacterium, and to demonstrate it in an observable way, a small admixture of oxygen to argon and a small circular solid permanent magnet of few hundreds of gauss pass below the Petri dish (the diameter of the magnet is less than the diameter of the Petri dish) to increase the inactivation efficiency. The tests were conducted as follows: live culture of *E. coli* spread over Petri dishes, two extra Petri dishes were kept as controls. For different times, five Petri dishes were exposed to Argon discharge, five Petri dishes were exposed to Argon- oxygen discharge for 1% O_2, five Petri dishes were exposed to Argon-oxygen discharge for 3% O_2, and finally five Petri dishes were exposed to Argon-oxygen discharge for 3% O_2, in the presence of magnetic field. In all of these tests, the plasma-treated samples were prepared as follows it has been previously suggested [12-14]. The bacterium (*E. coli*) was the organism of choice for our work. An overnight culture containing approximately 10^9 cell-forming unit, cfu/ml was prepared. One of the wells was kept as control while the other wells were exposed to the plasma discharge for two different treatment times. The plasma was in contact with the agar (the

Journal of Physics: Conference Series **431** (2013) 012014 doi:10.1088/1742-6596/431/1/012014

plasma discharge cover the dishes). After exposure, all the Petri dishes were incubated overnight at 37 ^{o}C.

3. Results and discussion

When gas breakdown take place, a sudden increase in the rate of ionization, particularly near the cathode mesh, was observed. Since positive ions move in the electric field much more slowly than the electrons (due to their masses), the electrons were swept rapidly far away from the mesh and a dense positive space charge was formed above the mesh [15]. When an object exposed to plasma, such as a Petri dish contains live culture of *E. coli*, the dish bombarded by both electrons and positive ions, but will initially collect much more electron current due to the higher mobility of the lighter electrons [16]. As a result, the Petri dish becomes negatively charged relative to the plasma bulk. At some negative potential, the electron and ion fluxes will become equal since more electrons are repelled from the Petri dish and positive ions are accelerated toward it to begin the inactivation process.

In this study, bacterial inactivation using ULFP, was studied under two the absence and the presence of magnetic field, and in the presence of oxygen.

3.1. Absence of magnetic field

It was noticed that by increasing the exposure time to Petri dishes from 4 min, at the beginning of the inactivation process to 20 min, slightly a complete process. By increasing oxygen ratios to argon from 1% to 3 %, the inactivation region appears as holes of the etched surface of *E- coli* culture and become wider, then contacted together to form more wider holes as shown in Figure (3, groups A, B, and C). The similarity in the surface morphologies implies that the etched surface processes are a self-similar structure in time scale.

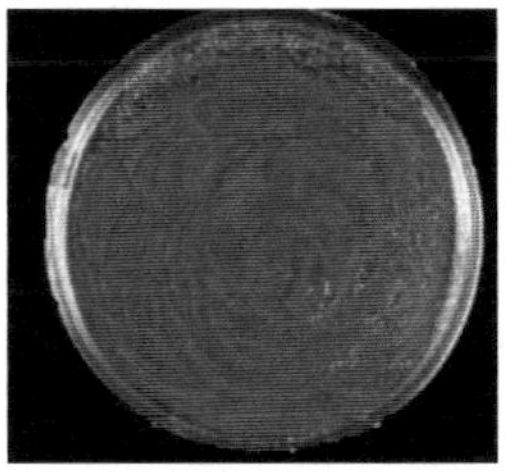

The Control

Exposure Time	**Group A Pure Argon**	**Group B Ar/1% O_2**	**Group C Ar/3% O_2**
16 min	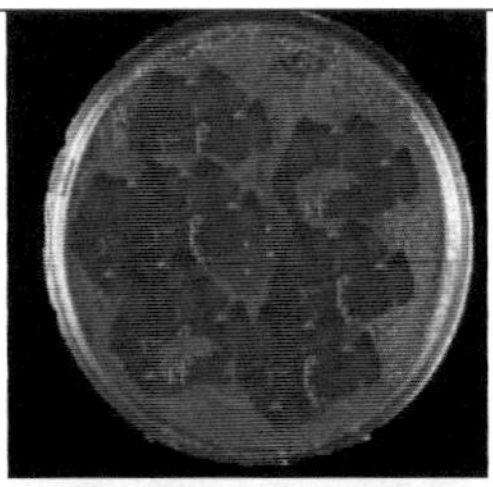	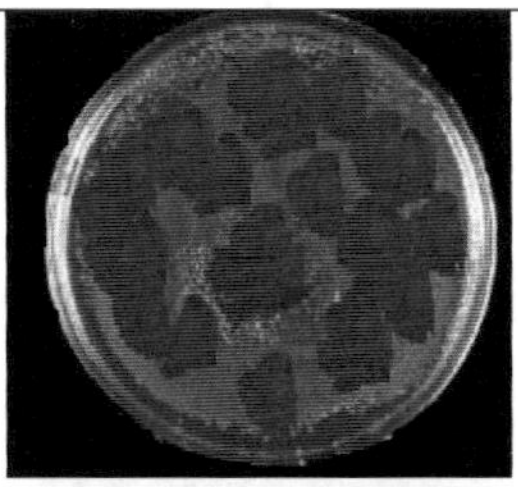	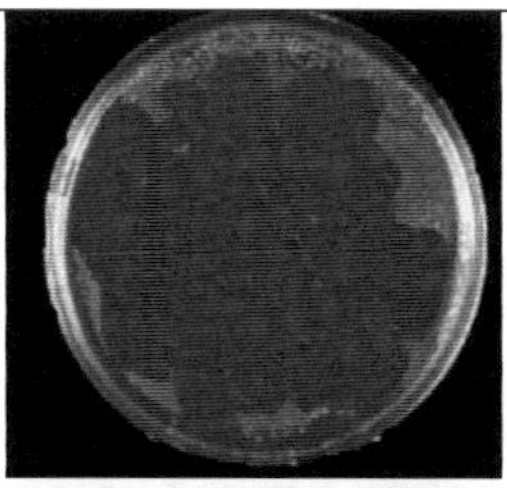
20 min	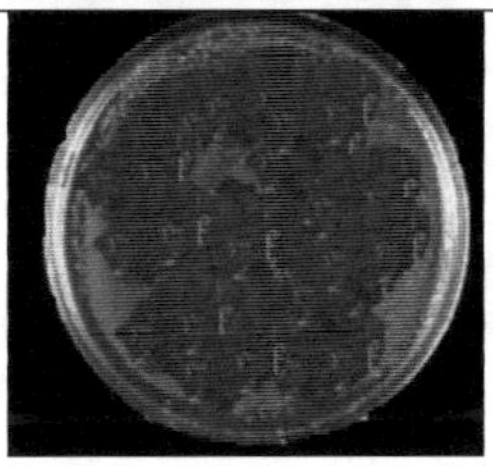	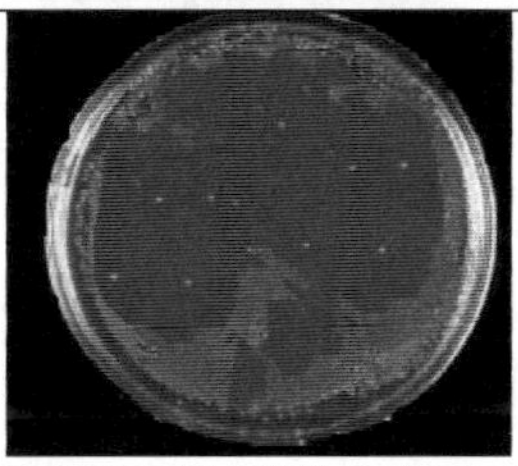	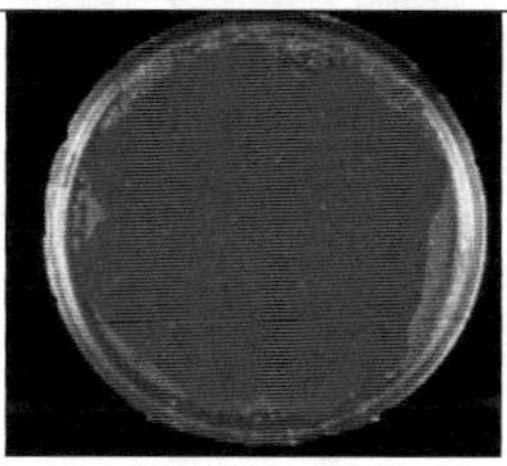

Figure 3. In the absence of magnetic field, inactivation process of *E. coli* bacteria increased and the exposure time decreased, as the addition values of O_2 increased.

3.2. Presence of magnetic field

When the same parameters for the exposure time and the adding of the oxygen ratio, in the absence of the magnetic field, a dark circles in the center showed no growth regions, as shown in Figures (4) for exposure time from 2 min to 10 min. Around these regions, a dense mat of bacterial growth covers the agar and masks the dark background. We noticed that the effect of plasma discharge on *E. coli* bacteria increased at small exposure time, approximately half the value at time needed for inactivation process in the absence of the magnetic field. From these photographs shown in Figures (4), one can conclude that a centered circular area of the inactivated region increases with increasing the treatment exposure time. Furthermore the area of the inactivated region is much greater when oxygen was added to argon using Ar/3% O_2.

Exposure time	**2 min**	**6 min**	**10 min**
Ar/3% O_2, with magnetic field under Petri dish	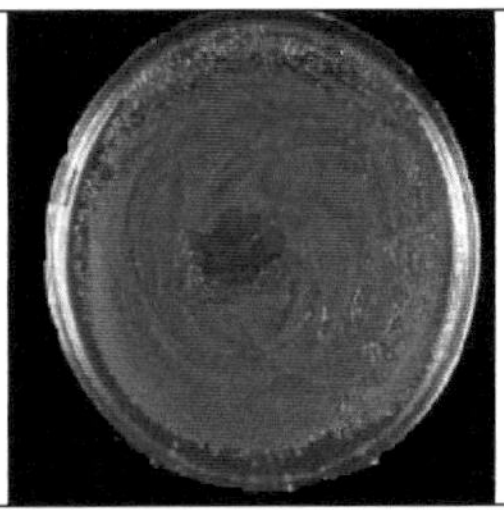	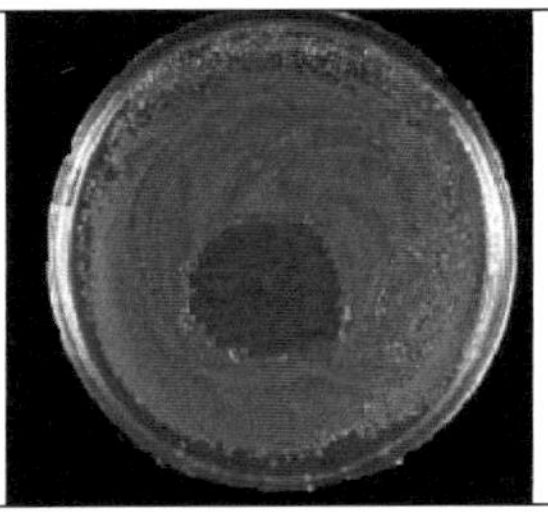	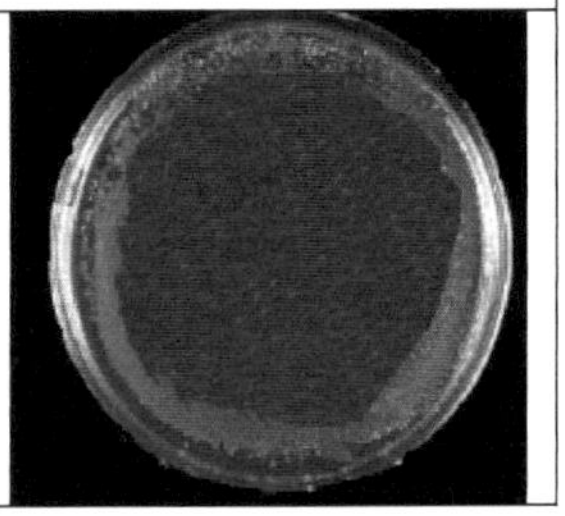

Figure 4. In the presence of magnetic field, inactivation process of *E. coli* bacteria increased and the exposure time decreased as the addition values of O_2 increased.

3.3. Survival curves

In this study the survival curves were focused on the effects of oxygen and magnetic field. Figure (5) showed the survival curves using Ar/0% O_2, Ar/1% O_2 and Ar/3% O_2, in the absence of the magnetic field. For pure argon, there are two phases, phase one gives the fact that radicals from the gaseous phase are involved and given the limited penetration depth of UV photons of plasma discharge. Due to radicals, electrons and positive ions, the destruction of isolated viable cells by UV photons dominates during this phase and phase two which is partly dependent on the erosion rate of the various materials (coats, debris, dead cells) covering still viable cells [17-19].

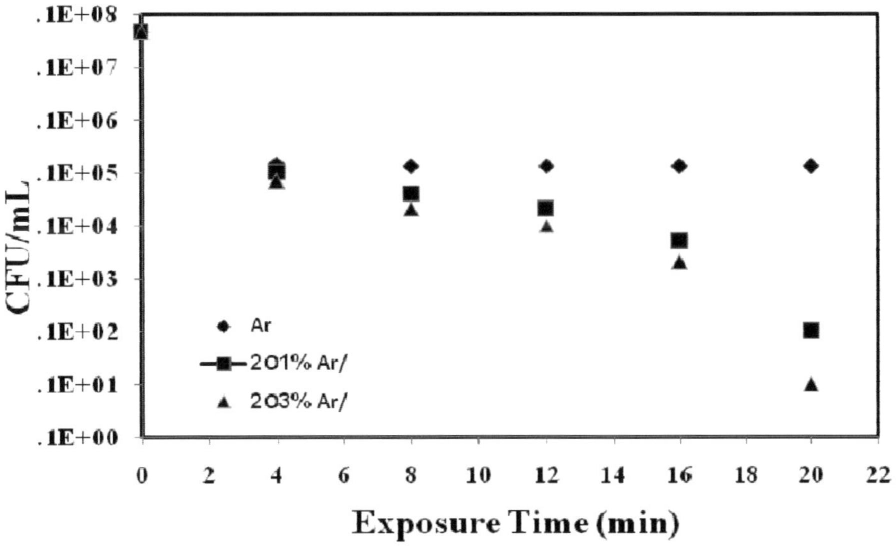

Figure 5. Survival curves of *E. coli* cells exposed to a flowing afterglow, adding O_2 to argon in the discharge without magnetic field.

The survival curve for Ar/1% O_2 and Ar/3% O_2, contains phase three beside the two previous phases with the same characteristics. Phase three starts when the viable cells that were not inactivated during phases 1 and 2 have been sufficiently cleared from debris for the UV photons to finally kill them. This explains why the D_3 time is very often close to D_1 as shown in Table (1) and why this third phase is observed immediately before sterilization is achieved in all cases. Bacterial spore inactivation is ultimately the result of the DNA destruction by UV photons [20-23].

Table 1. D- values changes due to the increasing of the concentration ratio of O_2%

D - Values (min)	D_1	D_2	D_3
Argon	2.1	18	-
Argon/1% O_2	2.1	6	2
Argon/3% O_2 in the absence of magnetic field	2.1	7	2.1
Argon/3% O_2 in the presence of magnetic field	1.2	4.3	1.5

The survival curves for argon/3% O_2 under the influence of magnetic field due to strong permentant magnet contain three phases as in the absence of the magnetic field, but the D values in the presence of magnetic field were less than those in the absence of magnetic field due to the low values of the exposure time as shown in Figure (6).

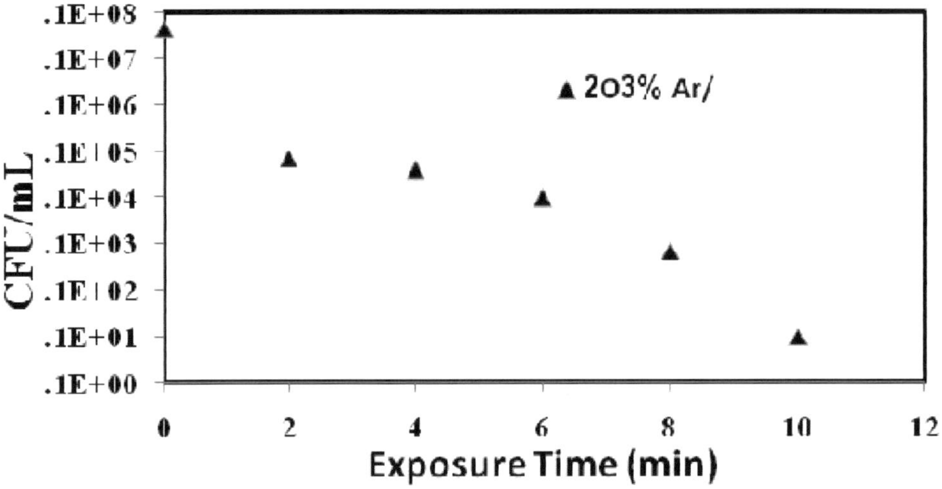

Figure 6. Survival curves of *E. coli* cells exposed to a flowing afterglow, adding 3%O_2 to argon in the discharge with magnetic field.

The longer treatment period and the higher oxygen concentrations increased the ability of low pressure plasmas to inactivate bacteria, as shown in Figure (7). Furthermore, the dramatic inhibitory effect of plasma discharge for the residual survival microbes ratio, in the presence of the magnetic field, was also measured (Figure 8).

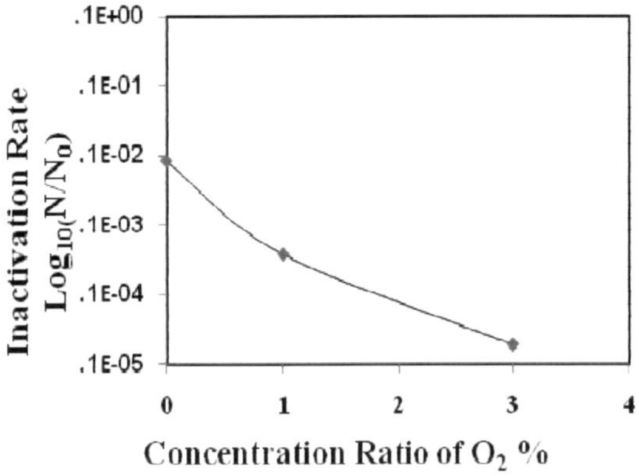

Figure 7. The bacterial inactivation rate increased as the concentration ratio of O_2 % increased.

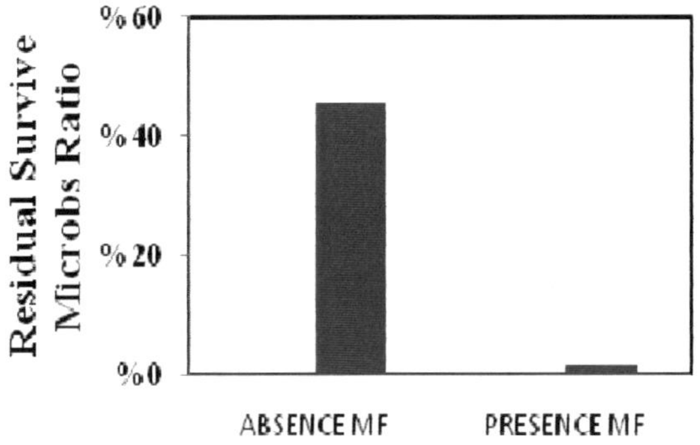

Figure 8. A dramatic inhibitory for the residual survival of microbe ratio in the presence of magnetic field.

Due to the magnetic field strength B (reach to 100 gauss), the Larmor radius [L_R] decrease, the diffusion coefficient across the field is reduced [L_R], then the electro temperature of electron decreases. Moreover, the density of the plasma increases [24-25] and become more dense over the *E. coli* microbes in the same region of the effected magnetic field. When using the magnetic field, the exposure time decreases approximately half the value comparing to those in the absence of the magnetic field.

4. Conclusions

It is concluded that there are many parameters affecting the survival curve shapes, such as the area of the inactivated region is much greater when oxygen was added to Argon using Ar/3% O_2, especially for the longer treatment time. Furthermore, the higher concentrations of oxygen increase the ability of low pressure plasmas to inactivate bacteria. It was found that he presence of the magnetic field increase the density of plasma over the sample, consequently increase the dramatic inhibitory effect of plasma discharge for the residual survival microbes ratio, followed by increase of the inactivation rate of the *E. coli* microbes. In addition decreasing of the exposure time required for inactivation process; The decrement in the electron temperature with a mixing ratio was mainly due to the large inelastic cross section of the mixed gases, and due to the increasing in the sheath around the grid wires affect the survival curves. It was also concluded that the electron temperature is a strong function of a mixing ratio. The cathode mesh (grid) in CCP (Capacitive Coupled Plasma) controlled the plasma parameters in region over the mesh and investigates plasma parameter variations as a function of a mixing ratio in Ar/O_2 plasma. The wires of the grid have a small distance between each other, was used to give ions with high energy. In the survival curves, phase three starts when the viable cells that were not inactivated during phases 1 and 2 have been sufficiently cleared from debris for the UV photons to finally kill them in the presence of the magnetic filed more than in the absence of the magnetic field. Furthermore sterilization is achieved in all cases, bacterial cells inactivation is ultimately the result of the DNA destruction by UV photons.

References

[1] Laroussi M, IEEE Trans. Plasma Sci, 2002 ,30, 1409
[2] Ahmed Rida Galaly, Journal of Modern Physics, 2012, 3, 1031
[3] Van Oost G, Hrabovsky M, Kopecky V, Konrad M, Hlina M, Kavka T, et al. , Vacuum 2013,88. 165
[4] Hyun Woo Lee, Seoul Hee Nam, Abdel-Aleam H. Mohamed, Gyoo Cheon Kim and Jae Koo Lee, Plasma Process. Polym. 2010, 7, 274
[5] Young Sik Seo, Abdel-Aleam H. Mohamed, Kyung Chul Woo, Hyun Wook Lee, Jae Koo Lee and Kyong Tai Kim, IEEE transactions on plasma science, 2010,10,1109,
[6] Kyenam Lee, Kwang-hyun Paek1, Won-Tae Ju1 and Yeonhee Lee, Journal of Microbiology, 2006, 44, 269.
[7] J. You, G. Y. Park, J. H. Kwon, J. H. Kim and H. Y. Chang, Appl. Phys. Lett., 2010, 96, 101504
[8] Ahmed Rida Galaly and F.F. El Akshar, International Journal of Multidisciplinary Research and Advances in Engineering (IJMRAE), Vol.5, No. I. January 2013.
[9] Ahmed Rida Galaly and F.F. El Akshar, International Journal of Engineering Research and Industrial Applications (IJERIA) which will be published in Vol.6, 1. February 2013.
[10] Ahmed Rida Galaly **,** M. A. Hassouba and U. M. Rashed**,** Plasma Physics Report Journal, 2013, 39, 3,1–8
[11] Ahmed Rida Galaly and F.F. El Akshar, " Comparison between experimental and theoretical consideration for three different schematics plasma reactors " 8th International Scientific Conference Environment, Development, and Bioinformatics, Faculty of Science, Al-Azhar University, 26 – 28 March 2012, Cairo, Egypt.
[12] Laroussi M,and Leipold, F, International Journal of Mass Spectrometry, 2004,233, 8,
[13] Laroussi M, D A Mendis and M Rosenberg, New Journal of Physics, 2003,5, 41
[14] Toader Emil, Plasma Sources Sci. Technol., 2004, 13, 646.
[15] Bibinov N, Halfmann H and Awakowicz P, Plasma Sources Sci. Technol, 2008, 17, 035004.
[16] Doggett B and Lunney J G, J. Appl. Phys., 2009, 105 ,033306.
[17] Zdenko Machala, Karol Hensel, and Yuri Akishev, "Plasma for Bio-Decontamination, Medicine and Food Security" Springer, (2011)

[18] Dragos Vicoveanu, Yasunori Ohtsu, and Hiroharu Fujita, Jpn. J. Appl. Phys. 2008, 47, 1130
[19] Shul, R.J and Pearton, S.J., "Handbook of Advanced Plasma Processing Techniques" Springer-Verlag, New York, Inc.(2000).
[20] Suwon Cho and M A Lieberman , Plasma Sources Science and Technology, 2003, 12,224
[21] M. Laroussi, *Plasma Proc. Polym.*, 2005, 2, (5), 391.
[22] I.O.Soloshenko, V.A.Khomich, V.V.Tsiolko et al, Proceedings of the 14th International Symposium on Plasma Chemistry, Prague, 19995, 2551-2556.
[23] I.A.Soloshenko, V.V.Tsiolko, and V.A.Khomich, Plasma Physics Report, 2000, 26,792.
[24] X. Lu, T. Ye, Y. G. Cao, Z. Y. Sun, Q. Xiong, Z. Y. Tang, Z. L. Xiong, J. Hu, Z. H. Jiang, and Y. Pan, *J. Appl. Phys.*, 2008, 104(5), 053309.
[25] Bohm D., in "The Characteristics of Electrical Discharge in Magnetic Fields, McGraw-Hill, New York, (1949).

Supercapacitor Electrodes from Activated Carbon Monoliths and Carbon Nanotubes

B N M Dolah[1], M A R Othman[1,*], M Deraman[1], N H Basri[1], R Farma[1,2], I A Talib[1] and M M Ishak[1]

[1] School of Applied Physics, Faculty of Science and Technology, Universiti Kebangsaan Malaysia, 43600 Bangi, Selangor, Malaysia
[2] Department of Physics, Faculty of Mathematics and Natural Sciences, University of Riau, 28293 Pekanbaru, Riau, Indonesia

E-mail: maro@ukm.my

Abstract. Binderless monoliths of supercapacitor electrodes were prepared by the carbonization (N_2) and activation (CO_2) of green monoliths (GMs). GMs were made from mixtures of self-adhesive carbon grains (SACG) of fibers from oil palm empty fruit bunches and a combination of 5 & 6% KOH and 0, 5 & 6% carbon nanotubes (CNTs) by weight. The electrodes from GMs containing CNTs were found to have lower specific BET surface area (S_{BET}). The electrochemical behavior of the supercapacitor fabricated using the prepared electrodes were investigated by electrochemical impedance spectroscopy (EIS) and galvanostatic charge-discharge (GCD). In general an addition of CNTs into the GMs reduces the equivalent series resistance (ESR) value of the cells. A cell fabricated using electrodes from GM with 5% CNT and 5% KOH was found to have the largest reduction of ESR value than that from the others GMs containing CNT. The cell has steeper Warburg's slope than that from its respective non-CNT GM, which reflect the smaller resistance for electrolyte ions to move into pores of electrodes despite these electrodes having largest reduction in specific BET surface area. The cell also has the smallest reduction of specific capacitance (C_{sp}) and maintains the specific power range despite a reduction in the specific energy range due to the CNT addition.

1. Introduction

Supercapacitor, ultracapacitor or electrochemical double layer capacitor (EDLC) are the terms used to a class of electrical components that bridge the gap between capacitors and batteries. These components have unique characteristics for higher energy storage than conventional capacitor and faster in charging and discharging than batteries, which make them promising candidate for power application. Despite this, there are two areas where the performance of the supercapacitor can be further improved; their specific energy is lower than batteries and their specific power is lower than conventional capacitor.

A supercapacitor stores their energy through the formation of EDLC from electrostatic attraction, pseudocapacitance from quick faradaic charge transfer reaction and combination of both EDLC and pseudocapacitance effect. The EDLC which form from electrolyte ion's charge separation at the solution/electrode interfaces depends strongly on the electrodes surface area. The electrodes with high porosity have large specific surface areas which can promote the EDLC formation. One of the prevailing sources of high porosity carbons is from biomass material because they are cheap and easily available in large quantities [1-11]. Some of these biomass material can be pre-carbonized to produce self-adhesive carbon grain (SACG) that can be converted into GMs form without the addition of any binding material [9, 12, 13]. These GMs have been converted into activated carbon pellet for application as electrode in supercapacitor [10, 11, 14]. However the electronic conductivity of activated carbon is low; in order of 10 to 100 S/m [15].

The electrical conductivity of the electrodes can be improved with the addition of CNTs into the green monoliths prior to electrode preparation because CNTs have a high electrical conductivity [16-18] in order of 10^3 (multi walled CNT [19]) to 10^8 (single walled CNT [20]) S/m. The purpose of adding the CNTs in the electrodes is to decrease the equivalent series resistance of the supercapacitor cells so that the specific power of the cells will be improved [21].

In the present study, the supercapacitor cells whose electrodes were prepared from three batches of mixtures with different combination of SACG, KOH and CNTs percentages were investigated by electrochemical impedance spectroscopy (EIS) and galvanostatic charge-discharge (GCD). The objective of this study is to determine their electrochemical behaviors, focusing on the correlation between the change in electrode porosity with specific capacitance, specific power, specific energy and ESR of the cells resulted from the present of CNTs in the electrodes.

2. Experimental details

2.1. Sample preparation

The activated carbon monoliths (ACMs) were prepared from fibers of empty fruit bunch (EFB) precursor by employing our previously reported method [22]. The EFB was first pre-carbonized (Furnace CTMSB46) at temperature of ~280 °C, followed by ball-milling process (ball-mill AC Motor BS 500-110) for 36 hours and sieving (Matest 24030 Brembate Sopra (BG)) to obtain SACG of particle size less than 106 microns. The composition for the mixtures of SACG, KOH, and CNTs, based on the weight percentages shown in table 1, were prepared and then converted into GMs using a press pelletizing machine (VISITEC 2009-Malaysia) with 8 tonnes of compressive force inside a mold (20 mm in diameter). Justification for selecting the composition shown in Table 1 was based on the results reported in reference [9].

Carbon monoliths (CMs) were then produced from GMs via carbonization process (Vulcan Box Furnace 3-1750) that was done separately according to their batches under N_2 gas with flow rate of 1.5 *l*/min [22, 23]. ACMs production were then continued with CO_2 activation (1 *l*/min flow rate) of the CMs at temperature of ~800 °C with 3 hour holding time [22]. The ACMs were polished to a desired thickness and washed with distilled water to achieve pH 7. The ACMs were used in symmetrical supercapacitor cells as electrodes with 0.4 mm in thickness. The cells were assembled using stainless steel 316L as current collector, 1 M sulfuric acid (H_2SO_4) as electrolyte, Teflon ring as a separator (0.1 mm) and cured resin as a housing. The cells fabricated were labeled according to the codes shown in table 1.

2.2. Physical and electrochemical characterizations

Measurement for the dimensions (Mitutoyo 193-253) and weight (Mettler Toledo AB204) of GMs, CMs, and ACMs produced from each batch were taken to determine the sample density. Studies on the microstructure of the electrodes fractured surface were carried out using Field Emission Scanning Electron Microscopy (FESEM) (Ziess SUPRA 55VP). The porosity properties of the ACMs were characterized using nitrogen adsorption-desorption isotherm technique (ASAP 2010, Micromeritics) at 77 K. The data collected from this technique were used to calculate the specific surface area (S_{BET}), micropore surface area (S_{micro}), mesopore surface area (S_{meso}), micropore volume (V_{micro}), mesopore volume (V_{meso}), total pore volume (V_{total}) and average pore diameter (D_p) of the ACMs by using the standard methods [24, 25].

Table 1. Composition of green, carbon and activated carbon monoliths and their labels.

Batch-Type	Green Monolith (% by weight)				Carbon Monolith	Activated Carbon Monolith
	SACG	KOH	CNT	Label		
1-1	95	5	0	GM1150	CM1150	ACM1150
1-2	90	5	5	GM1255	CM1255	ACM1255
2-1	95	5	0	GM2150	CM2150	ACM2150
2-3	89	5	6	GM2356	CM2356	ACM2356
3-4	94	6	0	GM3460	CM3460	ACM3460
3-5	89	6	5	GM3565	CM3565	ACM3565

Table 2. Weight and dimension of ACMs.

ACMs	Weight (g)	Tickness (mm)	Diameter (mm)	Density (g/cm^3)
ACM1150	0.26	1.49	14.13	1.13
ACM1255	0.29	1.56	14.92	1.07
ACM2150	0.30	1.91	14.67	0.93
ACM2356	0.33	2.02	15.53	0.86
ACM3460	0.32	1.69	14.55	1.14
ACM3565	0.33	1.78	15.27	1.01

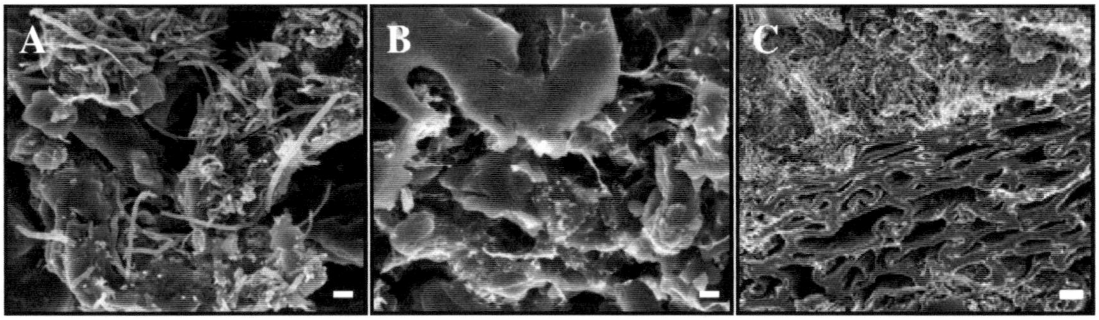

Figure 1. FESEM images of the fractured surface of (a) the ACM1150 with 20,000× magnifications and ACM1255 electrodes with (b) 20,000× and (c) 2,500× magnifications.

The electrochemical behavior of the supercapacitor cells fabricated was characterized by EIS and GCD (Solartron 1287). All the measurements were conducted at room temperature (25 °C). The EIS measurement was conducted over a frequency range of 500 kHz to 10 mHz, and the GCD measurement was conducted at a current density of 10 mA cm^{-2}.

3. Results and discussion

The weight and dimension of ACMs from SACG mixed with varying combination of KOH and CNT are shown in table 2. This table shows that the addition of CNTs reduced the density of the ACMs.

The porous properties of the samples with and without CNTs are clearly revealed by the presence of macropores in the FESEM images recorded at 20,000 and 2,500 magnification scales in figure 1. The shape, orientation and distribution of the ACMs grains look very similar for both samples, ACM1150 and ACM1255, in figures 1(a) and 1(b) respectively. The present of CNTs can be seen in the figure 1(b); there are unoccupied mesopores and macropores occupied by CNTs which are in agglomerated form. However, at low magnification of 2,500×, it is revealed that the CNTs is mostly occupied the outer pores and the pores within the grains remains unoccupied. This type of CNTs distribution creates spacing between grains where it lowers the electrodes density as in table 2.

It was found that the nitrogen adsorption-desorption isotherm data exhibited a typical type IV isotherm curve for all the ACMs samples. Figure 2 shows the comparisons of nitrogen absorption-desorption capacities for all batches. It indicates that ACMs from all batches were porous, consisting both meso and micro pores; with a high internal surface area. However the electrodes with CNTs have lower absorption-desorption capacities compared to electrodes without CNTs. Table 3 list the porosity parameter values obtained from the nitrogen absorption-desorption data. Generally, the additions of CNTs into the electrodes lower the S_{BET}. This behavior is shown by the value of S_{micro}, S_{meso}, V_{micro}, V_{meso} (except for Batch 3) and D_p for electrodes with and without CNTs as in Table 3. ACM1255 has the largest reduction of S_{BET}, S_{micro} and S_{meso} compared to the electrodes without CNTs (ACM1150). This is in agreement with other study that show decrease in S_{BET} when CNTs was added into activated carbon derived from seaweed [26]. The CNTs have very small specific surface area of about 82 m^2/g [27] but very high conductivity. Thus, by mixing small portion of CNTs into active material with high specific surface area, the good conductivity of CNTs can be inherited by the new composite.

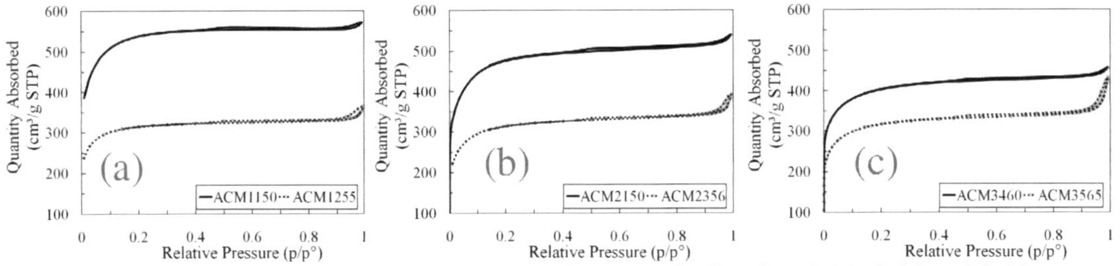

Figure 2. Nitrogen adsorption-desorption isotherm for Batches 1 (a), 2 (b) and 3 (c).

Table 3. The values of ACMs porosity parameters.

ACMs	Porosity Parameters					
	S_{BET} (m^2/g)	S_{meso} (m^2/g)	S_{micro} (m^2/g)	V_{meso} (cm^3/g)	V_{micro} (cm^3/g)	D_p (nm)
1150	1704	539	1165	0.16	0.62	2.1
1255	987	279	708	0.13	0.38	2.2
2150	1592	656	936	0.31	0.45	2.1
2356	1001	395	606	0.20	0.33	2.4
3460	1349	495	855	0.24	0.41	2.1
3565	1056	343	713	0.31	0.34	2.4

The EDLC behavior is associated with the diffusion of electrolyte ions into pores of the porous electrodes. Thus its diffusion depends on many factors such as electrolyte resistance, current collector intrinsic resistance, active material/electrolyte interface resistance, active material intrinsic resistance and contact resistance between particles [18]. The electrochemical behavior of the samples was determined from the EIS and GCD measurements.

The EIS data presented in a form of the Z" versus Z' is known as a Nyquist plot. A typical Nyquist plots for EDLC behavior consisting three significant part of a semi-circle curve at high frequency region, a Warburg line (slope ~ 45°) at intermediate frequency region and a nearly straight vertical line at low frequency region. This typical plot can be seen in figure 3, i.e. for the samples with and without CNTs, ACM1255 and ACM1150, respectively. The Nyquist plots of EIS data for the other batches show a very similar shape of Nyquist plot (not shown here). These data show that the cells using electrode with and without CNTs exhibit EDLC behavior. The diameter of the semi-circle curve represents the ESR of the cells, as shown in the inset of figure 3, which depends on the pore characteristic of the ACMs and ionic accessibility into the ACM's pores [28]. The Warburg's slope in the EIS represents ion diffusion resistance in the porous electrodes. The slope's 45° angle represent *-j* gradient in Nyquist plot. The low frequency region represents the cells capacitance dominant region.

The ESR values are lower for electrodes with CNTs as shown in figure 4. The largest decrease in ESR value was for ACM1255. The decrease in ESR for electrodes with CNTs can be attributed to the uniform network of highly conductive CNTs that effectively lower the active material intrinsic resistance and contact resistance between particles. This increases the polarization range in the electrodes far from the current collector metal. This is trivial in a 0.4 mm-thick electrode.

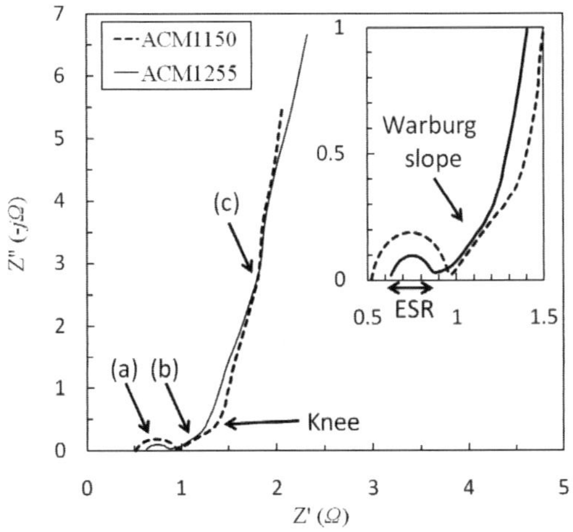

Figure 3. The ESR, Warburg slope and knee in Nyquist plot for ACM1150 and ACM1255. Shown are (a) the semi-circle curves at high frequency regions, (b) Warburg lines (slope ~45°) at intermediate frequency regions and (c) nearly straight vertical lines at low frequency regions.

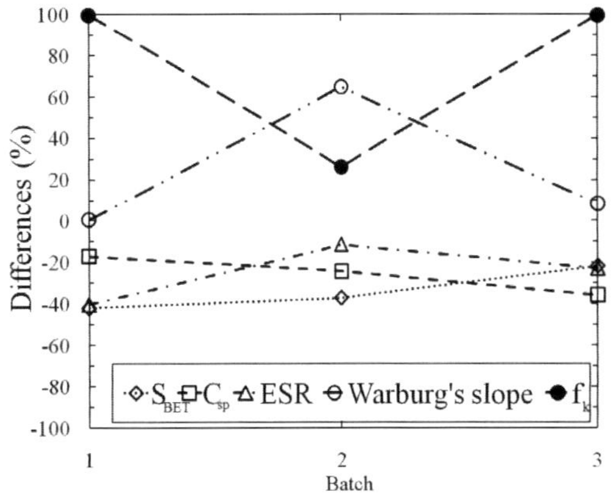

Figure 4. The differences of EIS parameters in percentage for electrodes with CNTs compared to electrodes without CNTs for all batches.

Figure 4 show that ACM2356 has the highest increase in the Warburg's slope gradient, indicating the effect of CNTs in the sample. Whereas there are almost no changes in the other two ACMs batches. The specific capacitance was calculated from the EIS data using $C_{sp} = -2/(2\pi m f|Z''|)$, where m is the mass of one electrode, f is the frequency and $|Z''|$ is the magnitude of imaginary impedance. C_{sp} at f = 10 mHz was found to decrease with the addition of CNTs. The smallest reduction of C_{sp} was from ACM1255 (-17 %) even though S_{BET} was reduced (-42 %) as shown in figure 4. This prove that the gain or reduction in C_{sp} is not a linearly proportional with S_{BET} but depends on the effectiveness of double layer formation [29]. Knee in the Nyquist plot is a point that joints the Warburg line and the nearly vertical line of capacitance dominant region at low frequency as in figure 3. Thus, a knee frequency, f_k, is a maximum frequency where the capacitive behavior is dominant. The higher f_k gained by electrodes containing CNTs as shown in figure 4 depict an improvement in their specific power [30]. However higher f_k shift the lowest frequency point (10 mHz) to higher impedance value (|Z|), which result decrease in C_{sp} of the electrodes.

Figure 5 shows a comparison of the GCD behavior of all batches over a potential range of 0–1 V and a current density of 10 mA cm^{-2}. Both the charge and discharge curves of all batches exhibit an almost linear behavior; a typical shape for carbon based supercapacitors. Both cells show a sharp voltage drop at the beginning of the discharge curves, which is associated with the ESR of the supercapacitor cells. The cells with CNTs charged and discharged quicker than the cells without CNTs as shown by smaller duration needed to perform the charge-discharge cycle, however due to they have smaller areas under the charge-discharge curve reflect their smaller capacitance values.

The values of the specific power (*P*) and specific energy (*E*) were calculated from the charge–discharge curves (figure 5) using the equations $P = Vi/m$ and $E = Vit/m$, respectively, where *i* is the discharge current, *V* is the voltage excluding the *iR* drop occurring at the beginning of the discharge, *t* is time and *m* is the mass [28]. Figure 6 shows Ragone plots that compared the variation of the *P* with the *E* of the cells with and without CNTs. All cells show a typical specific power-energy relationship and exhibit a very similar shape, with a specific energy that remains almost unchanged at low specific power and then gradually decreases before showing a relatively larger decrease in the region of higher

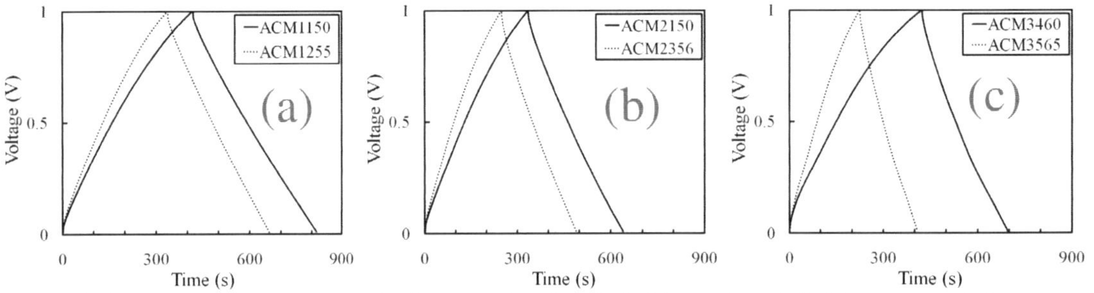

Figure 5.Charge–discharge curves for Batch 1 (a), Batch 2 (b) and Batch 3 (c), measured for current density at 10 mA cm^{-2}.

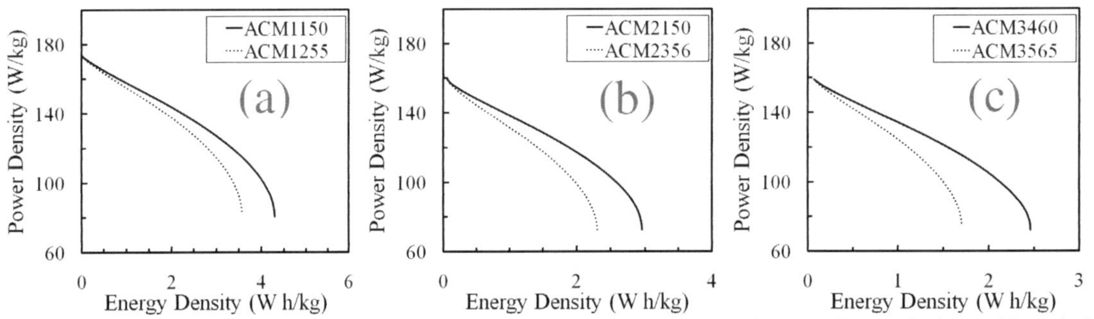

Figure 6. Ragone plots of specific power-energy correlations, for Batch 1 (a), Batch 2 (b) and Batch 3 (c).

specific power. Although the results show a very similar shape, the differences in their values clearly indicates the role of CNTs in improving specific power-energy relationship, with a maximum specific energy of 3.6 Wh kg^{-1} corresponding to a specific power of 82 W kg^{-1} over the electrodes without CNTs.

4. Conclusion

The supercapacitor cells fabricated from electrodes prepared by chemical and physical activation of GMs containing pre-carbonized biomass and mixtures of pre-carbonized biomass and small portion of CNTs, respectively, were investigated. It was found that the addition of small quantity (5 – 6 % by weight) of CNTs reduced the specific surface area and specific capacitance of the electrodes, but it can improve electronic conductivity of electrodes as revealed by the decrease in the ESR values of the cells. This improvement is associated with results that the cells with CNTs have smaller specific energy range but can retain their specific power range.

Acknowledgments

We acknowledged the grants from the Universiti Kebangsaan Malaysia (UKM-GUP-216-2011, UKM-DLP-2012-022, UKM-DLP-2012-023), and the support of CRIM (Centre for Research and Innovation Management). The author also thank to Mr. Saini Sain for helping with the laboratory work.

References

[1] Ismanto A E, Wang S, Soetaredjo F E and Ismadji S 2010 *Bioresource Technol.* **101** 3534-40
[2] Jisha M R, Hwang Y J, Shin J S, Nahm K S, Prem Kumar T, Karthikeyan K, Dhanikaivelu N, Kalpana D, Renganathan N G and Stephan A M 2009 *Mater. Chem. Phys.* **115** 33-9
[3] Liu L, Liu Z, Huang Z, Liu Z and Liu P 2006 *Carbon* **44** 1598-601
[4] Olivares-Marín M, Fernández J A, Lázaro M J, Fernández-González C, Macías-García A, Gómez-Serrano V, Stoeckli F and Centeno T A 2009 *Mater. Chem. Phys.* **114** 323-7
[5] Rufford T E, Hulicova-Jurcakova D, Khosla K, Zhu Z and Lu G Q 2010 *J. Power Sources* **195** 912-8
[6] Rufford T E, Hulicova-Jurcakova D, Zhu Z and Lu G Q 2008 *Electrochem. Commun.* **10** 1594-7
[7] Subramanian V, Luo C, Stephan A M, Nahm K S, Thomas S and Wei B 2007 *J. Phys. Chem. C* **111** 7527-31
[8] Wu F C, Tseng R L, Hu C C and Wang C C 2006 *J. Power Sources* **159** 1532-42
[9] Deraman M, Saad S K M, Ishak M M, Awitdrus, Taer E, Talib I A, Omar R and Jumali M H 2010 *AIP Conf. Proc.* **1284** 179-86
[10] Taer E, Deraman M, Talib I A, Umar A A, Oyama M and Yunus R M 2010 *Curr. Appl. Phys.* **10** 1071-5
[11] Taer E, Deraman M, Talib I A, Awitdrus, Hashmi S A and Umar A A 2011 *Int. J. Electrochem. Sc.* **6** 3301-15
[12] Deraman M, *et al.* 2010 *AIP Conf. Proc.* **1325** 50-4
[13] Awitdrus, Deraman M, Talib I A, Omar R, Jumali M H H, Taer E and Saman M M 2010 *Sains Malays.* **39** 83-6
[14] Awitdrus, Deraman M, Talib I A, Farma R, Omar R and Ishak M M 2012 *Adv. Mat. Res.* **501** 13-8
[15] El-Kady M F, Strong V, Dubin S and Kaner R B 2012 *Science* **335** 1326-30
[16] Deraman M, Ishak M M, Farma R, Awitdrus, Taer E, Talib I A and Omar R 2011 *AIP Conf. Proc.* **1415** 175-9
[17] Farma R, Deraman M, Omar R, Awitdrus, Ishak M M, Taer E and Talib I A 2011 *AIP Conf. Proc.* **1415** 180-4
[18] Portet C, Taberna P L, Simon P and Flahaut E 2005 *J. Power Sources* **139** 371-8
[19] Pham G T, Park Y B, Liang Z, Zhang C and Wang B 2008 *Compos. Part B-Eng.* **39** 209-16
[20] Yang Y L, Wang Y D, Ren Y, He C S, Deng J N, Nan J, Chen J G and Zuo L 2008 *Mater. Lett.* **62** 47-50
[21] Kötz R, Hahn M and Gallay R 2006 *J. Power Sources* **154** 550-5
[22] Deraman M, Omar R, Zakaria S, Mustapa I R, Talib M, Alias N and Jaafar R 2002 *J. Mater. Sci.* **37** 3329-35
[23] Deraman M, Omar R, Harun A G and Ismail M P 1998 *Journal of Materials Science Letters* **17** 2059-60
[24] Brunauer S, Emmett P H and Teller E 1938 *J. Am. Chem. Soc.* **60** 309-19
[25] McEnaney B 1987 *Carbon* **25** 69-75
[26] Raymundo-Piñero E, Cadek M, Wachtler M and Béguin F 2011 *Chemsuschem* **4** 943-9
[27] Lu W, Hartman R, Qu L and Dai L 2011 *J. Phy. Chem. Lett.* **2** 655-60
[28] Taer E, Deraman M, Talib I A, Hashmi S A and Umar A A 2011 *Electrochim. Acta* **56** 10217-22
[29] Obreja V V N 2008 *Physica E* **40** 2596-605
[30] Du C and Pan N 2006 *Nanothechnology* **17** 5314-8

Active Layer Spin Coating Speed Dependence of Inverted Organic Solar Cell Based on Eosin-Y-Coated ZnO Nanorod Arrays

R. T. Ginting[1], C. C. Yap[1,*], M. Yahaya[1], V. Fauzia[2] and M. M. Salleh
[1]School of Applied Physics, Faculty of Science and Technology, Universiti Kebangsaan Malaysia, 43600 UKM Bangi, Selangor, Malaysia
[2]Institute of Microengineering and Nanoelectronics (IMEN), Universiti Kebangsaan Malaysia, 43600 UKM Bangi, Selangor, Malaysia

*Email: ccyap@ukm.my

Abstract. The active layer spin coating speed dependence of the performance of inverted organic solar cells (OSCs) based on Eosin-Y-coated ZnOnanorods has been investigated. An active layer consisted of poly(2-methoxy-5-(2'-ethyl)-hexyloxy-p-phenylenevinylene) (MEH-PPV) as donor and phenyl-c61-butyric acid methyl ester (PCBM) as acceptor was employed, whereas ZnO nanorods were utilized as electron transporting layer. The active layer was deposited on top of Eosin-Y-coated ZnO nanorods with various spin coating speeds (1000-4000 rpm). Inverted OSCs with a structure of FTO/Eosin-Y-coated ZnO nanorods/MEH-PPV:PCBM /Ag were characterized through the current density-voltage (J-V) measurement under illumination intensity of 100 mW/cm^2. Based on the investigation, the short circuit current density (Jsc) and the power conversion efficiency (PCE) enhanced significantly, where as fill factor slightly increased with spin coating speed. The two-diode equivalent model was found to fit the experimental J-V curves very well. The optimum PCE of 1.18 ± 0.07 % was achieved at the highest spin coating speed of 4000 rpm, as a result of the decrement of diffusion current density (J_{diff}), recombination current density (J_{rec}), and ideality factor, thus further confirms the strong built-in electric field in thinner photoactive layer.

1. Introduction

Organic solar cells (OSCs) based on conjugated polymers and fullerene derivatives have the potential to offer low-cost processing using roll-to-roll printing techniques to fabricate large-area photovoltaic devices on flexible and light-weight substrates [1, 2]. Recently, inverted OSCs based on ZnOnanorod arrays (NRAs) have attracted a lot of attention due to the environmental stability and ease of processing [3]. Vertically aligned ZnO NRAs known as good electron transporting layer can be grown at low temperature by hydrothermal method, thus promising for low-cost fabrication process of inverted OSCs device.

Previously, the effects of MEH-PPV:PCBM blend active layer thickness in conventional bulk heterojunction solar cells structure of ITO/PEDOT:PSS/MEH-PPV:PCBM/Al have been investigated by controlling the solution concentration and spin coating speeds [4]. It has been found that open circuit voltage (V_{oc}) remains almost unchanged with spin coating speed, whereas short circuit current density (J_{sc}) gradually increases due to stronger built-in electric field which results in shorter pathways for photogenerated electron and holes to transport to the corresponding electrodes. However, when spin coating speed is higher than 5000 rpm (thinner film), both V_{oc} and J_{sc} reduce.

Chou et al. (2009) reported that the performance of inverted OSCs with structure of ITO/ZnO NRAs/P3HT:PCBM/Ag) can be improved by varying spin coating rate in order to lengthen the solidification time of polymer and thickness of active layer. It was noted that the decrease of spin coating rate results in thickened active layer and thus enhanced light harvesting. Although the probability of exciton recombination is relatively high in a thicker (~400 nm) active layer, these limitation can be overcome by effective infiltration of polymer into the ZnO NRAs which further promotes better charge separation [5]. Recently, the correlation between experimental current density-voltage (*J-V*) plots and theoretically *J-V* plots based on single and two-diode model approach for conventional OSCs with structure of ITO/PEDOT/MEH-PPV:PCBM/LiF/Al have been well studied [6]. It was found that two-diode model fits well to the experimental *J-V* curve compared with single diode model under illumination intensity of 100 mW cm^{-2}.

The aim of this study is to investigate the effect of active layer spin coating speed on the photovoltaic performance of inverted OSCs. The diode characteristics of inverted OSCs were investigated by extracting several important parameters from experimental current density-voltage (*J-V*) curve using two-diode equivalent model. The inverted device structure of fluorine doped tin oxide (FTO)/Eosin-Y-coated ZnO NRAs/MEH-PPV:PCBM/Ag were used[7]. Herein, a thin organic Eosin-Y dye layer was spin coated onto ZnO NRAs to improve the wettability of ZnO and enhance charge transfer efficiency at polymer/ZnO NRAs interface. Based on the results, it was found that the enhancement of Jsc with spin coating speed is due to the decrement of recombination current density (J_{rec}) and ideality factor. The spin coating speed of 4000 rpm leads to optimum device performance with power conversion efficiency (PCE) of 1.18 ± 0.07 %.

2. Methodology

The thin ZnO seed layer was made by the sol-gel spin coating process. The sol-gel precursor solution consisting of 0.2 M equimolar of zinc acetate dehydrate and diethanolamine (DEA) in ethanol, was spin coated on top of pre-cleaned FTO substrate. The sample was immediately annealed at 300 °C for 1 h in air. The ZnO NRAs with length of about ~250 nm were hydrothermally grown at 90 °C using oven for 45 min in equimolar solution of 40 mM zinc nitrate hexahydrate and hexamethylenetetramine (HMT). The ZnO NRAs-coated substrate was rinsed with deionized water and dried to remove residual organic material. A thin layer of 0.3 mM Eosin-Y organic dye was spin coated on top of ZnO NRAs. The MEH-PPV and PCBM(1:3 by weight) dissolved in chloroform were spin coated onto Eosin-Y-coated ZnO NRAs with different spin coating speed from 1000 to 4000 rpm in air. Finally, without further annealing the samples, silver electrodes (180 nm) were sputtered with a low deposition rate of 2.20 nm s^{-1} by introducing oxygen at flow rate of 5 cc/min. The effective active area of devices was defined as 0.07 cm^2 using mask.

The absorption spectra of blend films consisted of organic active layer and Eosin-Y-coated ZnO NRAs were investigated using a Dynamica Halo DB-20 UV-Vis spectrometer. The J-V characteristics of the devices were measured with Keithley 237 SMU under illumination at 100 mW/cm^2 from a solar simulator with AM 1.5G filter. The incident photon to current conversion efficiency (IPCE) was measured using Newport IPCE system at a chopping frequency of 10 Hz. All device preparation, fabrication and measurements were conducted under ambient atmosphere (without encapsulation).

3. Results and discussion

Figure 1 shows the optical absorption spectra of MEH-PPV:PCBM deposited on Eosin-Y-coated ZnO NRAs at different spin coating speeds. Absorption of MEH-PPV:PCBM at visible range of 400-580 nm decreased significantly as the spin coating speed increased from 1000 to 4000 rpm. Similar phenomena has been observed on the thickness of the active layer on top of the ZnO NRAs layer which reduced from ~375 nm to ~150 nm (measured by surface profilometer) with increasing spin coating speed.

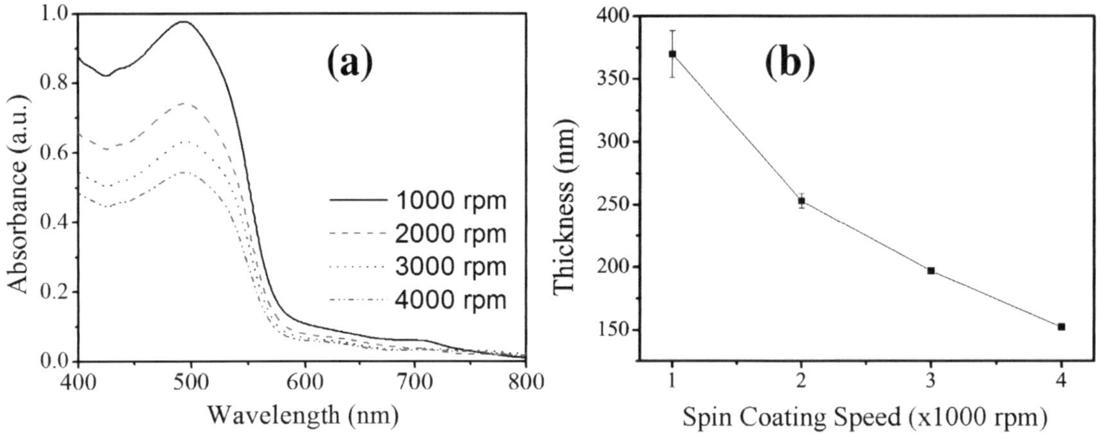

Figure 1.(a) Optical absorption spectra and (b) Active layer thickness of MEH-PPV:PCBM deposited on Eosin-Y-coated ZnO NRAs at different spin coating speeds.

Four devices for each parameter have been fabricated and were characterized under 100 mW cm^{-2} illumination. Figure 2a shows the *J-V* characteristics of MEH-PPV:PCBM devices with different spin coating speeds. The series resistance (R_s) and shunt resistance (R_{sh}) were derived from the inverse slope of *J-V* curve at V_{oc} and J_{sc}, respectively, as shown in Table 1. TheJ_{sc} improved significantly from 3.42 ± 0.24 to 5.61 ± 0.08 mA cm^{-2}with increasing spin coating speed. In contrast, the V_{oc} decreased slightly from 0.63 to 0.59 V with spin coating speed. This can be attributed to the increase of leakage current from ZnO NRAs to Ag electrode as suggested by the decrement of shunt resistance (R_{sh}) at higher spin coating speed.

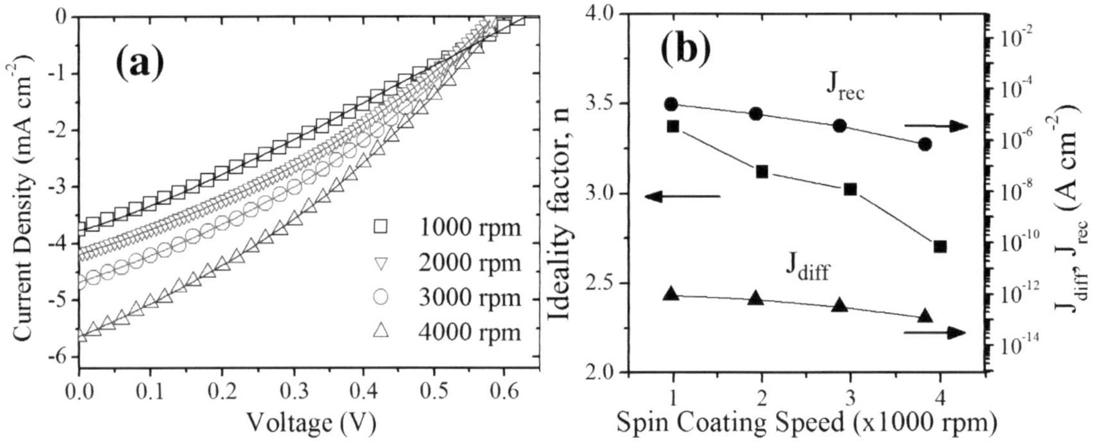

Figure 2. (a)Current density-voltage (*J-V*) characteristics (b) ideality factor (*n*), J_{diff} and J_{rec} of inverted OSCs device with different spin coating speeds under the illumination of simulated AM 1.5 G sunlight at 100 mW cm^{-2}.

Table 1. Summary of photovoltaic parameters of devices at various spin coating speeds.

Spin-coating speed (x1000 rpm)	J_{sc} (mA cm^{-2})	V_{oc} (V)	PCE (%)	FF	R_s (Ω cm^2)	R_{sh} (Ω cm^2)	IPCE (%)
1	3.42±0.24	0.63±0.03	0.67±0.01	0.31±0.03	127±29	288±58	21
2	4.11±0.11	0.61±0.01	0.84±0.03	0.34±0.02	79±5	252±13	27
3	4.52±0.19	0.59±0.01	0.92±0.02	0.35±0.01	63±5	240±11	30
4	5.61±0.08	0.59±0.01	1.18±0.07	0.36±0.02	45±6	223±18	38

The increase of J_{sc} with spin coating speed could be correlated with the active layer thickness. It is observed from Table 1 that series resistance (R_s) reduced dramatically by ~ 65 % from 127 ± 29 to 45 ± 6 Ω cm^2 with spinning speed. This indicates that the probability of photogenerated charge carriers, electron in particular, to transport from active layer to ZnO NRAs became higher with thinner active layer. The highest Jsc achieved at spin coating speed of 4000 rpm leads to optimum PCE of 1.18 ± 0.07 %. In order to further explain the dependence of photovoltaic performance on the active layer spin coating speed, the Eq. (1) based on two-diode equivalent model was used to fit the experimental *J-V* curve under light illumination [6].

$$\underbrace{J = J_{ph}}_{\text{Photocurrent}} - \underbrace{\frac{V+JR_s}{R_{sh}}}_{\text{Shunt current}} - \underbrace{J_{01}[exp(\frac{q}{kT}(V+JR_s) - 1]}_{\text{Diffusion current}} - \underbrace{J_{02}[exp\frac{q(V+JR_s)}{nkT} - 1]}_{\text{Recombination current}} \quad (1)$$

Where J is current density, J_{01}describes the saturation current density, which is related to the bulk recombination, diffusion and drift effect process [8] throughout the device and dominate at high forward bias voltage or denoted as diffusion current density, J_{diff}. The J_{02}corresponds to charge carrier recombination via deep levels in the space-charge region of the junction and also known as recombination current density or J_{rec}[9], whereas idealityfactor, n indicates the diode quality, and J_{ph} is the current generation upon light illumination. The two-diode equivalent model is an effective method to extract more detailed aspects in inverted OSCs where the parameters of the diode model can be related to the mechanisms of solar energy conversion [6].

It was found that the two-diode model fitted well to the experimental *J-V* (symbol) curves as shown in Figure 2a, wherethe line and symbol in Figure 2a represents the fitting line and experimental *J-V* curve, respectively. Figure 2b shows the extracted data parameters of ideality factor, J_{diff} and J_{rec} from fitted *J-V* curve of Figure 2a. The *Rs* and *Rsh* from Table 1 were used as initial or estimated values to fit the experimental *J-V* curves. From Figure 2b, J_{diff} and J_{rec} decreased significantly with respect to spin coating speed, where $J_{diff}$was several order smaller than J_{rec}. Generally, lowering the spin coating speed will result in a higher photocurrent due to enhancement of light absorption with thick active layer. However, the series resistance and charge recombination rate also increase with increasing active layer thickness. Therefore, further increase in the active layer thickness above optimum value normally leads to lower device efficiency [10]. In this inverted OSCs, the lowest spin coating speed showed an increase of charge carrier recombination in the space charge region (J_{rec}) by two-order magnitude as shown in Figure 2b. It could be due to the increase of charge recombination of free charge carriers by space charge effect, low charge carrier mobilities and increased series resistance of device [11, 12]. On the other hand, the decrement of ideality factor with increase of spin coating speed is also consistent with the decrement of J_{diff} and J_{rec} and also the reduction of series resistance of the devices. These can be due to the reduction of trap-assisted recombination in the space charge region [13], where the photogenerated electron and holes turned out to be effectively separated and transport to the corresponding electrodes. This further confirms the existence of strong built-in electric field in thinner active layer which shortened the pathways for transporting the charge carriers, resulting in the increment of J_{sc} and enhancement of device efficiencies. Based on the result, it is suggested that the decrement of J_{diff}, J_{rec}, and ideality factor (as shown in Figure 2b) with spin-coating speed, was mainly correlated to the photoactive layer thickness, since the growth of ZnO NRAs and other parameters remained the same for all devices.

Furthermore, the IPCE of the devices was measured with different active layer spin coating speeds as shown in Figure 3, whereas the IPCE values at wavelength closed to the absorption peak of MEH-PPV (500 nm) areshown in Table 1.For active layer spin coated at 4000 rpm, the device exhibited highest IPCE of ~38 %, which indicates better photocurrent generation efficiency compared to lower spin coating speed. The IPCE result is also consistent with the trend in J_{sc}which increased with spin coating speed.Despite the increase of optical absorption (shown in Fig. 1a) with spin coating speed, the opposite trend was observed from IPCE spectra. The IPCE spectrum of the device with the lowest spin coating speed (1000 rpm) exhibited low photon conversion efficiency at wavelength (500 nm) corresponding to absorption peak of MEH-PPV. For thicker active layer deposited at low spin coating speed, where the majority of photons with wavelength close to 500 nm were mostly absorbed in the region close to transparent conducting oxide (FTO)/Eosin-Y-coated ZnONRAs. Meanwhile, photons with wavelength away from 500 nm were more uniformly absorbed throughout the active layer. The distortion of IPCE spectrum of thick active layer suggests that photocurrent generation in the region close to Ag was more efficient than that in the region close to FTO at wavelength below 500 nm [14].

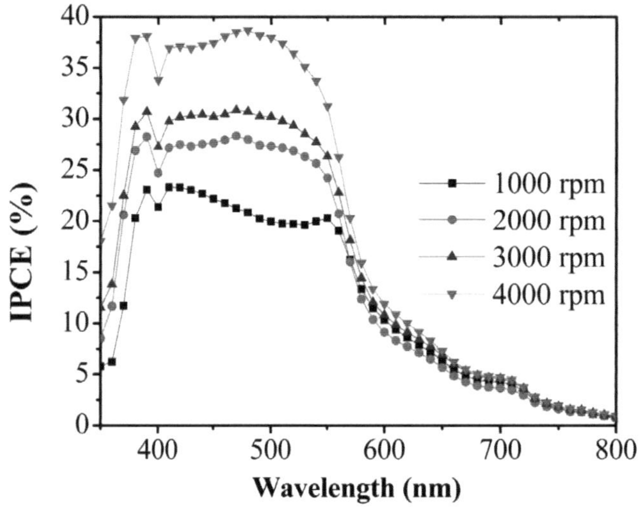

Figure 3. IPCE spectra of devices prepared with different spin coating speeds.

The air-stability of optimum inverted organic solar cells with structure of FTO/ZnONRAs/Eosin-Y/MEH-PPV:PCBM/Ag was measured at ambient air in room temperature over a week as shown in Fig. 4. Interestingly, after storing in air for 4 days, J_{sc}, V_{oc}, PCE, and FF increased significantly. It could be explained that oxidation of Ag electrode with time during exposure in air increased its effective work function to −5.0 eV [15] as shown in inset Figure 4. The decrease in series resistance from 45 to 34 Ω cm^2 provides evidence for the decrease of barrier for hole transport at the photoactive layer-silver electrode interface as a result of increase in the effective work function of Ag [16]. The optimized device exhibited maximum PCE of 1.47 % after 4 days.

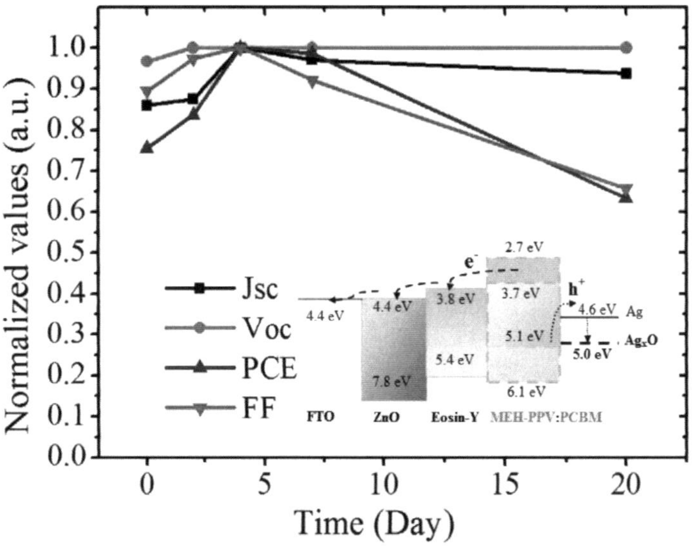

Figure 4. Plot of optimized inverted OSCs device parameters as a function of time in ambient air (without encapsulation). Inset shown the band diagram of inverted OSCs device structure FTO/Eosin-Y-coated ZnO NRAs/MEH-PPV:PCBM/Ag.

Similar behavior has been observed from previous report by Olson et al. (2007), where the performance of the device (ITO/ZnO/ZnO fibers/P3HT/Ag) increased significantly after storing in air for 3 days. The increase of Ag work function was also proven by the increase of V_{oc} by 0.2 V after 2 days and the V_{oc} maintained up to 20 days. Therefore, by having air stable organic solar cells devices, the encapsulation process can be performed under ambient conditions, thus substantially reducing fabrication complexity and leading to cost-effective organic solar cells [15].

4. Conclusions

The effects of active layer spin coating speed on the performance of inverted OSCs based on Eosin-Y-coated ZnO NRAs have been investigated. It was found that the two-diode equivalent model fitted well to the experimental *J-V* curves of inverted OSCs. The decrease of ideality factor is also consistent with the decrement of $J_{diff}$and J_{rec} and reduction of series resistance of the devices. These lead to the enhancement of $J_{sc}$and device efficiencies and further confirm the existence of strong built-in electric field in thinner photoactive layer. The optimum PCE of 1.18 ± 0.07 % was achieved at spin coating speed of 4000 rpm and was found to increase significantly after storing in air for 4 days which is probably due to the better ohmic contact between the active layer and metal contacts.

Acknowledgement

This work has been carried out with the financial support of Malaysian Ministry of Science, Technology and Innovation (MOSTI) under ScienceFund 03-01-02-SF0725. The authors would like to thank Mr. Idris Zulkifle from School of Applied Physics, UniversitiKebangsaan Malaysia for silver sputtering.

References

[1] Krebs F C, Gevorgyan S A and Alstrup J 2009 *J. Mater. Chem.***19** 5442-5451
[2] Krebs F C 2009 *Sol. Energy Mater. Sol. Cells* **93** 465-475
[3] Olson D C, Lee Y-J, White M S, Kopidakis N, Shaheen S E, Ginley D S, Voigt J A and Hsu J W P 2007 *J. Phys. Chem. C***111** 16640-16645
[4] Liu X-d, Zhang F-j, Xu Z, Zhao S-l, Song J-l, Li J-m, Song D-d and Wang Y-s 2009 *Spectros. Spect. Anal.***30** 5
[5] Chou C Y, Huang J S, Wu C H, Lee C Y and Lin C F 2009 *Sol. Energy Mater. Sol. Cells***93** 1608-1612
[6] Cheknane A, Hilal H S, Djeffal F, Benyoucef B and Charles J-P 2008 *Microelectr. J.***39** 1173-1180
[7] Ginting R T, Yap C C, Yahaya M and Salleh M M 2013 *Int. J. Photoenergy***2013** 8
[8] Araujo G L, Sánchez E and Martí M 1982 *Sol. Cells***5** 199-204
[9] Brabec C J, Hauch J A, Schilinsky P and Waldauf C 2005 *Mater. Res. Soc. B***30** 50
[10] Lee S-H, Kim D-H, Kim J-H, shim T-H and Park J-G 2009 *Synth. Met.***159** 1705-1709
[11] Lenes M, Koster L J A, Mihailetchi V D and Blom P W M 2006 *Appl. Phys. Lett.***88** 243502-243503
[12] Mihailetchi V D, Wildeman J and Blom P W M 2005 *Phys. Rev. Lett.***94** 126602
[13] Sah C-T, Noyce R N and Shockley W 1957 *Proc. of the Institute of Radio Engineering* vol 45 pp 1228-1243
[14] Thomas J K B, Yana V, Zhe L, Dinesh K, Richard H F and Christopher R M 2012 *J. Phys. D Appl. Phys.***45** 415101
[15] Zhang F, Xu X, Tang W, Zhang J, Zhuo Z, Wang J, Wang J, Xu Z and Wang Y 2011 *Sol. Energy Mater. Sol. Cells***95** 1785-1799
[16] Kim J B, Kim C S, Kim Y S and Loo Y-L 2009 *Appl. Phys. Lett.***95** 183301-183303

Performance Enhancement of Hybrid Solar Cells via Surface Modification with Diluted P3HT

E L Lim[1], C C Yap[1,*], M Yahaya[1], and M M Salleh[2]
[1]School of Applied Physics, Faculty of Science and Technology, Universiti Kebangsaan Malaysia, 43600 Bangi, Selangor, Malaysia
[2]Institute of Microengineering and Nanoelectronics (IMEN), Universiti Kebangsaan Malaysia, 43600 Bangi, Selangor, Malaysia

Email: *ccyap@ukm.my

Abstract. Surface modification of ZnO nanorods with diluted P3HT in inverted hybrid solar cell was studied. ZnO nanorods were grown on ZnO nanoparticles-coated FTO substrate via chemical bath deposition. 0.2 mM solution of Eosin-Y was spin-coated onto ZnO nanorods at a spinning speed of 1000 rpm for 60 s. Diluted P3HT (0.5 mg/ml) was then coated on ZnO nanorods by using two different methods, namely spin coating and immersion technique. The fabrication process was followed by depositing P3HT from a 25 mg/ml solution. A device without surface modification was also made for comparison. The structure of the devices was FTO / Eosin-Y-coated ZnO nanorods/P3HT/Ag. The optical properties of the samples were investigated and the devices performance was characterized by IV measurement under 100 mW/cm^2 simulated AM 1.5G sunlight. It was found that device coated with diluted P3HT via immersion method showed the highest efficiency, followed by without surface modification and that coated with diluted P3HT using spin coating method.

1. Introduction

Organic solar cell has drawn attention due to low cost processing, large area fabrication, and flexibility [1]. Poor photostability of active layers, phase segregation and low mobility of charge carriers limit the performance of the device [2, 3]. An inorganic semiconductor metal oxide (ZnO) is introduced to form organic-inorganic hybrid solar cell. The use of ZnO nanorods array as an electron acceptor is preferable due to its high electron mobility (115-155 $cm^2V^{-1}S^{-1}$) which can further improve electron transportation in bulk heterojunction solar cells[4, 5]. In cooperate with hole conducting donor polymer such as P3HT, hybrid solar cells have exhibited efficiencies ranging from 0.04 % to 0.5 %. [6, 7]. Generally, complete infiltration of polymer into porous spacing between surface of inorganic semiconductors leads to better power conversion efficiency. Among the strategies to encourage polymer infiltration into porous spacing are surface modification of the inorganic semiconductors and heat treatment[2, 8]. In this paper, the architecture of the device was FTO/ Eosin-Y-coated ZnO nanorods/ diluted P3HT/ dense P3HT/ Ag. ZnO nanorods array was modified with diluted P3HT by using two methods, namely immersion and spin coating method. The device modified by immersion method gave the highest efficiency, with short circuit current density, J_{sc} = 3.48 mA/cm^2 and open circuit voltage, V_{oc} = 0.34 V under illumination of a simulated AM 1.5 sunlight 100 mW/cm^2.

2. Experimental

0.2 M zinc acetate dehydrate and 0.2 M diethanolamine was dissolved in ethanol and stirred at 60°C for 30 min. The colloidal solution was spin coated onto cleaned FTO substrate at 3000 rpm for 30 s followed by preheating at 100°C for 15 min to remove residual solvents. Same process was repeated 3 times. Then, the samples were annealed at 300°C for 1 hour inside furnace to form dense ZnO nanoparticles thin film. ZnO nanorods arrays were grown using chemical bath deposition method. The seeded sample was immersed and tilted upside down against the wall into the aqueous solution of 0.04 M zinc nitrate hydrate and 0.04 M hexamethylenetetramine and heated inside laboratory oven at 90°C for 45 min. Subsequently, the substrate was taken out of the solution and rinsed with deionized water for several times to remove any residual salt and organic materials from the surface and dried in air. 0.2 mM of Eosin-Y dissolved in acetone was spin coated onto surface of ZnO nanorods at a rate of 1000 rpm for 60 s. P3HT was dissolved in chlorobenzene with concentration of 0.5 mg/ml and 25 mg/ml, respectively and stirred at 80°C for 1 hour to increase the solubility of P3HT. The diluted P3HT (0.5 mg/ml) was then coated on dye-modified ZnO nanorods by using two different methods, namely spin coating and immersion technique. By using spin coating method, diluted P3HT (0.5 mg/ml) was spin coated onto dye-modified ZnO nanorods at 1000 rpm for 60 s, whereby for immersion method, the dye-modified ZnO nanorods were immersed in diluted P3HT (0.5 mg/ml) for 15 min. The fabrication process was followed by depositing P3HT from a 25 mg/ml solution at a rate of 1000 rpm for 60 s. A device without surface modification was also made for comparison. The sample was annealed inside the vacuum oven at 100°C for 15 min under ~0.09 MPa to get better infiltration of P3HT. Then, ~250 nm of silver (Ag) anode was sputtered at a pressure of 4×10^{-3} Torr by flowing 3 sccm oxygen gas. The active area of the device was defined as 0.07 cm^2 by using mask.

The absorption spectra of the samples were investigated using Dynamica Halo DB-20 spectrophotometer. Current density-voltage characteristics of the devices were measured with Keithley 237 source measurement unit by illuminating the devices with solar simulator equipped with AM1.5G filter. The light intensity was adjusted to 100 mW/cm^2 by using a calibrated silicon solar cell. Time resolved photoluminescence spectroscopy was carried out with Edinburgh FLS920 spectrophotometer equipped with a picosecond laser diode (470 nm) as excitation source. Lastly, the incident photon to current efficiency (IPCE) was measured for each device using Newport IPCE system at chopping frequency of 10 Hz.

3. Results and Discussion

Figure 1 shows absorption spectra of Eosin-Y-coated ZnO nanorods array/P3HT blend films with and without diluted P3HT modification. Broad absorption ranging from 400 nm to 660 nm is correlated with π-π* band of P3HT [9]. Two vibronic peaks at 550 nm and 600 nm indicate an enhancement ordering of the polymer [10]. There was no distinguished variation in absorption spectra of unmodified and modified samples, suggesting photo generated exciton could not be effectively improved by diluted P3HT modification.

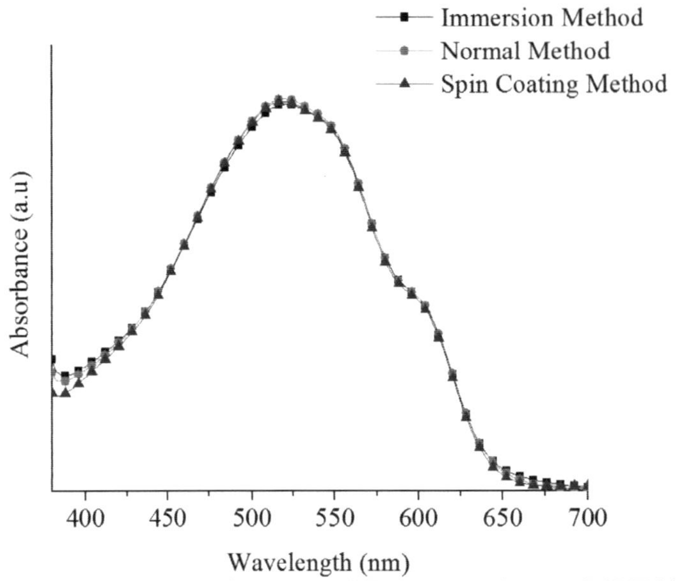

Figure 1. Absorption spectra of Eosin-Y-coated ZnO nanorods array/P3HT blend films with and without diluted P3HT modification.

Figure 2 shows the current density-voltage (J-V) curves of devices with and without diluted P3HT modification under simulated AM 1.5 G sunlight at 100 mW/cm^2. A significant improvement of the PCE was achieved in the device modified with diluted P3HT via immersion method, giving the highest PCE of 0.63%, with J_{sc} of 3.48 mA/cm^2, V_{oc} of 0.34 V and FF of 0.53. The corresponding device exhibited approximately 1.5 times increase in the PCE with respect to that device without diluted P3HT modification. However, the PCE of the device modified with diluted P3HT via spin coating method was smaller than that without modification. The photovoltaic performance of all devices is summarized in Table 1.

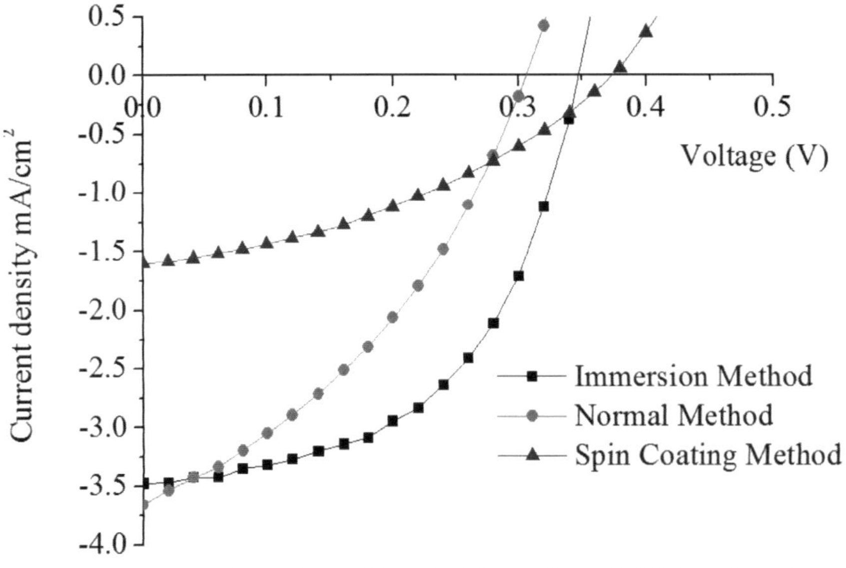

Figure 2. J-V curves of the devices with and without diluted P3HT modification under illumination of simulated AM 1.5G sunlight at 100mW/cm^2.

Generally, an ideal solar cell should have low series resistance and high shunt resistance [11]. Series resistance is mainly described by the contact and bulk resistance. It was found that the device with diluted P3HT modification via immersion method gave the lowest series resistance. It is strongly believed that low series resistance was a result of better polymer infiltration [12] and good interfacial contact [13]. Meanwhile, significant increase in shunt resistance after diluted P3HT modification via spin coating (856.27 Ωcm^2) and immersion methods (968.86 Ωcm^2) could be attributed to the suppression of leakage current from ZnO nanorods to Ag electrode [13, 14]. The decrease in series resistance and increase in shunt resistance lead to significant increase in FF and Voc, resulting in higher PCE of device modified with diluted P3HT via immersion method.

Table 1. Photovoltaic properties of devices

Method of diluted P3HT modificvation	J_{sc} (mA/cm^2)	V_{oc} (V)	FF	PCE (%)	R_{shunt} (Ωcm^2)	R_{series} (Ωcm^2)
Unmodified	3.66	0.30	0.38	0.42	184.94	36.02
Immersed	3.48	0.34	0.53	0.63	968.86	22.32
Spin	1.60	0.36	0.40	0.23	856.27	78.09

The role of diluted P3HT modification in the photovoltaic performance has been studied by using time resolved photoluminescence spectroscopy at wavelength of 690 nm corresponding to emission of P3HT. Figure 3 shows time resolved photoluminescence decay of the samples with and without diluted P3HT modification. The photoluminescence decay was fitted by a single-exponential decay function. Introducing diluted P3HT layer on the Eosin-Y-coated ZnO nanorods results in a shorter PL decay lifetime, which indicates enhancement in exciton dissociation at the interface of Eosin-Y-coated ZnO nanorods/P3HT [15]. The measured photoluminescence decay lifetime of sample with diluted P3HT modification via immersion method (402 ps) was smaller than that of sample without diluted P3HT modification (468 ps). However, the photoluminescene decay lifetime of the sample modified via spin coating method increased to 440 ps with respect to that modified via immersion method. This suggests that uniformity of thin P3HT layer coating via immersion method was better than that via spin coating method, resulting in higher excition dissociation rate. Even though the exciton dissociation rate of device modified via spin coating method was slightly higher than the unmodified device, the J_{sc} of the later was much higher. This suggests that the probability of holes to reach Ag from P3HT was lower in the device modified via spin coating method, as indicated by the higher series resistance.

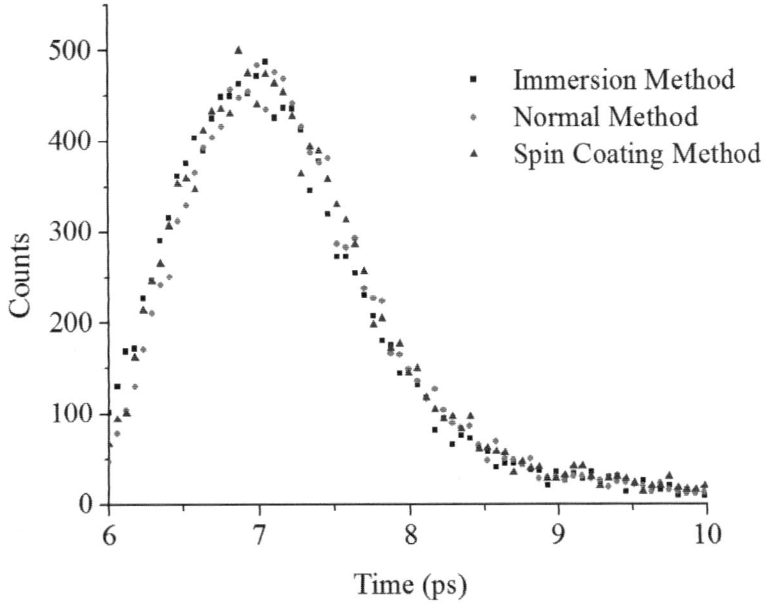

Figure 3. Time resolved PL curve of of ZnO nanorods array/P3HT blend films with and without diluted P3HT modification.

Figure 4 shows the IPCE spectra of the devices with and without diluted P3HT modification. The IPCE is defined as the ratio of the photogenerated charge carriers to the number of photons striking the active area of solar cell. All of the devices exhibited the highest IPCE at the wavelength closed to the absorption peak of P3HT and had similar shapes. There was a peak around 370 nm in the IPCE spectra corresponding to absorption of ZnO, indicating exciton generated on ZnO also contributed to the total photocurrent [16]. The IPCE result is consistent with the trend in J_{sc}, which follows the sequence of without diluted P3HT modification, diluted P3HT modification via immersion method and lastly diluted P3HT modification via spin coating method.

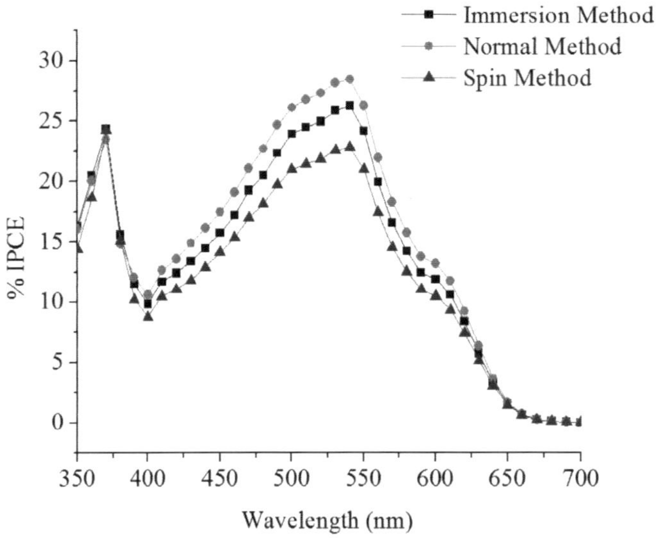

Figure 4. IPCE spectra of devices with and without diluted P3HT modification.

4. Conclusion

The effects of surface modification with diluted P3HT on the photovoltaic performance of Eosin-Y-coated ZnO nanorods/P3HT device have been investigated. The increase in shunt resistance of the devices modified with diluted P3HT via immersion method as a result of reduced leakage current leads to significant increase in FF and Voc as compared to unmodified device. The device, modified with diluted P3HT via immersion method, showed the highest PCE of 0.63%, with J_{sc} of 3.48 mA/cm^2, V_{oc} of 0.34 V and FF of 0.53. It has been shown that surface modification with diluted P3HT via immersion method could be one of the valuable approaches to boost the performance of inverted hybrid solar cell.

Acknowledgement

This work has been financially supported by Malaysian Ministry of Science, Technology and Innovation (MOSTI) under ScienceFund 03-01-02-SF0725. The authors would like to thank Mr. Idris Zulkifle from School of Applied Physics, Universiti Kebangsaan Malaysia for silver sputtering.

References

References

[1] Wu S, Li J, Lo S-C, Tai Q and Yan F 2012 *Org. Electron.* **13** 1569

[2] Ravirajan P 2004 *J. Appl. Phys.* **95** 1473

[3] Lu S, Sun S-S, Niu J, Zeng L, Liu H and Zhao X 2011 *J. Mater. Sci: Mater.Electron.* **23** 251

[4] Kanmani S S, Ramachandran K and Umapathy S 2012 *Int. J. Photoenergy* (doi:10.1155/2012/267824)
[5] Sekine N, Chou C-H, Kwan W L and Yang Y 2009 *Org. Electron.* **10** 1473
[6] Olson D C, Piris J, Collins R T, Shaheen S E and Ginley D S 2006 *Thin Solid Films* **496** 26
[7] Vaynzof Y, Kabra D, Zhao L, Ho P K H , Wee A T S and Friend R H 2010 *Appl. Phys. Lett.* **97** 033309
[8] Olson D C , Lee Y J, White M S, Kopidakis N, Shaheen S E, Ginley D S, Voigt J A and Hsu J W P 2007 *J. Phys. Chem. C* **111** 16640
[9] Ji L W, Shih W S, Fang T H, Wu C Z, Peng S M and Meen T H 2010 *J. Mater. Sci.* **45** 3266
[10] Chou C -Y, Huang J -S, Wu C -H, Lee C -Y and Lin C -F 2009 *Sol. Energy Mater. Sol. Cells* **93** 1608
[11] Zhang F, Xu X, Tang W, Zhang J, Zhuo Z, Wang J, Wang J, Xu Z and Wang Y 2011 *Sol. Energy Mater. Sol. Cells* **95** 1785
[12] Song J, Xu Z, Zhang F, Zhao S, Hu T, Li J, Liu X, Yue X and Wang Y 2009 *Sci. China Ser. G-Phys. Mech. Astron.* **52** 1606
[13] Saunders B R and Turner M L 2008 *Adv. Colloid Interfac.* **138** 1
[14] Huang J -S, Chou C -Y, Liu M -Y, Tsai K -H, Lin W -H and Lin C -F 2009 *Org. Electron.* **10** 1060
[15] Chen C -T, Hsu F -C, Kuan S -W and Chen Y -F 2011 *Sol. Energy Mater. Sol. Cells* **95** 740
[16] Ruankham P, Macaraig L, Sagawa T, Nakazumi H and Yoshikawa S 2011 *J. Phys. Chem. C* **115** (2011) 23809

Electrochemical Effect of Different Modified Glassy Carbon Electrodes on the Values of Diffusion Coefficient for Some Heavy Metal Ions

M. M. Radhi*[1], Y. K. A. Amir[2], S. H. Alwan[1] and T. W. Tee[3]
[1]Department of Radiological Techniques, College of Health and Medical Technology, Baghdad, Iraq
[2]Department of Chemistry College of science, Al-Mustanssiria University, Baghdad, Iraq
[3]Department of Chemistry, Faculty of Science, University Putra Malaysia, 43400, UPM, Serdang, Selangor, Malaysia

Email: mmradhi@yahoo.com

Abstract. Glassy carbon electrode (GCE) was modified with carbon nanotubes (CNT), C_{60} and activated carbon (AC) by mechanical attachment method and solution evaporation technique to preparation CNT/GCE, C_{60}/GCE and AC/GCE, these electrodes were modified in Li^+ solution via cyclic voltammetry (CV) potential cycling to preparing CNT/Li^+/GCE, C_{60}/Li^+/GCE and AC/Li^+/GCE. The sensing characteristics of the modified film electrodes, demonstrated in the application study for different heavy metal ions such as Hg^{2+}, Cd^{2+}, and Mn^{2+}. Cyclic voltammetric effect by chronoamperometry (CA) technique was investigated to determination the diffusion coefficient (D_f) values from Cottrell equation at these ions. Based on Cottrell equation (diffusion coefficient) of the redox current peaks of different heavy metal ions at different modified electrodes were studied to evaluate the sensing of these electrodes by the diffusion coefficient values. The modification of GCE with nano materials and Li^+ act an enhancement for the redox current peaks to observe that the diffusion process are high at CNT/Li^+/GCE, C_{60}/Li^+/GCE and AC/Li+/GCE, but it has low values at unmodified GCE.
Keywords: Voltammetry, chronoamperometry, Cottrell equation, CNT/Li^+/GCE, C_{60}/Li^+/GCE, AC/Li^+/GCE, diffusion coefficient

1. Introduction

In the electrochemistry of cyclic voltammetry for the modified electrodes have been used to determination of diffusion coefficient values by Cottrell equation [1,2] for different chemical compounds such as biochemical compounds [3-6], metal ions [7,8], blood [9,10], etc. Chronoamperometric experiments at modified electrodes are used to quantify both the concentration of $[Fe(CN)_6]^{3-/4-}$ in the ionic liquid and the diffusion coefficient was determined [11]. The electrochemically derived of the diffusion coefficient values of several species such as Ru(bpy) in a hydrated clay. The small diffusion coefficients (typical solution value = 10^{-5} cm^2/s) correspond well with what is known about the interaction of these species with clay surfaces. These small values indicate that the activation energy required to jump from site to site is quite large [12]. The chitosan films are permeable to both cationic $[Ru(NH_3)_6{}^{3+/2+}]$ and anionic $[Fe(CN)_6]^{3-/4-}$ redox couples, but anionic complexes are retained in the chitosan film. Electrochemical parameters, including apparent diffusion coefficients for the redox probes at the electrodeposited chitosan modified electrodes are presented and are comparable to values reported fore cast chitosan films [13].

Determination of the diffusion coefficient of $[S_2O_3]^{2-}$ in the solution using Cottrell equations by the chronoamperometric data obtained at the short times, Because at short times, double-layer charging current and electrode surface related Faradic current are not negligible and, in fact, are often comparable in magnitude to the desired signal. For an electroactive material with a diffusion coefficient, D, the current corresponding to the electrochemical reaction under diffusion control condition is described by Cottrell's law. Chronoamperometry can also be used for the evaluation of the catalytic rate constant. At relatively short times [14]. Hydrogen diffusion coefficients in Ni, Co, Mn, and Al metal hydride electrode as a function of depth of discharge (DOD) and 3.75, 0.65, 0.4, 0.2 temperature were evaluated with modified Warburg impedance which describes more precisely the practical diffusion behavior. It was found that hydrogen diffusion coefficient in this electrode increases with the increase in DOD at ambient temperature, and for this electrode with 50% DOD, hydrogen diffusion coefficient increases with the increase of temperature and the activation energy for 21 hydrogen diffusion in it is 35 kJ mol [15].

The charge-transfer diffusion coefficient for the new electrochromic mixed-valence-compound film of vanadium hexacyano ferrate (VHF) was determined for both its oxidation and reduction processes, in H_2SO_4 + M_2SO_4 aqueous solutions, using short-time transient chronoamperometry (CA). It was revealed that the overall charge transfer process in a VHF film was limited by the electron diffusion rate between adjacent redox sites within the film. So Dct is identified as an electron diffusion coefficient De, which has an average value of 1.4×10^{-9} cm^2s^{-1}. The calculated E-t data based on the interaction parameter theory, agree very well with the experimental curve [16]. The diffusion coefficient (D_f) of the Mn(II) ion in 0.1M KCl at lithium doped ITO electrode by using Chronoamperometry equal to 5.75×10^{-6} cm^2/s was determined. For diffusion control process, current versus time in accordance to Cottrell equation, a monotonic rising and current transient (current dependent of $t^{-1/2}$) should be observed [17].

In this work, CNT, C_{60} and activated carbon (AC) modified Mediators are fixed on the GCE surface using different methods such as CNT/Li^+/GCE, C_{60}/ Li^+/GCE and AC/Li^+/GCE to study the affect of diffusion coefficient (D_f) values, using Cottrell equation for different heavy metal ions Mn^{2+}, Hg^{2+}, and Cd^{2+}.

2. Materials and Methods

2.1. Materials

CNT (Fluka, 99%), C_{60} (Fluka, 98%) and activated carbon (AC) (Fluka 98%). Other chemicals and solvents were used of annular grade and as received from the manufacturer. Deionizer water was used for the preparation of aqueous solutions. All solutions were deairated with oxygen free nitrogen gas for 15 minutes prior to making the measurement.

2.2 Instruments

Electrochemical workstations of Bioanalytical System Inc. USA: Models BAS CV 50W with potetiostate driven by electroanalytical measuring software was connected to Personal computer to perform cyclic voltammetry (CV), chronocoulometry (CC) and chronoamperometry (CA), an Ag/AgCl (3M NaCl) and Platinum wire were used as a reference and counter electrode respectively [1].

2.3 Preparing of working electrodes

There are two methods for modification of working electrodes, mechanical attachment technique method (MA) was used by pressing of a clean GCE surface onto a few mg of CNT or AC powder placed on a filter paper. The other method is solution evaporation technique, which includes the application of a 2 μL of saturated C_{60} in acetonitrile solvent and subsequently dried by hot air blower before placing in voltammetric cell.

The working electrodes used in this study were GCE only and modified GCE with CNT and AC by mechanical attachment method (CNT/GCE and AC/GCE) [18, 19]. Another modified electrode with C_{60} has evaporated solution on the GCE (C_{60}/GCE) [20-21]. CNT/Li^+/GCE, C_{60}/Li^+/GCE and AC/Li^+/GCE were prepared by the doping of Li^+ ion on to CNT/GCE, C_{60}/GCE and AC/GCE via 10 potential cycling between +600 to -600mV in presence of 0.1M LiOH during cyclic voltammetry. A platinum wire (1mm diameter) counter electrode and an Ag/AgCl (3M NaCl) reference electrode were used in CV analysis.

3. Results and Discussion

3.1. Effect of different modified electrodes

The redox current peaks of Mn^{2+}, Hg^{2+} and Cd^{2+} were considerably enhanced and shifting towards the origin, when the CNT/Li^+/GCE was used in comparison with the C_{60}/Li^+/GCE, AC and GCE. The results confirm the electro-catalytic activity of CNT was also exerted on the redox current under the conditions of cyclic voltammetry. The degree of sensitivity/electro-catalytic response for the different electrodes increases in the order below at the Mn^{2+} and Cd^{2+}.

CNT/Li^+/GCE > C_{60}/Li^+/GCE >AC/ Li^+/GCE > GCE

But, it seems the modified AC/ Li^+/GCE has a high redox current peaks comparison with CNT/Li^+/GCE and C_{60}/Li^+/GCE in Hg^{2+} as show in Figure 1. The activated carbon with mercury ion has a high sensitivity for redox current peaks as a good sensor with doping in Lithium ion.

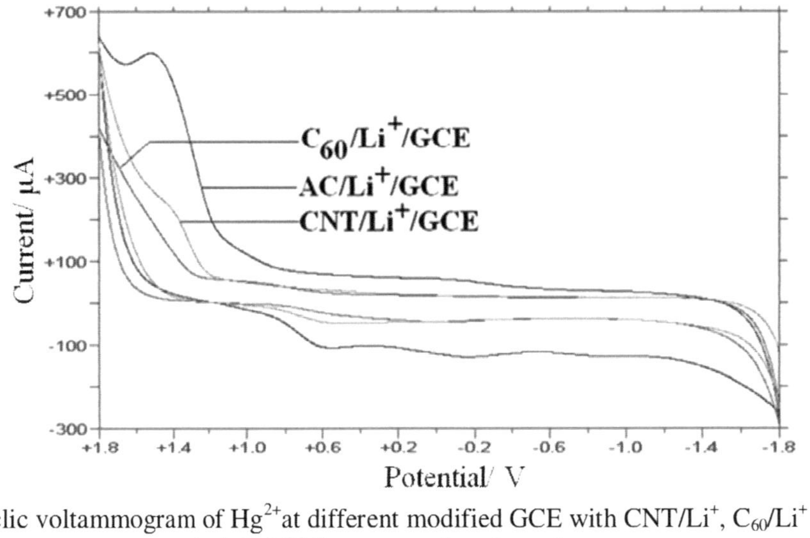

Figure 1. Cyclic voltammogram of Hg^{2+}at different modified GCE with CNT/Li^+, C_{60}/Li^+ and AC/ Li^+ in 0.1M KCl as supporting electrolyte

3.2. Affect the different modified electrodes on the values of diffusion coefficient (D_f)

Mass Transport in Electrochemistry In order to react a species at an electrode it needs to be transported from bulk to surface. Three principal mechanisms involves are; 1) the diffusion where the movement of molecules along a concentration gradient, from an area of high concentration to an area of low concentration; 2) migration transport of a charged species under the influence of an electric field; and 3) convection transport of species by hydrodynamic transport (e.g. natural thermal motion and/or stirring) [22].

3.3. Chronoamperometry from Cottrell equation

In Chronoamperometry (CA) the potential is stepped from an individual value E_i to E_t and the accompanying current is recorded as a function of time for an electrode in unstirred solution [1, 23-25]. The current decays as the electrolysis proceeds to deplete the solution near the electrode of electroinactive species. The current response is described by the Cottrell equation [22]:

$$I = nFAC(D/\pi t)^{1/2} \quad (1)$$

where

I = Current
n = Number of electron per molecule
F = Faraday constant
A = Electrode area
D = Diffusion coefficient of electroactive species
C = Concentration of electroactive species
T = Time

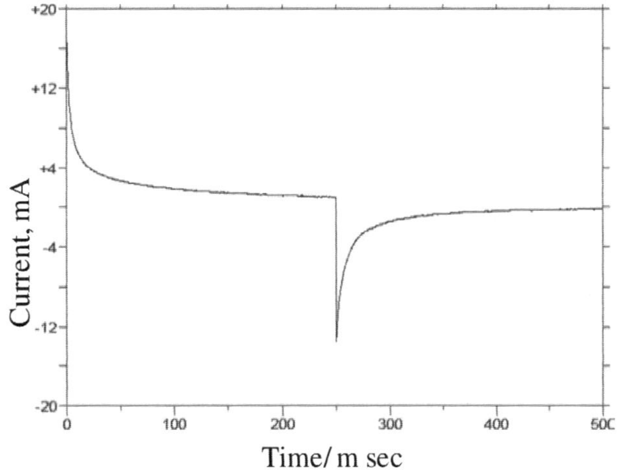

Figure 2. Chronoamperomogram or Cottrell plot obtained for the reduction of heavy metal ion in 0.1M KCl as supporting electrolyte using modified GCEs. Potential was scanned in a negative direction from -1800 to +1800mV with 250msec. pulse width versus Ag/AgCl.

From the slope of a plot of I vs $t^{-1/2}$ at different concentrations of Mn(II), Hg(II) and Cd(II) in 0.1 M KCl as supporting electrolyte at different working modified GCEs (CNT/Li^+/GCE, C_{60}/Li^+/GCE and AC/Li^+/GCE), the D_f values were estimated according to the Cottrell equation [23]. The slopes of the resulting straight lines were then calculating the average values of diffusion coefficient from Cottrell equation as shown in table 1. The results show that the electron transfer is a diffusion-controlled process and the obtained diffusion coefficient value is in good agreement with reported values of diffusion coefficient.

Electrode reaction rates and most double layer parameters are extensive quantities and have to be referred to the unit area of the interface. For diffusion control process, current versus time in accordance to Cottrell equation, a monotonic rising and current transient (current dependent of $t^{-1/2}$) should be observed. However, Figure 2 shows a non-monotonic rising current transient versus time dependent of power (-1/2) was observed instead indicating presence of a non diffusion controlled process. It supported the earlier finding of a complex surface process based on the scan rate study. From table 1, it seems that the values of D_f of heavy metals of Mn(II), Hg(II) and Cd(II) at the modified CNT/Li^+ electrodes have 1.53x10^{-7}, 3.25x10^{-6}, and 4.52x10^{-7} cm^2/sec respectively, also for C_{60}/Li^+/GCE, 5.12x10^{-5}, 5.12x10^{-6}, 3.12x10^{-6} cm^2/sec and for AC/Li^+/GCE, 6.55x10^{-4}, 1.75x10^{-6}, 1.501x10^{-6} respectively, these values are good diffusion coefficient comparing with the values at GCE 5.19x10^{-2}, 2.23x10^{-3}, and 1.55x10^{-3} respectively.

Table 1. Diffusion coefficient (D_f) values of different heavy metal ions at different modified electrodes

Electrodes	D_f, Mn(II), cm^2/sec	D_f, Hg(II), cm^2/sec	D_f, Cd(II), cm^2/sec
GCE	5.19x10^{-2}	2.23x10^{-3}	1.55x10^{-3}
CNT/Li^+/GCE	1.53x10^{-7}	3.25x10^{-6}	4.52x10^{-7}
C_{60}/Li^+/GCE	5.12x10^{-5}	5.12x10^{-6}	3.12x10-6
AC/Li^+/GCE	6.55x10^{-4}	1.75x10^{-6}	1.501x10^{-6}

3.4. Scanning Electron Microscopy (SEM)

Scanning Electron Microscopy (SEM) the fractured surfaces of the nanocomposites were studied using a JEOL attached with Oxford Inca Energy 300 EDXFEL scanning electron microscope operated at 20 to 30 kV. The scanning electron photographs were recorded at a magnification of 1000X to 6000X depending on the nature of the sample. SEM analysis was carried out to investigate microcrystal. Samples were dehydrated for 45 minutes before being coated with gold particle using SEM coating unit Baltic SC030 sputter Coater. SEM was used to examine the morphology of CNT and AC microcrystal by mechanical attached on a graphite electrode surface; C_{60} has evaporated on a graphite electrode surface. CNT/Li^+, C_{60}/Li^+ and AC/ Li^+ were prepared by the doping of Li^+ ion on the electrodes, SEM of these electrodes show in Figure 3.

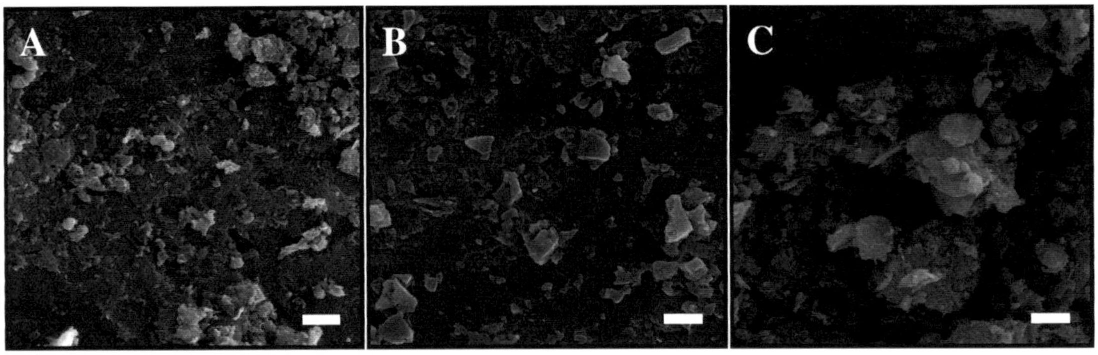

Figure 3. SEM of the a) CNT/Li^+ b) C_{60}/Li^+ and c) AC/Li^+ microcrystal attached to a graphite electrode surface via solvent cast on to 5 mm diameter basal plane graphite electrode and doping of Li^+ ion respectively. Scale bars are 2 μm.

4. Conclusion

CNT/Li^+/GCE, C_{60}/Li^+/GCE and AC/Li^+/GCE were fabricated successfully. These modified electrodes are useful in certain electrochemical applications. Diffusion coefficient (D_f) of electroactive species (heavy metal ions) such as Mn(II), Hg(II) and Cd(II) can be estimated from the electrochemical data from Cottrell equation. Furthermore, an understanding of the mechanisms of charge transport, control of the pore dimensions to enhance transport properties, and complete elucidation of the role of the electrolyte in ion-exchange reactions and charge conduction are required. As these factors are mapped out, the technique will become more generally applicable in the determination diffusion coefficients for heavy metal ions chemists. The electroactive surface of the modified electrodes by determination of diffusion coefficient (D_f) were calculated from cyclic voltammetric with chronoamperometric response. Under optimized conditions, good analytical performance was obtained, including suitable precision, excellent linear dynamic range and reproducibility when using the nonmaterial in working electrode to determination of diffusion coefficient.

References

[1] Instruction manual, CV 50W, version 2. Bioanalytical System, Inc. USA, Feb, 1996.

[2] Wee Tee Tan, Farhan Yusri and Zulkarnain Zainal, Electrochemical Reduction of Potassium Ferricyanide Mediated by Magnesium Diboride Modified Carbon Electrode, Sensors & Transducers Journal, Vol. 104, Issue 5, May 2009, pp. 119-127.

[3] Z. Nasri, E. Shams, Application of silica gel as an effective modifier for the voltammetric determination of dopamine in the presence of ascorbic acid and uric acid, Electrochim. Acta, 54 (2009) 7416-7421.

[4] Y. Bai, W. Yang, Y. Sun, C. Sun, Synthesis of flower-like CuO nanostructures as a sensitive sensor for catalysis, Sens. Actuators B, 134 (2008) 761–768.

[5] X. Wang, N. Yang, Q. Wan, Cyclic voltammetric response of nicotinic acid and nicotinamide on a polycrystalline gold electrode, Electrochim. Acta, 52 (2006) 361.

[6] P. Kalimuthu, S. A. John, Electropolymerized film of functionalized thiadiazole on glassy carbon electrode for the simultaneous determination of ascorbic acid, dopamine and uric acid, Bioelectrochemistry 77 (2009) 13.

[7] M. M. Radhi, W. T. Tan, M. Z. B. A. Rahman, A. B. Kassim, Voltammetric Detection of Mn(II) in Blood Sample at C_{60} and MWCNT Modified Glassy Carbon Electrodes, Am. J. Appl. Sci., 7 (2010) 439.

[8] M. M. Radhi, W. T. Tan, M. Z. B. A. Rahman, A. B. Kassim, Electrochemical Redox of Hg^{2+} Mediated by Activated Carbon Modified Glassy Carbon Electrode, Int. J. Electrochem. Sci., 5 (2010) 615.

[9] H. Liu, G. Wang, D. Chen, W. Zhang, C. Li, B. Fang, Sens. Actuators B, 128 (2008) 414.

[10] R. Zhang, X. Wang, C. Chen, Electrochemical Biosensing Platform Using Carbon Nanotube Activated Glassy Carbon Electrode, Electroanalysis, 19 (2007) 1623.

[11] Jay D. Wadhawan, Uwe Schro der, Andreas Neudeck, Shelley J. Wilkins a, Richard G. Compton, Frank Marken, Crestina S. Consorti, Roberto F. de Souza, Jaı rton Dupont, Ionic liquid modified electrodes. Unusual partitioning and diffusion effects of $[Fe(CN)_6]^{4-/3-}$ in droplet and thin layer deposits of 1-methyl-3-(2,6-(S)-dimethylocten-2-yl)-imidazolium tetrafluoroborate, Journal of Electroanalytical Chemistry 493 (2000) 75–8

[12] Alanah Fitch, Clay-modified electrodes: a review, Clays and Clay Minerals, Vol, 38, No. 4, 391-400, 1990

[13] Rebecca A. Zangmeister, Jung J. Park , Gary W. Rubloff , Michael J. Tarlov, Electrochemical study of chitosan films deposited from solution at reducing potentials, Electrochimica Acta 51 (2006) 5324–5333.

[14] Reza Emamali Sabzi, Ali Hassanzadeh, Parvaneh Heravi and Khosrow Ghasemlu, Al Electrode Modified by Au Atoms as a Novel Electrode for Electrocatalytic Oxidation of Thiosulfate, Journal of the Chinese Chemical Society, 2007, 54, 977-982.

[15] Xianxia Yuan, Naixin Xu, Determination of hydrogen diffusion coefficient in metal hydride electrode by modified Warburg impedance, Journal of Alloys and Compounds 329 (2001) 115–120.

[16] Dong Shaojun, Li Fengbin, Researches on chemically modifiedelectrodes: Part XVI. Electron diffusion coefficient in vanadium hexacyanoferrate film, Journal of Electroanalytical Chemistry and Interfacial Electrochemistry,Volume 217, Issue 1, 23 January 1987, Pages 49–63

[17] Paul M. S. Monk, Fundamental of Electroanalytical Chemistry, John Wiley and Sons Ltd, England (2001)

[18] W. T. Tan, G. K. Ng, and A. M. Bond, Electrochemical oxidation of microcrystalline tetrthiafulvalene(TTF) at an electrode-solid-aqueous(KBr) interface, Malysian Journal of Chemistry, 2, 2000, pp. 34-42.

[19] F. Scholz, and B. Lange, Abrasive stripping voltammetry-an electrochemical solid state spectroscopy of wide applicability, Trends in Analytical Chemistry, 11, 1992, pp. 359-367.

[20] F. Scholz and B. Lange, Abrasive stripping voltammetry - an electrochemical solid state spectroscopy of wide applicability.Trends in Analytical Chemistry, 11 (1992) 359.

[21] M. M. Radhi, W. T. Tan, M. Z. B Ab Rahman, and A. B. Kassim, Electrochemical Reduction of Mn (II) Mediated by C_{60}/Li^+ Modified Glassy Carbon Electrode, Int. J. Electrochem. Sci., 5 (2010) 254-266.

[22] A. J. Bard and L. R. Faulkner, Electrochemical Methods, Fundamentals and Applications, Wiley, New York, (2001).

[23] Gynthia G.Zoski, Handbook of Electrochemistry, 1st. ed. 2007, Elsevier, US.

[24] Yeo May Ching, Tan Wee Tee, Zulkarnain Zainal, Electrochemical Studies of Mn(II) Mediated by Li+ Doped Indium Tin Oxide(ITO) Electrode, Int. J. Electrochem. Sci., 6 (2011) 5305 – 5313.

[25] Ganchimeg Perenlei, Tan Wee Tee, Nor Azah Yusof, Goh Joo Kheng, voltammetric detection of potassium Ferricyanide mediated by multi-walled carbon nanotube/Titanium dioxide composite modified glassy carbon electrode, Int. J. Electrochem. Sci., 6 (2011) 520 – 531.

Quantum Dots Sensitized Solar Cell: Effect of CdSe Nanoparticles Purification Procedure of QD Sensitized Photoanodes

K. A Yaacob, M. N. Ishak and N. N Alias

School of Materials and Mineral Resources Engineering, Universiti Sains Malaysia, Engineering Campus, 14300 Nibong Tebal, Pulau Pinang, MALAYSIA

Email: mraisha@eng.usm.my

Abstract. In this research the effect of purification of CdSe nanoparticles for application in quantum dots sensitized solar cells (QDSSC) photoanodes are studied. The CdSe nanoparticles are attached to the titanium dioxide surface using a linker based approached (CdSe nanoparticles disperse in toluene) and direct mode attachment (CdSe re-disperse in dichloromethane (DCM)). Colloidal CdSe nanoparticles with estimated size of 3.0 nm were synthesized by hot injection method in trioctylphosphine oxide (TOPO) as stabilizing solvent. Prior to the sensitization, the CdSe nanoparticles were purified using a common purification step involving the alternate cycles of precipitation / redispersion in non-polar solvent and polar solvent. With increasing the number of purification, the concentrations of CdSe nanoparticles attached to the titanium dioxide were also increased; from 2.47 x 10^{15} dots/cc for 3x wash CdSe nanoparticles to 3.70 x 10^{15} dots/cc for 4x wash CdSe nanoparticles. Polysulfide electrolyte and Cu_2S counterelectrodes were used to assemble a complete QDSSC. The highest efficiency of 0.05% was obtained from 4x wash CdSe nanoparticles; V_{oc}=0.2V, J_{sc}=0.34 mA/cm^2 and FF= 0.07).

1. Introduction

The solar cells with semiconductor nanoparticle have attracted interest in recent years. This generation of solar cells is moving toward for reducing the cost with better efficiency. Furthermore semiconductor nanoparticles have some unique features such as large surface-to-volume ratio, fast charge transport to surface, and controllable morphology, ease for large area processing, and potential for inexpensive fabrication on flexible substrates. Moreover by using nanoparticles one can enhance charge injection into wide band gap semiconductor.

One of the most significant recent advances in photovoltaic was the development of the dye sensitized solar cell (DSSC) or the Gratzel cell. The photoanode in such cells is constituted by a mesoporous TiO_2 layer sensitized to the visible radiation by an adsorbed dye. Recently, instead of using dyes the sensitization of TiO_2 mesoporous electrodes can be achieved through modification of the oxide with nanoparticles of semiconducting materials so called quantum dot sensitized solar cell. The use of nanoparticles enables band gap tuning by controlled the size of the nanoparticle size, which allows adjusting the light absorption. There are several methods used to fabricate QDSSC, such as chemical bath deposition (CBD) [1,2] and successive ionic layer adsorption and reaction (SILAR)[3,4]. These two methods give a high coverage of nanoparticles with good anchorage to the substrate. But there is a limitation of these methods, where it is difficult to control the size of the nanoparticles and a broad size distribution of the deposited nanoparticles is obtained. This problem can be overcome by fabricating the QDSSC using pre-synthesized colloidal quantum dots for which it is much easier to prepare monodispersed samples with the required size[5-7]. Linker based approaches are the most common method used to incorporate these quantum dots into the mesoporous TiO_2 electrode. First the TiO_2 electrode is coated with bifunctional molecule linker followed by immersion of the electrode in quantum dots solution for long deposition times (24 – 96 hr) in order to achieve high surface coverage. However, linker based approached has been plagued by the instability between quantum dot to linker and linker to substrate interaction, which make this method have low repeatability. An alternative method to attach the QD onto TiO_2 mesoporous without using molecular linker termed as direct attachment method has been discover[7]. In this method CdSe nanoparticles capped with TOPO dissolve in dichloromethane (CH_2Cl_2) and directly absorbed onto TiO_2 mesoporous as light harvesting materials for converting light to current in solar cell.

Despite the attachment method, purification of CdSe nanoparticles will also increase the attachment of quantum dots on TiO_2 mesoporous films. Parkinson and co-workers have reported that the unwashed TOPO capped CdSe nanoparticles are agglomerated on rutile (110) TiO_2 substrate functionalised with 3-mercaptopropionic acid (MPA) [8]. They recommended that the excess free ligands in the CdSe nanoparticles solution such as TOPO, cadmium acetate, selenium, hexadecylamine (HDA) and tetra-decylphosphonic acid (TDPA) absorbed competitively to TiO_2 which avoid the attachment of quantum dots to the TiO_2 and also promoting the agglomeration of the quantum dots. Riley's group reported that three washing cycle increased the quantum coverage on the TiO_2 and less aggromeration [9]. Whereas, Kamat's group reported that five washing cycle was optimal to remove excess TOPO and retained some of capping ligands to avoid flocculation in solvent while also allow the quantum dots to interact with MPA on TiO_2[10].

In this paper we study two method of attachment, the linker based approaches and the direct attachment method, to link the quantum dots and the TiO_2, and study the effect of purification procedure of the quantum dots sensitized photoanodes. We used phosphorous NMR to analyses the present of TOPO in CdSe nanoparticle solution before and after purification. The *J-V* measurement was performed to measure the efficiency of the complete cell.

2. Experimental Methods

2.1 Synthesis of CdSe nanoparticles.

CdSe nanoparticles were synthesized using hot injection method, both precursors and coordinating solvents were heated together and nanoparticles formed and growth under non-isothermal conditions. Firstly the precursor of TOPSe and TOPCd solutions were prepared in glove box under dry atmosphere (argon). TOPSe was prepared by dissolving selenium (100 mesh with 99.99% trace metals basis, Aldrich) into trioctylphosphine (TOP) (90%, Aldrich). The cadmium stock solution, TOPCd was prepared by dissolving cadmium acetate (99.99% trace metals basis, Aldrich) into TOP. The coordinating solvent is prepared by mixing of trioctylphosphine oxide TOPO (99%, Sigma), hexadecylamine (90%, Aldrich) and tetra-decylphosphonic acid (97%, Aldrich). The mixture were dried, degassed under vacuum and refluxed at 110°C for 1 hour 30 minutes. TOPSe was then injected into the reaction mixture and the temperature increased to 300°C. TOPCd was then rapidly injected under vigorous stirring resulting in the nucleation of CdSe nanoparticles. Once the desired nanoparticle size was reached the reaction was quenched in 40 ml of toluene at room temperature.

2.2 Purification of CdSe nanoparticles solution

Figure 1 illustrated the procedure used to purify CdSe nanoparticles. In the ratio of 40 ml methanol: 40 ml CdSe solution the nanoparticles were precipitated and isolated by centrifugation (at 5000 rpm for 10 minutes). After centrifugation two layers of solution form with the colorless solution on top and the orange powder in the bottom. The colorless solution was poured out and the orange powder re-dispersed in 40 ml dichloromethane or toluene. This was labeled x1 purification cycle. The remaining solution was continued for next precipitation cycle, the ratio of 30 ml of CdSe nanoparticles with 30 ml of methanol wasused. The same centrifugation was done and the purified nanoparticle is re-dispersed in 30 ml dichloromethane or toluene and labeled x2 purification cycle. This stage was repeated for the desired number of precipitation. For each cycles, the UV-Vis spectroscopy and phosphorous NMR analysis was done to study the effect of precipitating of CdSe nanoparticles.

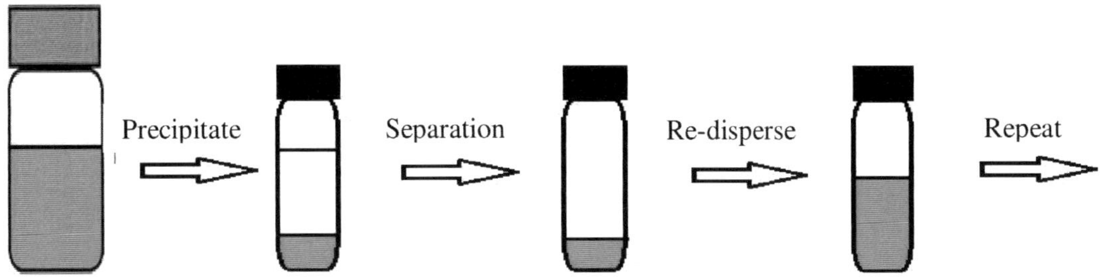

Figure 1: Schematic of precipitation procedure.

2.3 Preparation of CdSe-TiO_2 films.

Titanium dioxide (TiO_2) paste (Dyesol) was applied to conducting side of the fluorine doped tin oxide (FTO) using a glass rod and quick downward sweeping motions by Doctor Blade technique. Films were subsequently annealed in furnace at 450°C for 30 minutes, in order to remove organic molecule use in the paste and to achieve good electrical contact between the TiO_2 nanoparticles.

2.4 Molecular Linker Approach.

The annealed FTO/TiO_2 films were re-heated to 110 °C for 15 minutes to remove any moisture in TiO_2 films. The hot FTO/TiO_2 films were then directly immersed into acetonitrile (99.5%, Aldrich) solution with 3-mercaptopropionic acid (MPA) (99%, Aldrich) in the ratio of 10:1 for >12 hours. Finally, the TiO_2 films were washed with toluene and transferred into a CdSe toluene solution with different number of precipitation. The electrodes were immersed in the solution for >2 days, rinsed with toluene and left to dry in air.

2.5 Direct Absorption with DCM.

The TiO_2 films were re-heated to 110 °C for 15 minutes to remove any moisture in TiO_2 films. The hot films were then directly immersed into CdSe nanoparticles re-dispersed in dichloromethane with different number of precipitation for 2 days.

2.6 QDSSC preparation

In this research we used the Cu_2S as counter electrode. The Cu_2S counter electrodes were prepare by immersing brass into HCl solution at 70°C for 10 min and later drop a polysulfide on the brass surface for another 10 min. The composition of polysulfide electrolyte was prepared by mixing 1 M Na_2S, 1M S, and 0.1M NaOH solution in DI (deionized) water. The solar cells were prepared by assembling of the photoanode and counter electrode using a Scotch tape as a space and with a polysulfide as an electrolyte. For QDSSC, the Cu_2S and polysulfide were employed in order to avoid corrosion to the quantum dots [11].

2.7 Characterization

UV-vis absorption spectra of CdSe nanoparticles were recorded with Cary 50 Conc with wavelength range from 190 nm – 1100 nm. Phosphorous Nuclear Magnetic Resonance (^{31}P NMR) analysis wasperformed using Bruker Avance 400 MHz to measure the amount of phosphorus contained in CdSe nanoparticles before and after precipitation. The current density-voltage (J-V) curves for solar cell characterization were obtained using FRA-equipped PGSTAT 30 from Autolab. The cells were illuminated using a 150 Watt table lamp.

3. Results and Discussions

3.1 Optical Analysis of CdSe nanoparticles

UV-visible spectroscopy was used to monitor the CdSe nanoparticle solution throughout the quantum dots purification. Figure 2 shows the unpurified and purified CdSe nanoparticles from x0 to x4 purification cycle. The number of purifications does not affect the optical properties of nanoparticles. The positions of peak still remain at the same for unpurified CdSe nanopareticle solution throughout the purification which is 560 nm, which give the estimation diameter of ca. 3 nm. The quantum dots diameter was calculated by fitting to the mathematical formula describe by Peng *et al.* below [12]:

$$D = (1.6122 \times 10^{-9})\lambda^4 - (2.6575 \times 10^{-6})\lambda^3 + (1.6242 \times 10^{-3})\lambda^2 - (0.4277)\lambda + (41.57) \quad (1)$$

where$D$ is the size of a given nanoparticle sample (in nm). λ is the wavelength of the first excitonic absorption peak of the corresponding sample (in nm).

From the calculated diameter, the extinction coefficient per mole of particles, ε, can be determined by:

$$\varepsilon = 5857\,(D)^{2.65} \quad (2)$$

From figure 2b, the concentration of CdSe nanoparticles were slightly reduced from x1 purification cycle to x4 purification cycle. For x1 purification cycle and x2 purification cycle, at the higher wave length (650 – 800 nm) shows a raised baseline which is due scattering effect, compared to third and fourth precipitation. This may be caused by the present of excess impurities such as TOPO, HDA, TOPSe in the CdSe nanoparticles solution. From this we can say that the purification procedure remove the excess TOPO in the CdSe nanoparticles solution.

The concentration in unit dot/cc of the CdSe nanoparticles can be calculated from absorbance by using equation 3:

$$C = \frac{A \times N_A}{\varepsilon \times L \times 1000} \quad (3)$$

where A is the absorbance value at the peak position of the first excitonic absorption peak. C is the molar concentration (dot/cc) of the nanoparticles. L is the path length (1 cm) of the radiation beam used for recording the absorption spectrum. N_A is the Avogadro number.

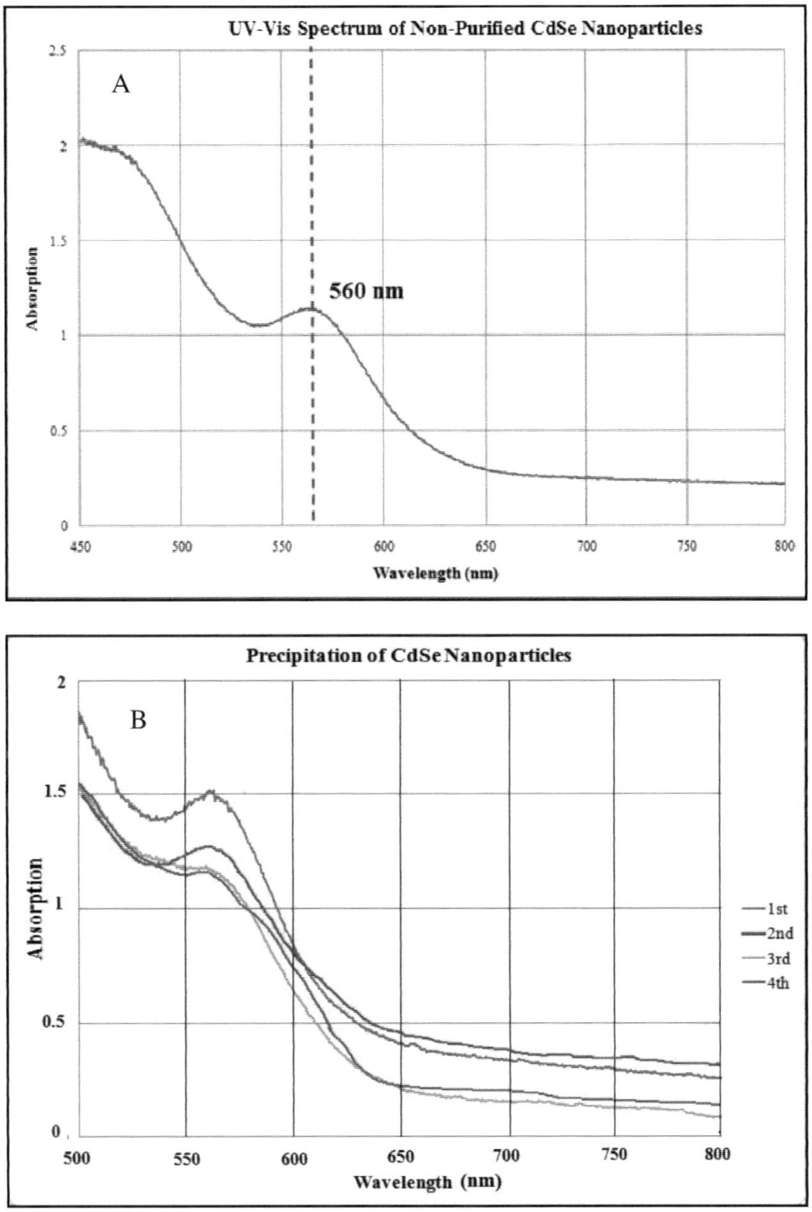

Figure 2. UV-Visible spectra of absorption of (a) non-purified (in toluene) and (b) purified CdSe nanoparticles (re-disperse in DCM).

3.2 Purification of CdSe

Phosphorous NMR spectroscopy is used to identify the phousphorous containing ligands and distinguishing between bound and free ligands in the CdSe nanoparticles solution. The TOPO will form as ligands around CdSe nanoparticles after synthesis. In the phosphorous NMR spectra, if the peak is broadened shows that ligand is bound to the surface. From the result obtained in Figure 3 show that a thin narrow peak, which indicate that ligand are not bound to the surface of CdSe nanoparticles. In this work, TOPO and TOPSe were first measure in order to confirm the peak position for TOPO and TOPSe. The results show that TOPO exist around 49 ppm and TOPSe exist around 37 ppm, similar results were reported by Morris-Cohen *e.t al* [13]. As can be seen in Figure 3, for unpurified

CdSe nanoparticles solution show the TOPSe peak at 37 ppm and no sign of TOPO exist in the solution. This is because, the as-synthesis CdSe nanoparticles were leave in dark for 3 days in order to allow the unreacted TOPO to solidified and sediment at the bottom of the vial. In this work we only collect the CdSe nanoparticles above the solidified TOPO for the phosphorous spectroscopy analysis. After the x4 purification cycle, the amounts of these two impurities were eliminated from the CdSe nanoparticles solution. Figure 3indicated that for x4 purification cycle, the presence of TOPSe and TOPO were not detected using phosphorous NMR measurement, which showed that the impurities were removed during precipitating process.

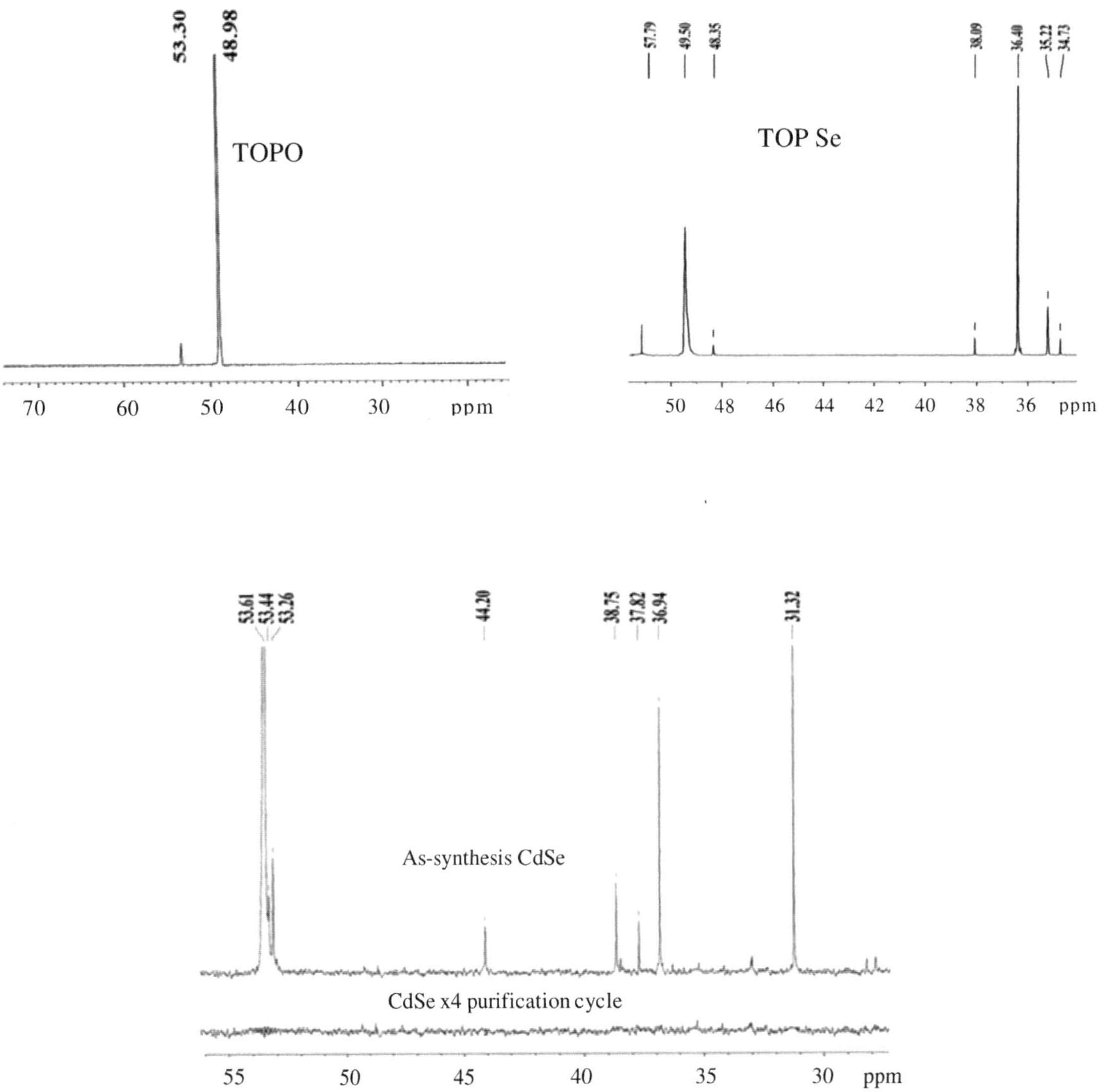

Figure 3. ^{31}P NMR spectra of TOPO, TOPSe precursor, as-synthesis CdSe and CdSe x4 purification cycle.

3.3 Attachment of CdSe nanoparticle to the TiO_2 mesoporous structure

In this work, we have attempt two methods of attachment the quantum dots to TiO_2 mesoporous structure, which were linker attachment method (CdSe re-dispersed in toluene) and direct absorption method (CdSe re-dispersed in dichloromethane). Figure 4 shows the attachment of CdSe nanoparticles onto TiO_2 films using direct absorption method and no attachment was observed for molecular linker approach. The absorption of CdSe nanoparticles on TiO_2 mesoporous films in toluene was difficult to achieve compared to direct absorption[14]. This is because the CdSe nanoparticles in toluene have better stability where the bonding between particles is much stronger if comparing to CdSe nanoparticles re-dispersed in DCM. As we can see in the inset of figure 4 after x1 and x2 purification, the TiO_2 films were yellowish in color, which is the original color of the TiO_2 films before sensitized with CdSe nanoparticle. In contrast, for x3 and x4 purification shows areddish color due to the CdSe nanoparticle has attached onto TiO_2 film. For x3 purification the CdSe nanoparticles deposited is not uniform as compare to the x4 purification. Besides, x4 purification show a highest coverage of CdSe nanoparticles as compare to x3 purification of CdSe nanoparticle coverage onto TiO_2, which show that the surface coverage of CdSe nanoparticles increase when the number of purification increased.

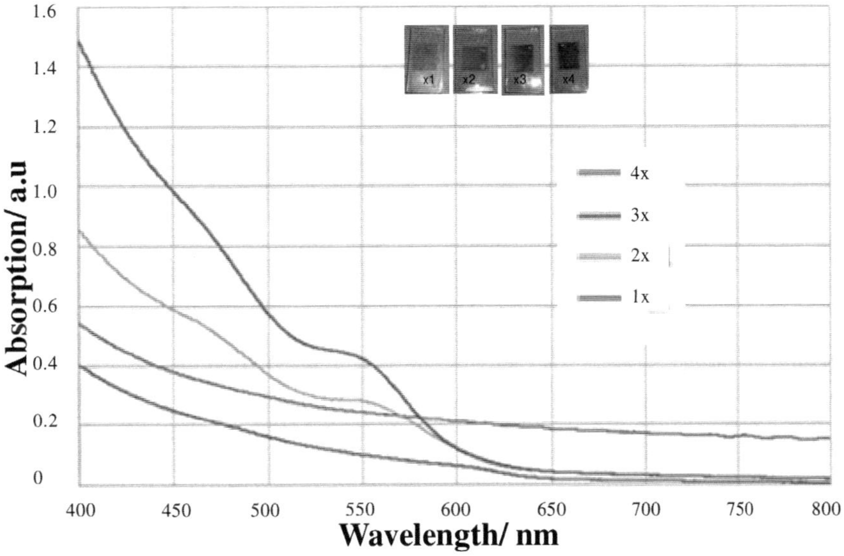

Figure 4. UV-Visible spectrum of deposited CdSe nanoparticles on TiO_2 for different number of precipitations. The inset shows photos of CdSe nanoparticles attachment to TiO_2 for different precipitations.

UV-vis absorption spectra of attachment of CdSe nanoparticles on TiO_2 films from x1 to x 4 purification cycles are represents in figure 4. For x3 and x4 purification cycles solution, the same excitonic peak corresponding to CdSe nanoparticles before the attachment occurs at the same wavelength (560 nm). However, for x1and x2 purification the peak is not observed due to no or small amount of CdSe nanoparticles that attached to TiO_2 film. As mention earlier, the impurities will reduce the performance of nanoparticle during application. This is because some of the TOPO on CdSe nanoparticles will be removed and leaved the Cd^{2+} enriched surface of CdSe nanoparticles. We suggest that this Cd^{2+} enriched surface of CdSe nanoparticles will physisorbed to the negatively charge of OH^- of TiO_2. Increasing the no. of purificationwill remove more TOPO and leaved more Cd^{2+} side on the surface of CdSe nanoparticle. High Cd^{2+} enriched surface generated, more CdSe nanoparticle will attached to the TiO_2. Using equation 1, 2 and 3, the concentration of CdSe nanoparticles on TiO_2 film was calculated and shows in Table 1. Concentration for x1 and x2 purification cycle were unable to calculated due to no absorption peak were observe in the UV spectroscopy data in Figure 4b. From the result, by increasing number of purification, the concentration of CdSe nanoparticles on TiO_2 film was also increase, which means more CdSe nanoparticles were attached to TiO_2 film due to removal of TOPO.

Table 1. The concentration of CdSe nanoparticles sensitized on TiO_2 film with different number of purification.

Number of Purification Cycle	Concentration (dot/cc)
3	2.47×10^{15}
4	3.70×10^{15}

3.4 J-V measurement of complete solar cell

The current density-voltage (*J-V*) curves was recorded to evaluate the solar energy conversion of full assemble of quantum dots sensitized solar cells. Under illumination of 150 Watt lamp, Figure 5 shows the results of *J-V* curves for x3 and x4 purification cycles of CdSe nanoparticles. The efficiency (η) and fill factor (*FF*) of the fabricated solar cell can be calculated using equations below:

$$FF = \frac{P\,max}{V\,oc\,.\,Jsc} \quad (4)$$

$$\eta = \frac{V\,oc\,.\,Jsc.FF}{Pinc} 100\% \quad (5)$$

Where P_{max} is the maximum power density of the solar cell, V_o is the open circuit voltage, J_{sc} is the short circuit current density and P_{inc} is the incident light power density.

After analyzing the solar cell parameters obtained for different purification cycles reported in figure 5. Theopen circuit voltage, V_{oc}, short circuit voltage, J_{sc}, fill factor and efficiency of our quantum dot sensitized solar cells for x3 purification cycle are -0.0063 V, 0.0156 mA/cm^2, 0.3 and 0.003% respectively and for x4 purification cycle are -0.2 V,3.4mA/cm^2, 0.07 and 0.05 % respectively. The efficiency, η, value increase with number of purification cycle due to the high coverage of CdSe nanoparticles at high purification cycle. This indicated that at high coverage more electrons were transfer to the substrate, which increases the efficiency.

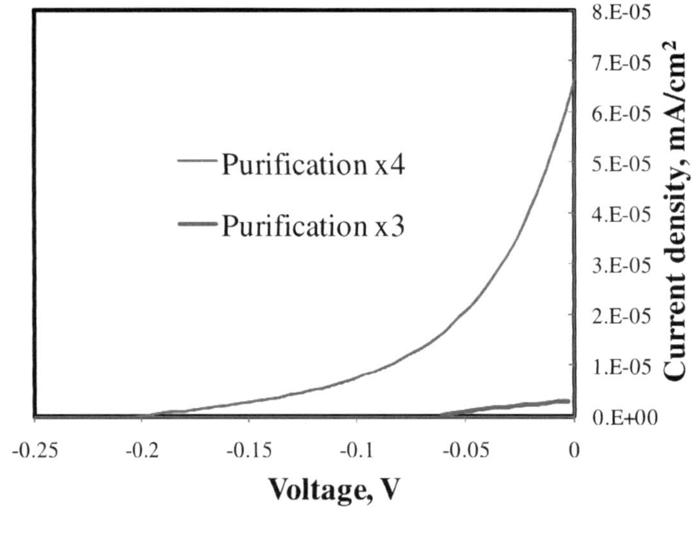

Figure 5. The J-V curves of the CdSe/TiO_2 system for x3 purification cycles and x4 purification cycle.

4. Conclusions

To conclude, we have report the effect of purification toward sensitization of CdSe nanoparticles onto TiO_2 mesoporous films using direct absorption method. The results show that increasing number of purification will increase the attachment of CdSe nanoparticles onto TiO_2 mesoporous structure, which indicated by the greater colouration of TiO_2 mesoporous films. This is due to the removal of the free ligands (TOPO and TOPSe) from the CdSe nanoparticles solution after the purification, which is supported by phosphorous NMR result. The devicesperformances were also observe, obtaining better result for x4 purification cycle. In future we will conduct a more detail characterization of the purified solution and the performance of quantum dots sensitized photoanodes.

Acknowledgement
The author would like to thank the Universiti Sains Malaysia for the financial support from USM Short Term Grant.

References

[1] Gorer, S., Hodes, G., (1994) J. Phys. Chem. 98 5338
[2] Niitsoo, O., Sarkar, S.K., Pejoux, C., Rühle, S., Cahen, D., Hodes, G., (2006), J. Photochem. Photobiol. A, 181 306
[3] Diguna, L. J, Shen, Q., Lobayshi, J. Toyoda, T., (2007), Appl. Phys. Lett., 91 023116
[4] Nicolau, Y. F., (1985) Appl. Surf. Sci. 22/3 1061
[5] Robel I., Subramanian V., Kuno M., and Kamat P.V., (2006). J. Am. Chem. Soc., 128, 2385
[6] López-Luque, T, Wolcott, A., Xu, L.P., Chen, S., Wen, Z., Li, J., Rosa E De La, Zhang, J. Z. (2008) J. Phys. Chem. C 112 1282
[7] Giménez, S., Mora-Séro, I., Macor, L., Guijarro, N., Lana-Villarreal, T., Gómez, R., Diguna, L. J., Shen, Q., Toyoda, T., Bisquert, J., (2009) Nanotechnology 20 295204
[8] Sambur, J.B., Novet, T., Parkinson, B.A., (2010) Langmuir, 26, 4839
[9] King, L.A., Riley, D.J., (2012) J. Phys. Chem. C, 116, 3349
[10] Pernik, D.R., Tvrdy, K., Ragich, J.G., Kamat, P.V., (2011) J. Phys. Chem. C, 115, 13511
[11] Gónzalez-Pedro, V., Xu, X., Mora-Seró, I., Bisquert, J., (2010) ACS Nano, 4, 5783
[12] Yu, W.W., Qu, L., Guo, W., Peng, X., (2003) Chem. Metter. 15, 2854-2860
[13] Morris-Cohen A.J., Donakowski M.D., Knowles K.E., Weiss E.A., (2010). J. Phys. Chem. C, 114, 897
[14] Lee H.J., Yum J.H., Leventis H.C., Zakeeruddin S.M., Haque S.A., Chen P., Seok S.I., Gratzel M., Nazeeruddin Md.K., (2008) J. Phys. Chem. B, 112, 11600

Journal of Physics: Conference Series **431** (2013) 012020 doi:10.1088/1742-6596/431/1/012020

Immobilization of Silver-based Bimetallic Nanoparticles on Titania-Support for Photocatalysis

J N Kuhn[a], E Qayyum [a], V A Castillo [a], K Warrington [a], and M A Barakat [b,c1]

[a]Department of of Chemical & Biomedical Engin., University of South Florida, Tampa, Fl, USA
[b] Department of Environmental Sciences , King Abdulaziz University, Saudi Arabia
[c] Central Metallurgical R & D Institute, Cairo, Egypt

Email: mabarakat@gmail.com

Abstract. Photocatalysts have potential for many applications including environmental remediation and selective catalytic transformations requiring decreased energy consumption. This study presents a systematic approach for the synthesis, characterization, and evaluation of titania-supported silver-based bimetallic nanoparticles. Titania was selected as a support due to its ability to form electron-hole pairs in the presence of appropriate radiation. This process is expected to be enhanced in the presence of silver-based bimetallic through the visible light absorption of its surface plasmon resonance. This work has been described the synthesis of silver-based bimetallic nanoparticles using colloid chemistry and the subsequent immobilization onto titania to form composite photocatalytic materials. The photocatalysts are characterized by electron microscopy, nitrogen physisorption, and x-ray diffraction. The work will also describe computational modeling of the silver-based particles surface plasmon resonance features. The results will be applied to photocatalytic environmental remediation.

1. Introduction

Photocatalysis is an emerging area for a range of applications including energy and environmental remediation [1-5]. The current application is aimed at composite photocatalysts for the energy efficient environmental remediation of hydrocarbon pollutants. This topic is of great interest to global society. The current effort is the use of surface plasmon resonance of silver-based nanoparticles as a method for converting visible sunlight into energy for activating titanium dioxide. Surface Plasmon resonance (SPR) phenomena have been identified as means for absorbing visible light for silver and gold particles. The absorption wavelength of the material changes with particle morphology, which is generally altered by modifying the particle morphology. Common morphology parameters are particle size, shape, and aspect ratio.

The synergistic combination occurs by having the plasmonic material assist titanium dioxide in generating electron-hole pairs for the decomposition of the targeted pollutants in water. Recent work [4] has shown 10 times rate enhancement for the photocatalytic decomposition of model organic molecules through the synergistic combination of silver (visible light adsorption) and titania (photocatalytic agent for splitting water spurring hydrocarbon decomposition). A further expected enhancement is expected by combining the plasmonic material with a catalytically active material.

Corresponding Author: mabarakat@gmail.com

Platinum is the leading and applied candidate for many catalytic reactions [6-8] and, thus, it is used a model catalyst material in the present study. The use of bimetallic particles may allow for the control of visible light absorption so that then entire range of visible light can be used for the photocatalytic hydrocarbon decomposition reactions. The demonstration of this control has occurred in previous studies of bimetallic particles. The surface plasmon resonance peak is a strong function of alloying in silver-copper nanoparticles and its position can be controlled. That is, it can be shifted towards absorption in the visible spectrum [9-10]. By modifying the composition and the atomic arrangement we can tune both the breadth and the location of the peak of the SPR spectrum of silver-copper alloy nanoparticles and thereby absorb a larger portion of the sunlight.

Recently, Christopher *et al.* [4] have shown that plasmonic silver nanocubes, characterized by excellent absorption of visible light through the creation of resonant surface plasmons can utilize concurrently thermal energy and a low-intensity photon flux (on the order of solar intensity) to drive catalytic oxidation reactions at significantly lower temperatures than those associated with conventional thermal processes. They demonstrated for the first time that metallic nanostructures can couple efficiently thermal energy and low-intensity visible photons to drive commercially important catalytic reactions; these results open new avenues towards the design of more energy-efficient catalytic processes with a significant fraction of energy input provided in the form of sunlight; the results suggest that, potentially, the long-term stability of catalysts and product selectivity could be enhanced because the reactions can be operated at low temperatures. Although the Ag nanocubes drastically enhance the absorption of visible light by the creation of resonant surface plasmons, the SPR peak is near the UV region and hence, there is lower utilization of the visible light spectrum of the sunlight.

Advances in nanotechnology now permit avenues to control the size, shape, and aspect ratio of the plasmonic materials. Great developments in this field have occurred in the recent years [11-14]. In particular, a great devotion of effort has focused on key catalytic metals which include platinum and rhodium [15-19]. In the present study, the size control of platinum-silver nanoparticles is demonstrated. The combination of a plasmonic metal (silver) and a catalytic metal (platinum) is proposed a powerful approach for developing the next-level materials for photocatalytic applications.

2. Materials and Methods

The synthesis of all particles follows the same general procedure and is outlined in each of the following sections. The overall process is also described in Figure 1.

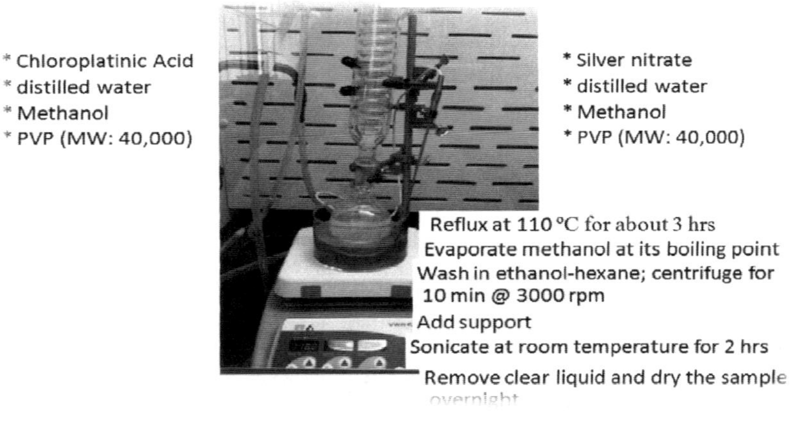

Figure 1. Synthesis of platinum, silver, and platinum-silver nanoparticles.

2.1. Synthesis of Pt Nanoparticles

Platinum nanoparticles, approximately 3 nm in size, were synthesized using previously established mechanisms [17, 20]. Solvents used in this synthesis were reagent grade chemicals from Sigma-Aldrich. Chloroplatinic acid (H_2PtCl_6; Sigma-Aldrich; mass of 62.15 mg) was combined with 20 ml of distilled water. Methanol (180 ml) was added to the fully dissolved acid solution in a round bottom flask. Subsequently, polyvinylpyrrolidone (PVP MW: 40,000 g/mol; Sigma-Aldrich; mass of 133 mg) was added and the solution, refluxed for three hours at 110°C. The solution turned dark brown during this process. Following the synthesis, the solvent was evaporated from the particles at 55°C and the particles were re-dispersed in ethanol. The particles were washed with hexane and re-dispersed in ethanol three times. For each washing, 3 ml of the ethanol – particle solution was combined with 9 ml of hexane and centrifuged (VWR clinical 100 centrifuge) at 3000 rpm for 10 minutes.

2.2. Synthesis of Ag Nanoparticles

The above procedure for Pt nanoparticles was repeated using the equivalent Ag precursor molar concentration as to that of Pt. The precursor was $AgNO_3$ (Sigma-Aldrich). Following reduction, the solution turned an orange-yellow type color due to the surface plasmon resonance phenomena of Ag nanoparticles.

2.3. Synthesis of Ag-Pt Bimetallic Nanoparticles

The above procedure for Pt nanoparticles was repeated using the equivalent total metal concentration. The molar ratio of Pt: Ag was 1: 1 in precursor solution. The precursor was $AgNO_3$ (Sigma-Aldrich). The solution turned dark brown during the reduction process.

2.4. Synthesis of Ag-Pt Mixture Nanoparticles

Monometallic Pt (section 1) and Ag (section 2) nanoparticles were physically mixed as colloidal solutions after washing. The molar ratio of Pt: Ag was 1 : 1 in the resulting mixture.

2.5. Immobilization of Nanoparticles onto Titania Supports

The particles for each above synthesis were immobilized onto titania (Alfa-Aesar). The particles were supported by mild sonication for 3 hours at a metal: titanium dioxide mass ratio of 0.3%**:** 99.7%. Following sonication, the composite materials were dried with mild heating on a hot plate (T ~ 60 °C) and then in a drying over (T ~ 90 °C).

2.6. Characterizations

Particle images were obtained using a FEI Morgagni 268D microscope. In addition to imaging, particle size distributions were obtained by measuring the size of 100 particles for each sample using the imageJ freeware, which was downloaded from the NIST site. Nitrogen physisorption was performed on a Quantachrome Autosorb I. The surface area determined using the standard BET analysis.

3. Results and Discussions

The particle imaging results are shown in Figures 2, 3, and 4, respectively, for platinum, silver, and platinum-silver bimetallic particles. In each case, the left sub-figure shows the particles with a scale bar of 500 nm and the right sub-figure shows the particles with a 200 nm scale bar.

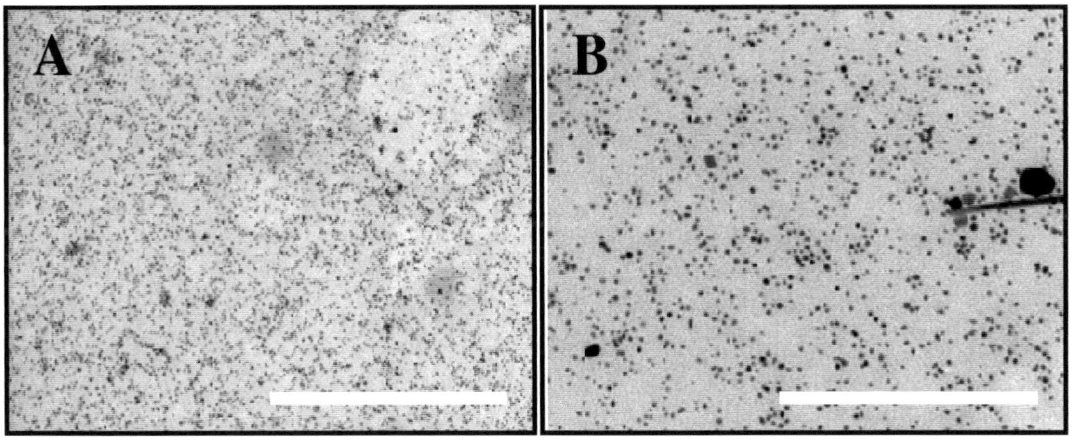

Figure 2. TEM images of platinum nanoparticles (Scale bar (A) is 500 nm and (B) is 200 nm).

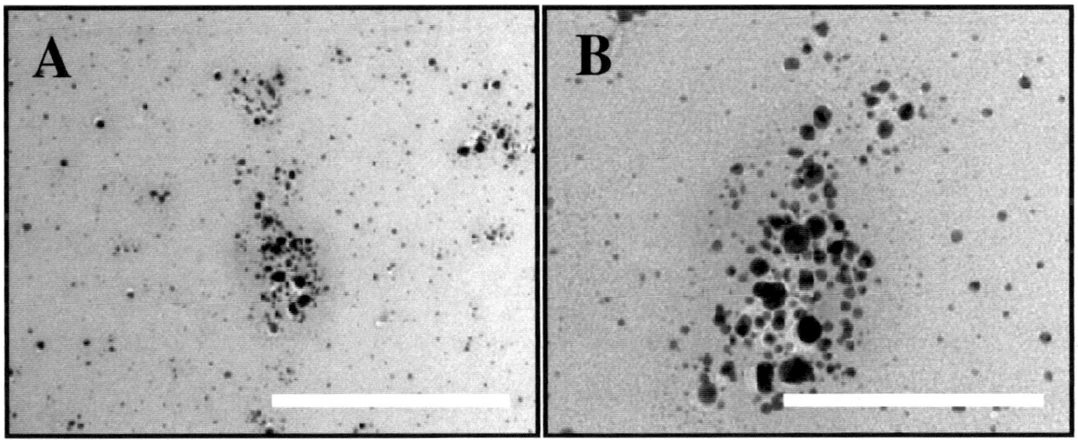

Figure 3. TEM of silver nanoparticles (Scale bar (A) is 500 nm and (B) is 200 nm).

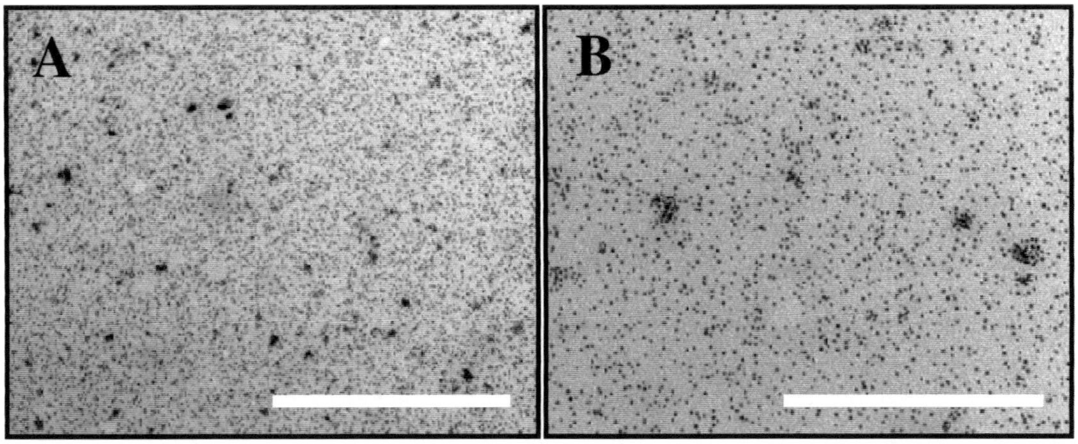

Figure 4. TEM of silver-platinum bimetallic Nanoparticles (Scale bar (A) is 500 nm and (B) is 200 nm).

The statistics of the particle sizes are compiled in Figure 5. The platinum particles (Figure 2) possessed a range of sizes between ~1.5 and 4 nm with an average near 3 nm. The size distributions are consistent with previous work on this synthesis system for platinum [15, 17]. As with the previous work, washing is very important to cleaning the metal surface from surfactant. The three cycles of ethanol-hexane washing removed more than 95% of excess carbon [15]. Following the synthesis, the platinum particles gave a black color, which indicates that it does not absorb visible light (expected for platinum). The size distributions of these particles (and all particles in this work) yield, using a common model in metal particle catalysis [8], a surface-to-bulk ratios or dispersions of 25 to 33%. That is, between a quarter and a third of the particles are available for catalysis because they are located at the surface of the particles.

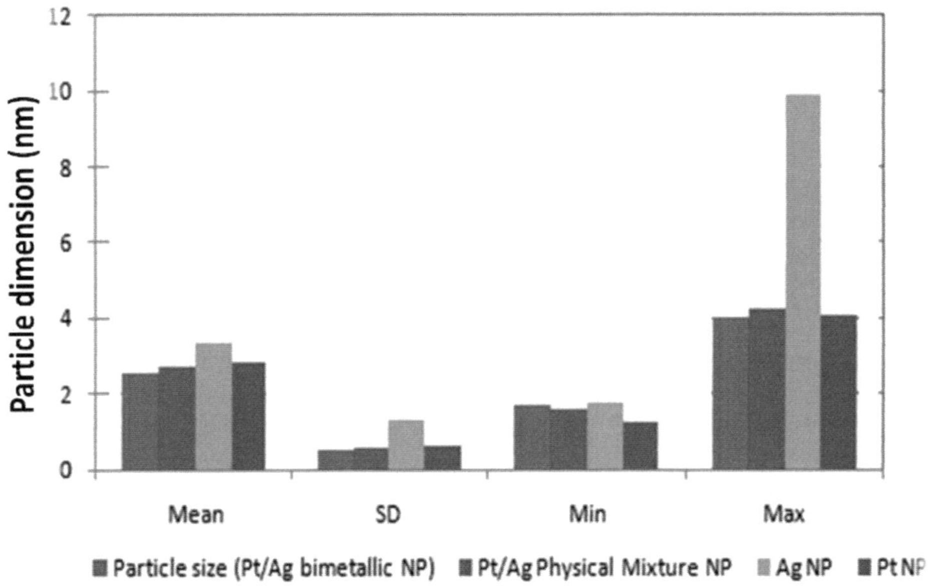

Figure 5. Size distribution statistics obtained from counting 100 synthesized nanoparticles. SD is the standard deviation.

With slight modification, the synthesis approach was adapted to silver. Following the synthesis, the silver particles in solution yield an orange color, which would be consistent with a SPR peak near 590 nm. The silver particles (Figure 3) were slightly larger than the monometallic platinum particles. The bulk portion of the particles was of similar size. The size distribution was skewed to a large particle size because a small minority of particles existed near a particle size of about 10 nm. Interestingly, the bimetallic platinum-silver particles (Figure 4) were more similar in size to platinum than to silver. This relationship implies that the platinum particles may act as nucleation sites for the subsequent growth of silver on its exterior. Advanced characterization studies are currently being performed to better assess the location of silver and platinum in the particles. The particles also yielded a black color, which indicates the platinum dominates silver's optical properties. A physical mixture of monometallic platinum and silver particles is also included. As expected, the results are in very good agreement with the parent compounds.

The four sets of particles described were immobilized on titania for further characterization. The surface area analyses for the particles supported on titania are presented in Figure 6. For comparison, the same particles supported a commercial silica (Cab-O-Sil) is presented. The results showed that the addition of the particles to supports, in general, caused a slight increase in the total surface. The effect is observed with great change when silver is added to the support. These particles are being investigated for photocatalytic conversion of model pollutants and the results are presented in a separate paper at this conference.

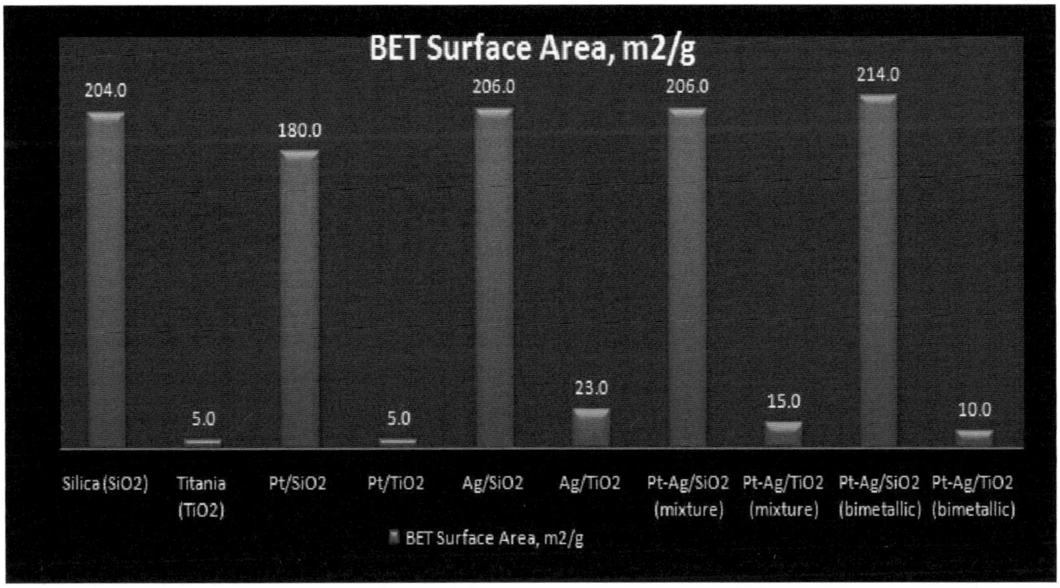

Figure 6. Specific Surface areas of silver, platinum, bimetallic silver-platinum, mixtures of silver and platinum particles supported on titania. A comparison to silica supports is included for comparison.

4. Conclusions

Photocatalysis powered by light through surface plasmon resonance (SPR) is a growing approach for the best use of the world's resources whether it is through energy or environmental applications. The properties of the light-absorbing metals can be enhanced by the synthesis of composite materials in which a semi-conductor is in intimate contact with the (SPR). The current work directly addresses this important issue by the advanced synthesis of SPR metals (silver and silver-platinum) and immobilization on titania. The silver, platinum, and silver-platinum particles were synthesized with good control of the particle size. After effective washing of the surfactant, the particles were successfully immobilized onto a photoactive titania for use in a range of catalytic and photocatalytic applications. The materials developed in this work are active for photocatalysis applications and this topic is the focus of a separate paper [21].

Acknowledgements

The authors gratefully acknowledge King Abdulaziz University (KAU, Saudi Arabia) for funding this work through the cooperation agreement with University of South Florida, USA (KAU-USF).

References

[1] Coutinho, C. A.; Gupta, V. K. Photocatalytic Degradation of Methyl Orange Using Polymer-Titania Microcomposites. *Journal Colloid and Interface Science* **2009**, 333, 457-464.

[2] Hou, W.; Liu, Z.; Pavaskar, P.; Hung, W. H.; Cronin, S. B. Plasmonic enhancement of photocatalytic decomposition of methyl orange under visible light. *Journal of Catalysis* **2011**, 277, 149–153.

[3] Christopher, P.; Ingram, D. B.; Linic, S. Enhancing Photochemical Activity of Semiconductor Nanoparticles with Optically Active Ag Nanostructures: Photochemistry Mediated by Ag

Surface Plasmons. *Journal of Physical Chemistry C* **2010**, 114, 9173–9177.

[4] Christopher, P.; Xin, H.; Linic, S. Visible-light-enhanced catalytic oxidation reactions on plasmonic silver nanostructures. *Nature Chemistry* **2011**, 1-6.

[5] Yen, C.-W.; El-Sayed, M. A. Plasmonic Field Effect on the Hexacyanoferrate (III)-Thiosulfate Electron Transfer Catalytic Reaction on Gold Nanoparticles: Electromagnetic or Thermal? *J. Phys. Chem. C* **2009**, 113, 19585–19590.

[6] Somorjai, G. A.; Park, J. Y. Molecular Factors of Catalytic Selectivity. *Angewandte Chemie International Edition* **2008**, 47, 9212-9228.

[7] Somorjai, G. A.; Li, Y. Introduction to surface chemistry and catalysis. *New York* **2010**, 2nd ed. , John Wiley & Sons, Inc.

[8] Bartholomew, C. H.; Farrauto, R. J. Fundamentals of Industrial Catalytic Processes. *Hoboken, NJ* **2006**, John Wiley & Sons, 2nd ed.

[9] Chowdhury, S.; Bhethanabotla, V. R.; Sen, R. Effect of Ag-Cu Alloy Nanoparticle Composition on Luminescence Enhancement/Quenching. *Journal of Physical Chemistry C* **2009**, 113, 13016–13022.

[10] Chowdhury, S.; Bhethanabotla, V. R.; Sen, R. Silver-copper alloy nanoparticles for metal enhanced luminescence. *APPLIED PHYSICS LETTERS* **2009**, 95, 131115-131113.

[11] Tian, N. Z., Z.-Y.; Sun, S.-G.; Ding, Y.; Wang, Z. L. Synthesis of Tetrahexahedral Platinum Nanocrystals with High-Index Facets and High Electro-Oxidation Activity. *Science* **2007**, 316, 732-735.

[12] Wang, C.; Daimon, H.; Onodera, T.; Koda, T.; Sun, S. A General Approach to the Size- and Shape-Controlled Synthesis of Platinum Nanoparticles and Their Catalytic Reduction of Oxygen. *Angw. Chem. Int. Ed.* **2008**, 47, 3588-3591.

[13] Burda, C.; Chen, X.; Narayanan, R.; El-Sayed, M. A. Chemistry and Properties of Nanocrystals of Different Shapes. *Chem. Rev.* **2005**, 105, 1025-1102.

[14] Narayanan, R.; El-Sayed, M. A. Some Aspects of Colloidal Nanoparticle Stability, Catalytic Activity, and Recycling Potential. *Top. Catal.* **2008**, 47, 15-21.

[15] Blavo, S., O.; Baldyga, L.; Sanchez, M. S.; Kuhn, J. N. The Effect of Stabilizing Agent on Platinum Nanoparticles and Implications Towards the Oxygen Reduction Reaction. *Journal of ASTM International* **2011**, 8, 1-9.

[16] Huang, W.; Kuhn, J. N.; Tsung, C.-K.; Zhang, Y.; Habas, S. E.; Yang, P.; Somorjai, G. A. Dendrimer Templated Synthesis of One Nanometer Rh and Pt Particles Supported on Mesoporous Silica: Catalytic Activity for Ethylene and Pyrrole Hydrogenation. *Nano Letters* **2008**, 8, 2027-2034.

[17] Kuhn, J. N.; Huang, W.; Tsung, C.-K.; Zhang, Y.; Somorjai, G. A. Structure Sensitivity of Carbon−Nitrogen Ring Opening: Impact of Platinum Particle Size from below 1 to 5 nm upon Pyrrole Hydrogenation Product Selectivity over Monodisperse Platinum Nanoparticles Loaded onto Mesoporous Silica *Journal of American Chemical Society* **2008**, 130, 14026-14027.

[18] Tsung, C.-K.; Kuhn, J. N.; Huang, W.; Aliaga, C.; Hung, L.-I.; Somorjai, G. A.; Yang, P. Sub-10 nm Platinum Nanocrystals with Size and Shape Control: Catalytic Study for Ethylene and Pyrrole Hydrogenation. *Journal of American Chemical Society* **2009**, 131, 5817-5822.

[19] Zhang, Y.; Grass, M. E.; Kuhn, J. N.; Tao, F.; Habas, S. E.; Huang, W.; Yang, P.; Somorjai, G. A. Highly Selective Synthesis of Catalytically Active Monodisperse Rhodium Nanocubes. *Journal of American Chemical Society* **2008**, 130, 5868-5869.

[20] Teranishi, T.; Hosoe, M.; Tanaka, T.; Miyake, M. Size Control of Monodispersed Pt Nanoparticles and Their 2D Organization by Electrophoretic Deposition. *J. Phys. Chem. B* **1999**, 103, 3818-3827.

[21] Barakat, M. A.; Al-Hutailah, R. I.; Hashim, M. H.; Qayyum, E.; and Kuhn, J.N. Titania-Supported Silver-based Bimetallic Nanoparticles as Photocatalysts, Environmental Science and Pollution Research (accepted).

Silicon Nitride Passivation of Silicon Nanowires Solar Cell

E S M Ashour[1]*, M Y Sulaiman[2], N Amin[3], and Z Ibrahim[1]
[1]Faculty of science and technology, UKM, Malaysia, [2]Solar Energy Research Institute (SERI),UKM, Malaysia, [3]Faculty of engineering and built environment, UKM, Malaysia.

Email: emansad12@gmail.com

Abstract. Vertical aligned silicon nanowires were synthesized using chemical etching of silicon wafer. Influence of a silicon nitride layer on the top of the silicon nanowires solar cell has been investigated. The optical properties of a Si NWs array with and without silicon nitride passivation layer are examined in terms of optical reflection property. In the presence of silicon nitride layer, 1% reflection ratio in the spectral range (250-1000nm) is achieved .In addition, the solar cell characteristics have been significantly improved, which exhibits high short circuit current as well high efficiency. Based on the current –voltage measurements and morphology results, we show that the silicon nitride layer can passivate the defects generated by wet etching processes.

1. Introduction

Silicon nanowires (SiNWs) attract significant attention because of their potential applications in many fields like sensors, transistors, lithium batteries, diodes and photovoltaic's [1-5]. Particularly it can be applied on silicon solar cells as an antireflection coating, due to low average reflectance values [6-7]. Several synthesise methods have been used to fabricate SiNWs including chemical vapour deposition [8], laser ablation [9], thermal evaporation, and solution methods [10-12]. Among these synthesis methods, wet chemical etching has been frequently used to prepare SiNWs. The electrochemical wet etching is advantageous for achieving SiNWs with controlled diameter, length, spacing and density, avoiding high –cost and low throughput conventional lithographic processes [13]. Recently,it was found that the silicon nanowires antireflection coating (ARC) prepared by electrochemical etching has achieved nearly perfect antireflection coating [14]. The superior antireflection property of the nanowires surface attributed to three reasons: huge surface area of SiNWs , rough surface morphology which leads to strong light scattering as well absorption, and graded refractive index profile between air and SiNWs that closely implies a multilayer antireflection coating [6,14-15]. Some other properties of SiNWs, as for example crystal ordination, good doping level, and excellent uniformity, imply utilization of SiNWs in silicon solar cells.

Despite all these features maximum obtained efficiency of planar solar cells used SiNWs ARC does not exceed 10%. This low efficiency can be attributed by many factors. One of factors is that the surface recombination velocity strongly increases when using SiNWs ARC, due to the large surface area [16-17]. Therefore it is necessary to develop a passivation way of the SiNWs surface and minimizing the surface states [18]. The most common and practical method of passivation is to grow high quality of nanowires silicon nitride as it is commonly done in standard solar cell processing [19]. This will also be beneficial for the development of the antireflection property of the SiNWs. In this

Journal of Physics: Conference Series **431** (2013) 012021 doi:10.1088/1742-6596/431/1/012021

work, we study the effect of SiNx layer on the surface passivation of the SiNWs ARC. Measurement of the open circuit voltage, V_{oc} and short circuit current I_{sc} gave brief information on the surface passivation quality.

2. Experimental details

Growth of aligned SiNW arrays was carried out on p-type (100) silicon (0-1Ωcm) wafers with 6 inch diameter. The etching of 3×3 cm ^{2} cutted samples were carried out in a Teflon beaker containing 5 M HF /0.02 M AgNO3 solution, at room temperature for different etching time. Prior to the etching, the samples were sequentially cleaned with acetone, ethanol and de-ionized water for 5 minutes each. Followed by cleaning with boiling piranha solution (H2SO4:H2O2 = 3:1 by volume, for 60 min) to remove any organic containments. The samples were then rinsed thoroughly with de-ionized water followed by dipping in 10% HF solution to remove any surface oxides. After that the cleaned silicon wafers were immersed in the etching solution .Hence, the tree like silver pattern wrapping the silicon samples were detached by NH3OH /H2O2 (3:1) solution. Finally, samples were rinsed with deionized water and blown dry in air.

After synthesis of SiNWs on p-type c-Si wafer, phosphorous diffusion was performed on the front surface using liquid POCl3 diffusion source at 900C. The samples were then dipped in 5% HF solution to remove the parasitic layer. Plasma –assisted nitride was deposited on SiNWs samples for 1,2,3,4 and 5 minutes. All plasma assisted nitridization was performed at a power of 100W and a temperature of 150°C, a pressure of 1 Torr and NH3/SiH4 gas flow ratio equals to 3. The back and front sides were deposited then using E-gun, followed by baking individual sides of the wafers. Finally, dried samples were co-fired in belt furnace to achieve an Ohmic contact at both ends. The optical and morphological characteristics of the SiNWs synthesized on c-Si wafer were estimated by spectrophotometer and field emission scanning electron microscopy (FE-SEM). The current–voltage (I–V) properties of the fabricated SiNWs solar cell were investigated using solar simulator (AM 1.5, 100 mA/cm2)

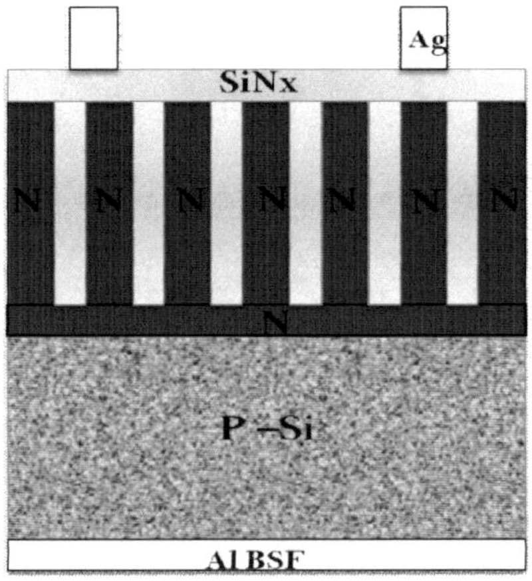

Figure 1. The schematic of the SiNx/ SiNWs-array solar cell

3. Results and discussion

3.1. Optical properties of Silicon nanowires layer

Etching time is the most essential parameter used to optimize the antireflection properties of silicon nanowires. In our case, silicon nanowires were etched for different duration time from 20 to 80 min. With the increase of etching time, the length of nanowires increased from few nanometres to micrometers. Referring to figure 2, it has been found that the length of SiNWs increases linearly with etching time for duration (0-80 min). The estimated average etching rate is 40 nm min^1 for HF and $AgNO_3$ concentration of 5.25 and 0.02 M, respectively at room temperature. The best reflectance is obtained when the etching time is 80 min, presents an effective reflectivity of 2.5 % (between 250 and1000nm). R value of our nanowires at different times ranging from 2.5 % to 3.85 % in the 250–1000 nm wavelength range. Referring to previous work [20] it was confirmed that the recombination effect is strongly dependent, increases with an increase in the etching time due to large SiNWs density as well larger surface area. For this reason our solar cell used SiNWs etched for 25 min with 3.5 % reflectivity. Figure 3 represents the spectral reflectivity of the used SiNWs along with polished silicon wafer.

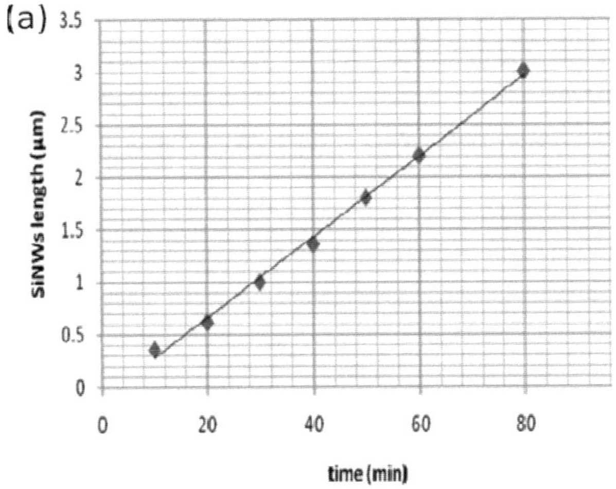

Figure 2. (a) Silicon nanowires length as a function of etching time

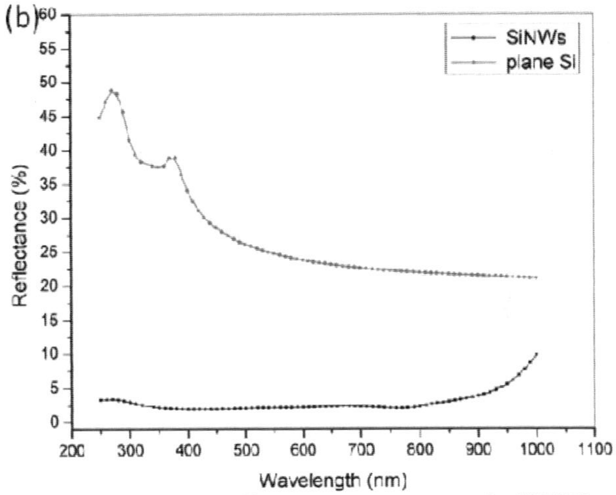

Figure 2. (b) Reflectance spectra of SiNW arrays and polished Si wafer.

3.2 Morphology of SiNWs/SiNx

The surface and cross-sectional views of the SiNWs with SiNx passivation layer were analyzed using field-emission scanning electron microscopy (FESEM), was shown Figure 3. The length of the NWs was found to be ca. 700 nm and the diameter is in the range of 50-100nm. By increasing the deposition time of SiNx, it can be seen that SiNWs diameters start to increase from 50- 100nm to 100- 130 nm. This increase in SiNWs diameters indicates radial SiNx coating of the nanowires which could be considered as a uniform passivation way. Following the FESEM photos of SiNx deposition from 1 to 5 min, it is clear that the SiNx passivation starts from bottom to up, fills the cavity between Nanowires and then start to be deposited on the surface. This can be approved by the SiNx thickness data achieved by ellipsometer measurement shown in figure 4. By comparing ellipsometer results for two samples; SiNWs and polished silicon coated with SiNx for 5 minutes, the thickness of SiNx decreases from 70 nm for polished silicon to 32 nm of SiNx for SiNWs. The decreasing in thickness confirmed the whole passivation of the SiNWs. After SiNx deposition we try to examine the reflectivity for different time durations Figure 4 , we found out that best reflectivity when we deposit SiN for 3 minutes , as we got 1% in the range 250-100nm. This result can be understood in conjunction with shallow passivation effect. Using the SiNx thickness measurements and have optimized deposition duration for SiNx (3min) with approximate thicknesses of 27 nm and spectral reflectivity of 1% in the range 250-1000nm.

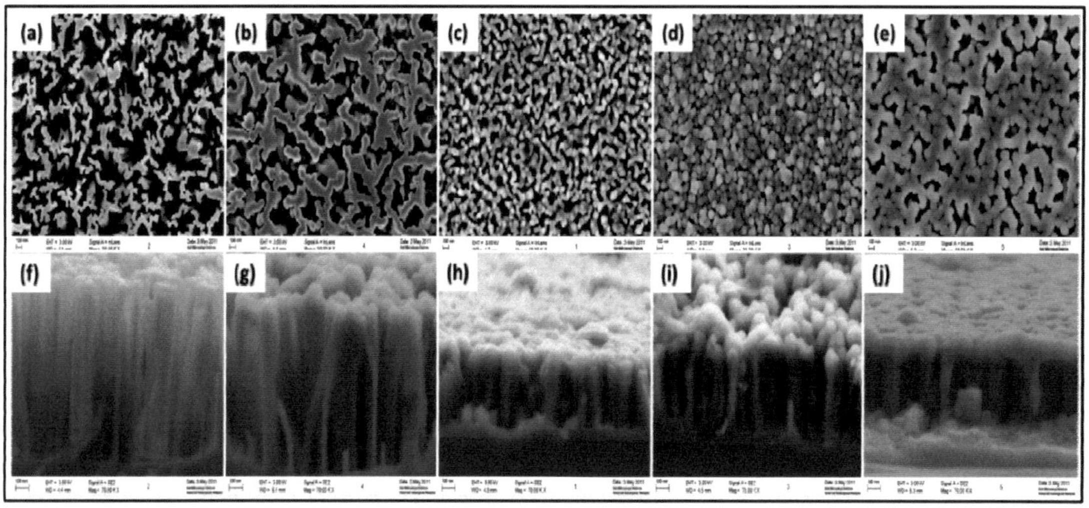

Figure 3. Plane- and cross-sectional SEM images of SiNx deposited on SiNWs for different durations (a,f) 1min, (b,g) 2min, (c,h) 3min, (d,i) 4 min, (e,j) 5 min.

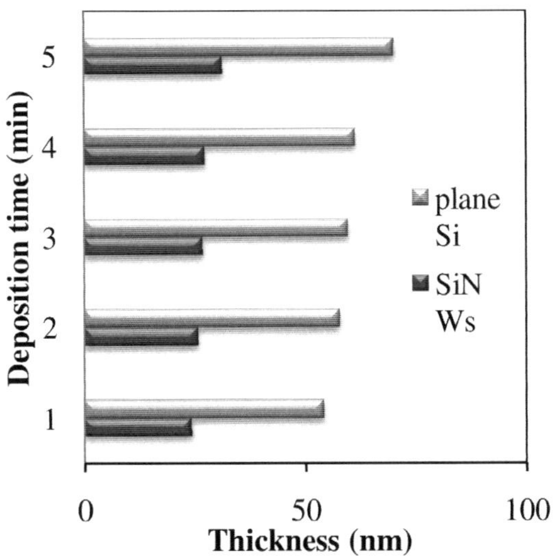

Figure 4. (a) silicon nitride coating thickness as a function of deposition time for plane silicon and SiNWs

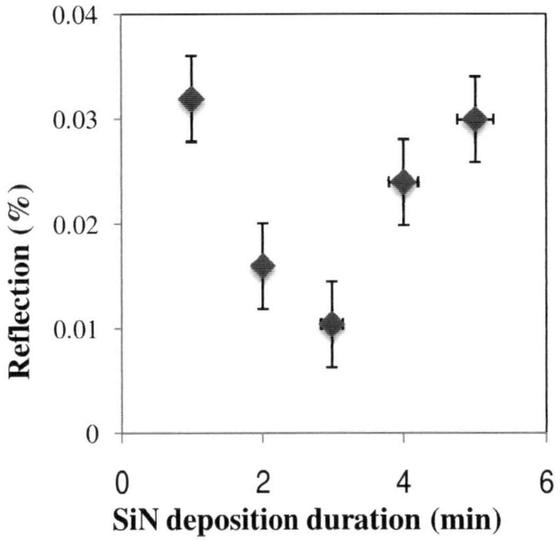

Figure 4. (b) SiNWs/SiNx reflection as a function of SiNx deposition durations.

3.3 Surface passivation results

Figure 5 show that a low reflectivity can be obtained using by optimization of SiNWs/SiNx structure and length leading to an increase in the short circuit current by 15%. The values show that the silicon nitride partly improves the surface passivation, leading to an increase of Voc, from 450 mV to 490mV around 8 %enhancement. . This passivation enhancement is related to the high content of hydrogen in the SiNx:H layers: H diffuses within the thin SiNWs layer and passivates the recombination centre elimination of the recombination centre results in better collection of electron hole pairs leads to large Voc which reflects relatively low surface recombination velocity of the Si/electrode junction and good bulk properties of NWs. However, low enhancement fraction of Voc reflects that the recombination rate and surface state density is not decreased well due to the high surface-to-volume ratio.

The calculated FF of SiNx passivated SiNWs solar cell improved by 6%, reaching 65.7%. This improvement can be attributed to decreasing contact area between electrode and SiNWs. However the original FF of SiNWs non passivated solar cell still low. This low magnitude is more related to the main problem facing SiNWs solar cell that is the the electrode contact (2). Hopefully by solving metal contact problem the fill factor can reach better results. Our SiNx passivated SiNWs solar cell exhibits an improved efficiency by 23%, an open circuit voltage by 7%, a short-circuit current by 15% and a fill factor by 6 %, as compared Control planar solar cells without SiNx, respectively. However, the recombination rate and surface state density is not decreased well due to the high surface-to-volume ratio which needs more optimization of passivation techniques as well passivation materials. Thus, solar cells with SiNx passivated SiNWs are better than the control samples without any SiNx, which verify the importance of passivation.

.

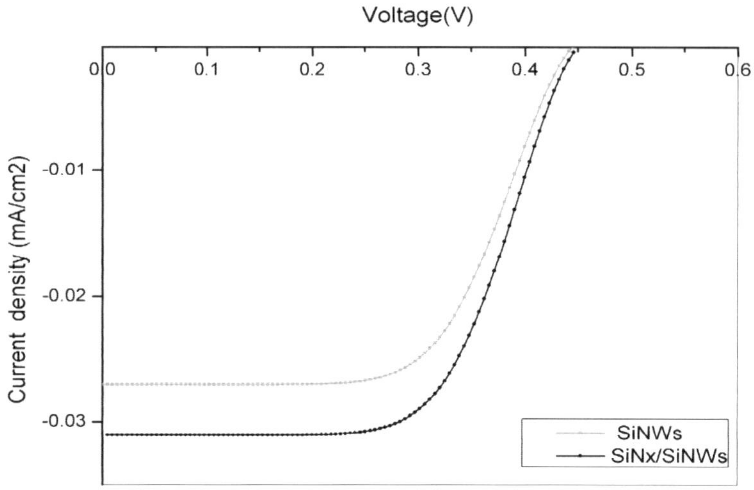

Figure 5. IV curve of SiNWs and SiNx /SiNWs solar cells

4. Conclusion

In order to reduce recombination effect in the SiNWs ARC solar cells, a SiNx film has been deposited to passivate the SiNWs array. The reflectivity can be effectively reduced up to 1% in 250-1000 nm range, leading to an increase in the short-circuit current density in the SiNWs ARC solar cells. The combination of SiNWs ARC with dielectrics layers allows increasing the surface passivation quality (Sun-Voc increase of ~ 8%). We obtain thus better reflectivity results than with a SiNx layer, with a minimisation of the surface degradation. However, obtaining high efficiency of SiNWs ARC solar cell needs optimizing the surface passivation and improving the electrode-contact property.

Acknowledgments

The authors thank Malaysian Government for funding this project.

References

[1] S. Huia, J. Zhang, X. Chena, H. Xua, D. Maa, Y. Liua and Bairui Taoa, *Sensors and Actuators B: Chemical*, 2011,155, 2, 592.
[2] Z. Tymieniecki, M. Li, C. Fobelets, K. Durrani, Z.A.K, *Electron Device Letters, IEEE*, 2010, 31,8, 860.
[3] Y. Huang, X. Duan, and C. M. Lieber, *Small 2005,***1**,1,142 C. K. CHAN,H. PENG ,G. LIU ,K. M. WRATH,X. F. ZHANG, *nature nanotechnology*, 2008, 3, 31
[4] J. Y. Jung , Z. Guo, S.W. Jee, H.D.Um, K.-T. Park, J.H. Lee, *Nanotechnology (IEEE-NANO), 2010*, 1166
[5] D. Kumar, S. K. Srivastava, P.K. Singh, M. Husain and V. Kumar**,** *Solar Energy Materials and Solar Cells*, 2011, 95,1,215.
[6] K. Peng, Y. Xu, Y. Wu, Y. Yan, S.T.Lee, and J. Zhu, *small* ,2005,11,1, 1062.
[7] S. Kodambaka, J. Tersoff, M. C. Reuter, F. M. Ross, *Phys. Rev. Lett.* 2006,96, 096105.
[8] Y. F. Zhang, Y. F. Tang, N. Wang, C. S. Lee, I. Bello, S. T. Lee, *Appl. Phys. Lett.* 1998, 72, 1835
[9] Junjie Niu, Jian Sha and Deren Yang, *Physica E: Low-dimensional Systems and Nanostructures*, 2004, 23, 1-2, 131
[10] J. D. Holmes, K. P. Johnston, R. C. Doty, B. A. Korgel, *Science* 2000, 287,1471.

[11] Z. Huang, H. Fang, J. Zhu, *Adv. Mater*. 2007, 19, 744.
[12] Z. Huang, N. Geyer, P. Werner, J. de Boor, U.Gösele, *Advanced Materials,* 2011, 23,2, 285.
[13] Y.A. Dai,H.C. Chang, K.Y. Lai,a C.A.Lin, R.J. Chung,G.R.Lina and J. H. He, *J. Mater. Chem*., 2010, 20, 10924.
[14] X. Li, J. Li, T. Chen, B. K. Tay ,J. Wang and H.Yu, *Nanoscale Res Lett* ,2010, 5,1721.
[15] H. Yue, R. Jia, C. Chen, W. Ding, Y. Meng, D. Wu, D.Wu, W. Chen, X. Liu, Z. Jin, W. Wang, and T.Ye, *J. Vac. Sci. Technol. B*, 2011, 29,3, 031208-1
[16] H. Fang, X. Li, S. Song, Y.Xu and J. Zhu, *Nanotechnology*, 2008, 19, 255703
[17] Y. Dan,K. Seo, K. Takei,J.H. Meza, A.Javey,and K.B. Crozier, Nano Lett. 2011, 11, 2527
[18] M. Lipiski, P. Ziba, P. Panek, S. Jonas, S. Kluska, H. Czternastek, A. Szyszka, B. Paszkiewicz,, *Proceedings of the 29th International Conference of IMAPS*, Poland, 2005, 203
[19] H. Li, R. Jia, C. Chen, Z. Xing, W. Ding, Y. Meng, D. Wu, X. Liu, and T. Ye, *Applied Physics Letter* , 2011,98, 151116
[20] H. Um, J.Y.Jung, H.S. Seo, K. T. Park, S.W. Jee, S. A. Moiz and J.H. Lee, 2010,49, 04DN02

The Capacitance and Temperature Effects of the SiC- and Si-Based MEMS Pressure Sensor

N Marsi[1], B Y Majlis[1], F Mohd-Yasin[2], A A Hamzah[1]

[1]Institute of Microengineering and Nanoelectronics (IMEN), Universiti Kebangsaan Malaysia (UKM), Bangi, Malaysia
[2]Queensland Micro- and Nanotechnology Centre (QMNC), Griffith University, Brisbane, QLD, Australia

Email: burhan@vlsi.eng.ukm.my

Abstract. This project develops the pressure sensor for monitoring the extreme conditions inside the gas turbine engine. The capacitive-based instead of piezoresistive-based pressure sensor is employed to avoid temperature drift. The deflecting (top) plate and the fixed (bottom) plate generate the capacitance, which is proportional to the applied input pressure and temperature. Two thin film materials of four different sizes are employed for the top plate, namely cubic silicon carbide (3C-SiC) and silicon (Si). Their performances in term of the sensitivity and linearity of the capacitance versus pressure are simulated at the temperature of 27°C, 500°C, 700°C and 1000°C. The results show that both materials display linear characteristics for temperature up to 500°C, although SiC-based sensor shows higher sensitivity. However, when the temperatures are increased to 700°C and 1000°C, the Si- based pressure sensor starts to malfunction at 50 MPa. However, the SiC-based pressure sensor continues to demonstrate high sensitivity and linearity at such high temperature and pressure. This paper validates the need of employing silicon carbide instead of silicon for sensing of extreme environments.

1. Introduction

Capacitive sensor is used in many applications such as gas turbine engine sensor [1]. It parallel plate structure contains flexible diaphragm as the top plate and fixed plate in the bottom of the structure, as shown in Figure 1 [2]. The external pressure, P will deflect the top diaphragm. A sealed cavity is required in between the plates as an insulator. Two parameters namely the distance between the plate, d and the permittivity, ε play an important role in determining the value of the generated capacitance, C. Capacitive sensor is also useful for measuring material properties since different materials have different values of dielectric constant as well as dielectric loss.

In the extreme environment, the selection of materials plays a very important role. The 3C-SiC material have been developed as a potential replacement to silicon due its superior thermal and electrical properties [3]. This paper will show that some of these properties, as shown in Table 1 [4] can affect the deflection values of the diaphragm. Our simulated results will demonstrate the advantages of 3C-SiC over Si diaphragms in improving the sensitivity, linearity and capacitance values of the pressure sensor at such extreme conditions [5].

Table 1. Material properties of tested materials [4].

Material properties	Silicon	3C-SiC
Young's modulus (GPa)	1.69	4.70
Poisson ratio (v)	0.30	0.22
Density (kg/μm^{-3})	2.5 x 10^{-15}	3.2 x 10^{-15}
TCE Integral Form (1/K)	2.5 x 10^{-6}	2.3 x 10^{-6}
Thermal Conductivity (pW/μmK)	1.48 x 10^{+8}	5.0 x 10^{+8}
Specific Heat (PJ/kgK)	7.12 x 10^{14}	1.34 x 10^{15}
Dielectric constant	9.72	11.68

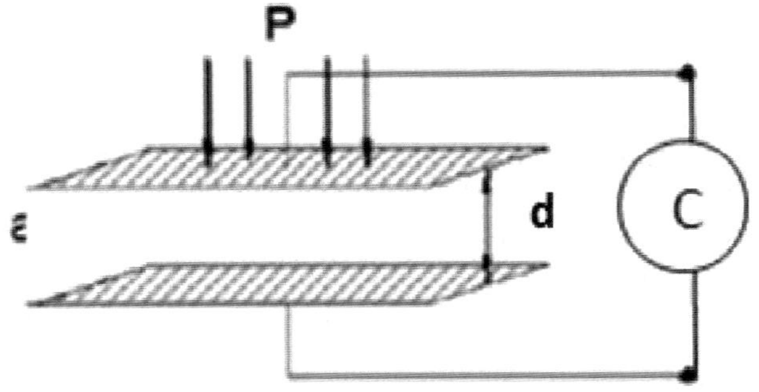

Figure 1. Schematic of capacitive pressure sensor [2]

2. Theory

2.1. Theory of Deflection and Capacitance

2.1.1. Deflection versus differential pressure under different temperatures. The edged-clamped diaphragm is shown in Figure 2. The gap of the capacitive pressure sensor is given by d, consider a length of diaphragm plat b, and width, a. Young modulus is denoted as E, Poisson's ratio is denoted as v, and the elastic rigidity is denoted as D. The application of a constant and uniform pressure P, on the center of diaphragm surfaces at point of coordinates (x, y), and thus the deflection $W(x, y, P)$ is determined. Diaphragm bending is the elastic type and deflection $W(x, y, P)$ remains small relative to the thickness h.

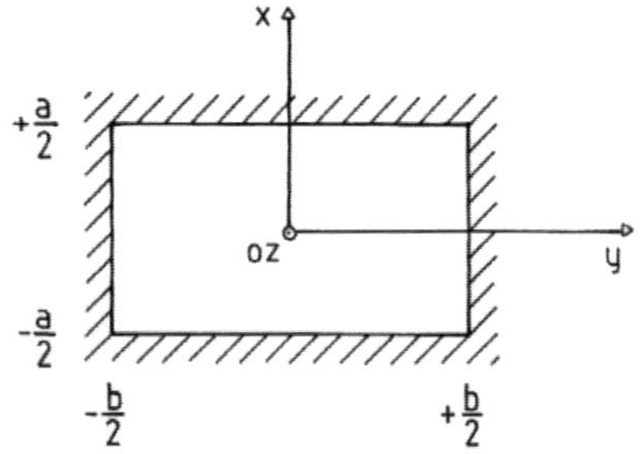

Figure 2. Deflection based on center coordination

In these positions, it is shown in equation (1) that the deflection of the diaphragm is governed by the following differential system [6]:

$$\frac{\delta^4 W}{\delta x^4} + 2\alpha \frac{\delta^4 W}{\delta x^2 \delta y^2} + \frac{\delta^4 W}{\delta y^4} = \frac{P}{D_0 h^3} \quad (1)$$

The partial derivates of the equilibrium position with length and width based on centre coordination as shown in Figure 2 are evaluated by equation (2):

$$W\left(x = \pm\frac{a}{2}; y\right) = 0\,; W\left(x; y = \pm\frac{b}{2} = 0\right)\,; \frac{\delta W}{\delta x}\left(x = \pm\frac{a}{2}; y\right) = 0\,; \frac{\delta W}{\delta y}\left(x; y = \pm\frac{b}{2}\right) = 0 \quad (2)$$

Where,

$$\alpha = v + \frac{2E(1 - v^2)}{E} \;\; ; \;\; D_0 = \frac{E}{[12(1 - v^2)]} \quad (3)$$

To attain a general formulation independent of the diaphragm symmetrical dimensions, a new integration domain is introduced in all cases, in this study is using a square of unit side:

$$X = \frac{2x}{a}, -1 \le X \le +1 \quad (4)$$

$$Y = \frac{2y}{b}, -1 \le Y \le +1 \quad (5)$$

The systems obtained using the Galerkin method indicates to the following approximately analytical solution [7]:

$$W(x, y, P) = W(0,0, P).F(X, Y, r) \quad (6)$$

Where F is given in equation (8). The analytical solutions substituting equation (3) into equation (1) associated equation (6) response for diaphragm centre deflection $W(x = 0, y = 0)$ calculated by:

$$W(0,0, P) = \frac{k(r)S^2}{16 D_0 h^3} P \quad (7)$$

Where;

$$F(X, Y, r) = [(1 - X^2)(1 - Y^2]^2 \times \sum_{i=0}^{n} \sum_{j=0}^{n} k_{ij} X^i Y^j \quad (8)$$

n, is an even positive integer
$i, j = 0, 2, 4, 6, \ldots, n$
S = area of the plate = ab
$r = b/a, k(r)$and $k_{ij}(r)$ are shape factors

Therefore, as the response for center deflection of diaphragm $W(0,0, P)$ approaches the area, S, shape factor, $k(r)$, thickness of diaphragm, h and elastic rigidity, D_0 material of the diaphragm under various pressure load, P, thus the entire top plate experience this displacement towards. Changes in the dimensional of the diaphragm plate and fixed plate, thus also affect the deflection value.

2.1.2. Capacitance between two parallel plates. A variable capacitor is recognized with the diaphragm (flexible plate) and a fixed plate. In the absence of applied pressure ($P = 0$), the distance between plates equals d. Capacitance calculation is given by the following equation:

$$C = \iint \frac{\varepsilon_0 \varepsilon dxdy}{h - W(X,Y)} \tag{9}$$

$$C = \frac{\varepsilon_0 \varepsilon S}{d} \tag{10}$$

Where, ε_0 stands for vacuum permittivity, ε stand for materials permittivity and $S = ab$ is the plate area.

It has been seen in Figure 3 that the application of a uniform pressure ($P \neq 0$) on diaphragm surfaces, at point (X,Y) causes a deflection $W(X,Y,P)$. At this point, the distance between plates becomes $d - W(X,Y,P)$. Capacitive pressure sensor response is then given by [8]:

$$C(P) = C(0) \int_0^1 \int_0^1 \frac{dXdY}{1 - \frac{W(X,Y,P)}{d}} \tag{11}$$

Where, $W(X,Y,P) < d$

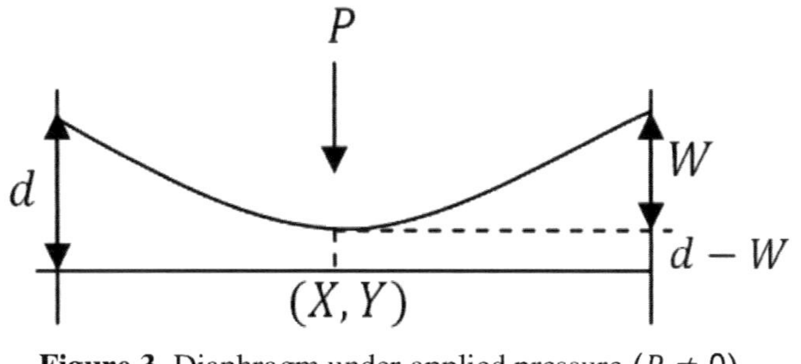

Figure 3. Diaphragm under applied pressure ($P \neq 0$)

Given the axial symmetry of the diaphragm, condition $W(0,0,P) < d$ is equivalent to $W(0,0,P) = d$. For $W(0,0,P) = d$, a short circuit occurs between plates. The sensor becomes inoperative. In the following, the specific value of P, for which $W(0,0,P) = d$ is noted P_{sc}. The analytical expression of P_{sc} is easily deduced from:

$$P_{sc}(r) = 16D_0 . \frac{dh^3}{S^2} . \frac{1}{k(r)} \tag{12}$$

From (6), (7) and (12), expression (11) can be written under the following normalized form:

$$C(P) = C(0) \int_0^1 \int_0^1 \frac{dXdY}{1 - \frac{P}{P_{sc}(r)} F(X,Y,r)} \tag{13}$$

Therefore, as the deflection $W(X,Y,P)$ approaches the gap distance, d, the capacitance, C and therefore the capacitance change ΔC increases drastically. Changes in the deflection of the diaphragm affects the distance of the diaphragm plate and fixed plate, thus also affects the capacitance value.

2.1.3. Effects of capacitance – temperatures. The capacitances are affected with differential temperature measurement has been performed based in temperature efficient offset (TCO) by using this equation [9]:

$$TCO = \frac{\Delta C_{offset,max}}{C_{max} - C_{min}}, \frac{10^6}{T_{max} - T_{min}} \frac{ppm}{°C} \quad (14)$$

Where,

$\Delta C_{offset,max}$	maximum offset capacitance for all temperatures;
C_{max} and C_{min}	maximum and minimum capacitances at the minimum temperature after offset compensation
T_{max} and T_{min}	maximum and minimum temperatures

In this case, TCO is the value of the output signal at a reference capacitance and the offset at the reference temperature.

2.2. Simulation model design

The diaphragm of capacitive pressure sensor can be modelled as a square plate that is deflected by a differential pressures and temperatures, as shown in Figure 4. CoventorWare ver.2008 simulation software is used to design, simulate and modify the performance of MEMS capacitive pressure sensor. It has three main components: Architect, Designer and Analyzer [10]. Architect is used to design the schematic of the capacitive pressure sensor. Designer is used to build the 3D design of the diaphragm. Analyzer is used to analyze the diaphragm deflection with given pressure and temperature.

Si and 3C-SiC materials with their respected electrical and mechanical properties are used to simulate the movable diaphragm (top plate). The capacitive pressure sensor is simulated at the initial voltage of 10 volts. Four diaphragm sizes are employed, namely 200 μm^2, 250 um^2, 300 um^2 and 350um^2. The separation between the diaphragm and fixed plate is 10 μm with the diaphragm thickness of 1.0 μm, and the range of applied pressures is between 0 to 100 MPa. The design steps include selecting the substrate layer and wet etching of the backside of the substrate to create the membrane. The next process is mesh creation for the device that is essential to allow the Analyzer to do the pressure and temperature analysis. Both are performed using CoventorWare mechanical solver MemMech. Figure 5 and Figure 6 shows the diaphragm deformation and maximum stress under a series of different pressures, respectively. In addition, the mechanical tool in MemMech module is employed to perform the deflections at various pressures, and the output of the capacitance changes between the diaphragm and the fixed plate, as shown in Figure 7.

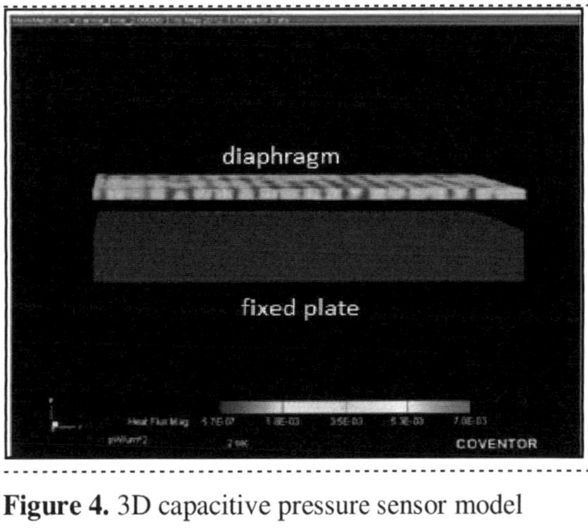

Figure 4. 3D capacitive pressure sensor model

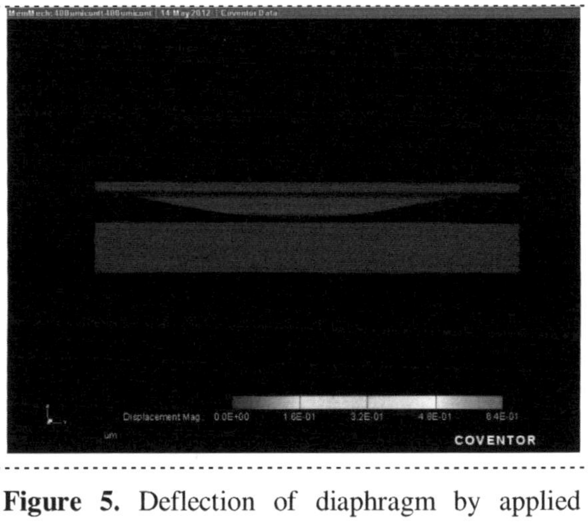

Figure 5. Deflection of diaphragm by applied pressure

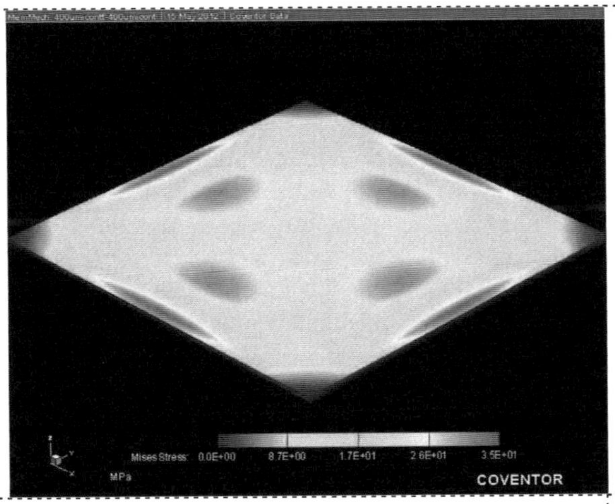

Figure 6. Maximum stress of diaphragm by applied pressure

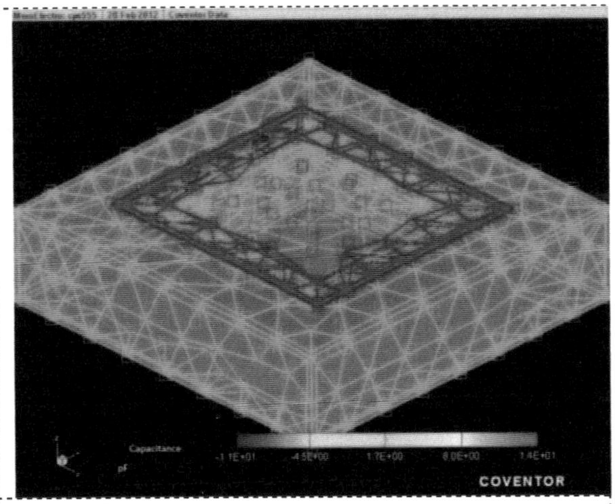

Figure 7. Capacitance results by CoventorWare ver.2008 software

Figure 8 shows the 3D model of the diaphragm using mapped bricks mesh generated by Designer [11]. There are three steps involved. Firstly is the substrate step, secondly is the silicon planar fill with an anistropic front side wet etch, and thirdly Si/SiC top diaphragm by using stack material modelling action. The MemCap solver produces an elastic electrostatic solution by solving for the charge and capacitance interaction between two plates. It computes the total number of charges and presents the final solution with charge distribution calculated for all the panels in the model as proposed by equation (13).

Figure 8. 3D diaphragm model using mapped brick mesh with relevant boundary conditions.

3. Results and discussion

Figure 9 and Figure 10 show the capacitance versus pressure plots of 3C-SiC- and Si-based pressure sensor at room temperature and 500 °C, respectively. Square diaphragms of four different sizes are used. Three main observations are recorded. First, both 3C-SiC- and Si-based sensors display linear relationship between the capacitance and the pressure. In addition, SiC-based sensor produces higher capacitances and thus is more sensitive towards the change in pressure. Second, the diaphragm with the largest are produces the largest capacitance value, as expected from equation (10). Third, it is interesting to note that starting at 40 MPa, the capacitance versus pressure curve of SiC-based sensor demonstrate slightly different path from the Si-based sensor.

In this work, the linearity is represented by the R square value. At room temperature, the 350 um^2 square shape diaphragms produce maximum capacitance value of 52.126 pF with linearity of 0.957 for 3C-SiC, compare to Si that produces 51.067 pF with linearity of 0.949. At 500°C, the superior performance of SiC-based material is noticeable. The SiC-based sensor achieves the linearity of 0.985 with maximum capacitance value of 55.99 pF, while the Si-based sensor has the linearity of 0.985 with the maximum capacitance value of 54.46 pF.

In general, it can be concluded that both 3C-SiC and Si0based pressure sensor maintain their expected performance at the applied pressure of 100 MPa and the temperature of up to 500 °C. Both have enough mechanical strength to withstand the external load without failure [12].

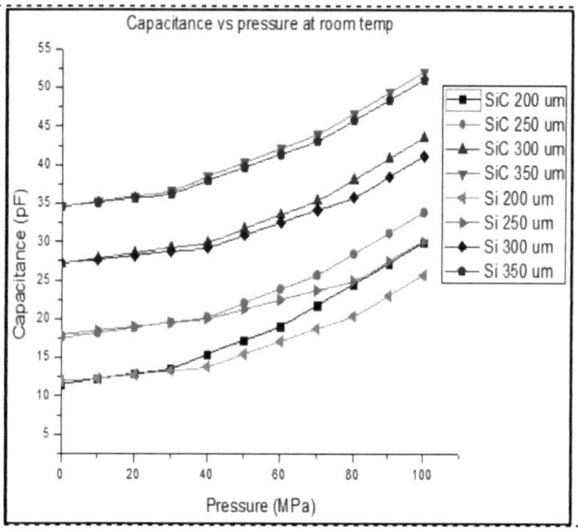

Figure 9. Capacitance versus pressure of 3C-SiC and Si at room temperature (27 °C)

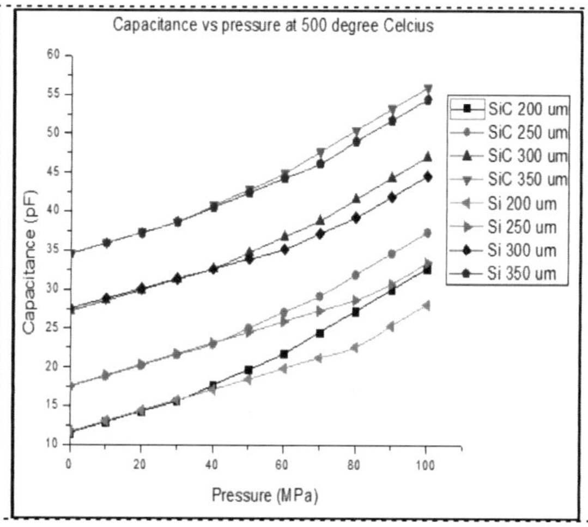

Figure 10. Capacitance versus pressure of 3C-SiC and Si at 500 °C

Figure 11 and Figure 12 shows the capacitance versus pressure plots of both SiC- and Si-based pressure sensor at 700 °C and 1000 °C, with four sizes of diaphragm. One important observation is the fact that the SiC- based pressure sensors continue its linear operation at such high temperature, while the Si-based pressure sensor start to fail at 50 MPa. In addition, the SiC-based pressure sensors continue its excellent linear behaviour. At 700 °C, the 200 μm^2 diaphragm has a linearity of 0.994 while Si-based sensor has a linearity of 0.814. At 1000°C, the 350 μm^2 diaphragm produces the linearity of 0.996 for SiC, while achieving the linearity of 0.393 for the Si. Based on the data from Figure 11 and Figure 12, SiC-based pressure sensor could sustain its performances at the higher temperature of 700 °C and 1000 °C, while SI-based pressure sensor fails to operate at these extreme temperatures. It is possible that this is due to the fragile properties of Si material due higher load and thermal cyclic on the diaphragm, which aggravate the cracks initiated by defects or scratches through the [13]-[14].

Based on these data, there seemed to be an effect of capacitance due to influence the alpha (α) based in the equation $a = v + 2E(1 - v^2)/E$, which is the parameter that is linked to Young Modulus, shear modulus and poison ratio that make the distinction in the capacitive performance between Si and 3C-SiC based pressure sensor. More work needs to be performed to confirm this relationship.

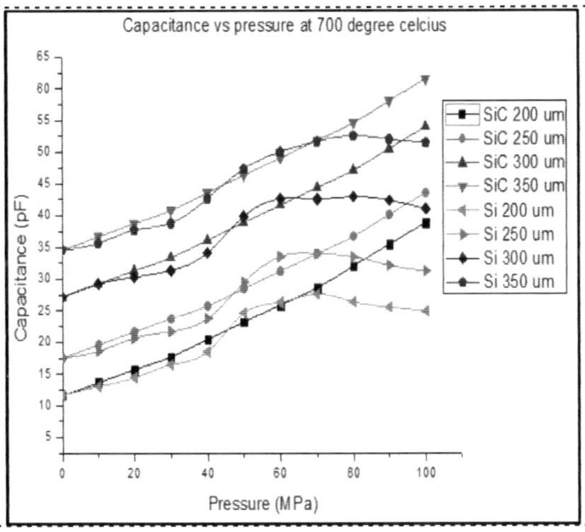

Figure 11. Capacitance versus pressure of 3C-SiC and Si at 700 °C

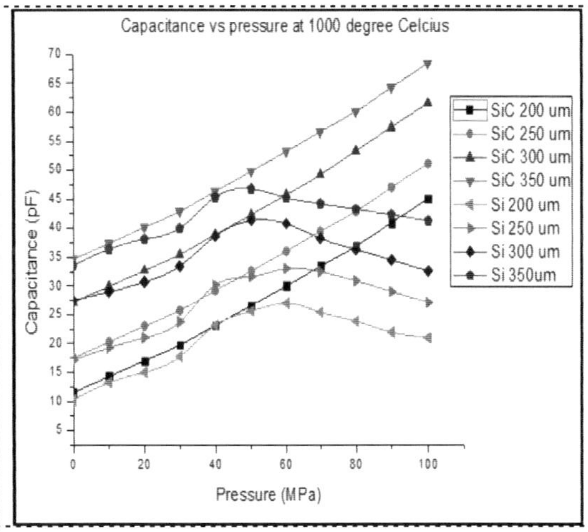

Figure 12. Capacitance versus pressure of 3C-SiC and Si at 1000 °C

4. Conclusion

This paper studies the capacitance and temperature effects of the capacitive pressure sensor when two thin film materials are employed, namely cubic silicon carbide (3C-SiC) and silicon (Si), SiC has been known to supersede Si in both electrical and mechanical properties at extreme environment, and the results from this paper supports this prevailing theory. The simulation data shows that the SiC-based pressure sensor shows higher sensitivity and linearity compares to Si-based sensor.

References

[1] Mehregany M and Zorman A 1999 SiC MEMS: Opportunities and Challenges for Applications in Harsh Environments *Thin Solid Film Vol. 355-356* pp. 355-359

[2] Ganji B A and Majlis B Y 2004 Modeling of Diaphragm for Micromachined Consider Microphones *Proc. IEEE International Conferences on Semiconductor Electronics* pp. 15-21

[3] Marsi N, Majlis B Y, Hamzah A A and Mohd-Yasin F 2012 Comparison of Mechanical Deflection and Mechanical Stress of 3C-SiC- and Si-based Pressure Sensor Diaphragms for Extreme Environment *Proc. IEEE International Conferences in Semiconductor Electronics pp. 186-190*

[4] Pratap R and Arunkumar A 2007 Material Selection for MEMS Devices *Indian Journal of Pure and Applied Physics* pp 358-367

[5] He F, Huang Q A and Qin M 2007 A Silicon Directly Bonded Capacitive Absolute Pressure Sensor *Sensors and Actuator vol 135* pp 507-514

[6] Lee K W and Wise K D 1982 SENSIM: A Simulation Program for Solid-State Pressure Sensors *IEEE Trans Electrin Devices* ED-29 pp 34-41

[7] Blasquez G, Naciri Y, Blondel N, Moussa B and Pons P 1987 Statistic Response of Maniature Capacitive Pressure Sensors with Square or Rectangular Silicon Diaphragm *Revue Physical Application 22* pp 505-510

[8] Hamzah A A, Majlis B Y, and Ahmad I 2004 Deflection Analysis of Epotaxially Deposited Polysilicon Encapsulation for MEMS Devices *Proc IEEE International Conferences on Semiconductor Electronics* pp 611-614

[9] Chuan C W, Bishnu P G, David J M and Carlos H M 2000 Contamination-Intensitive Differential Capacitive Pressure Sensors *Journal of Microelectromechanical System vol 9(4)* pp 538-543

[10] Ali M Y, Mehfuz R, Khan A A and Ismail A F 2012 Micromachining Techniques for

Fabrication Micro and Nano Structure *Tool Based Micromanufaturing InTech* pp 131-158
[11] Young M 1989 The technical Writer's *Handbook Mill Valley* CA: University Science
[12] Stephen E S, Manfred L, Thomas D, Gerhard P and Philip G N 1995 Thermal Capacitance Spectroscopy of Epitaxial 3C and 6H-SiC pn Junction Diodes Grown Side by Side n a 6H-SiC Substrate *Applied Physical Letters vol 66(26)* pp 3612-3614
[13] Ko W H, Bao M H and Hong Y D A 1982 High Sensitivity Integrated Circuit Capacitive Pressure Transducer IEEE *Trans Electron Devices ED-29* pp 47-56
[14] Buyong M R, Aziz N A and Majlis B Y 2011 MEMS Very Low Capacitive Pressure Sensor Based on CMOS Process *Sains Malaysiana vol 40(3)* pp 259-266

Analysis of Nanoparticle Additive Couple Stress Fluids in Three-layered Journal Bearing

T V V L N Rao[1], S Sufian[2] and N M Mohamed[3]
[1]Department of Mechanical Engineering, [2]Department of Chemical Engineering,
[3]Department of Fundamental & Applied Sciences, Universiti Teknologi PETRONAS,
31750 Tronoh, Perak Darul Ridzuan, MALAYSIA

[1]Email: tvvlnrao@gmail.com

Abstract. The present theoretical study investigates the load capacity and friction coefficient in a three-layered journal bearing lubricated with nanoparticle additive couple stress fluids. The couple stresses effects are analyzed based on Stokes micro-continuum theory. The nondimensional pressure and shear stress expressions are derived using modified Reynolds equation. The nondimensional load capacity increases and the coefficient of friction decreases using nanoparticle additive lubricants with couple stress effects. The three-layered journal bearing performance characteristics are improved with increase in both (i) surface adsorbent fluid film layer thickness and (ii) dynamic viscosity ratio of surface to core layer.

1. Introduction

Analysis of load capacity and coefficient of friction in hydrodynamic lubrication are significantly influenced by structure and properties of fluid film. Szeri [1] offered a new approach to the friction reduction in hydrodynamic bearing using a composite-film journal bearing. The composite film configuration consists of immiscible high-viscosity and low-viscosity layers adjacent to bearing and journal surfaces respectively. Tichy [2] developed a rheological model applicable for thin films that considers high viscosity layer adhered to solid surfaces. Meurisse and Espejel [3] derived generalized Reynolds equation for surface layer model in thin film lubrication. Higher load capacity and lower friction coefficient of journal bearing are obtained with increase in viscosity and thickness of surface layer.

Lubricants with additives are used in hydrodynamic lubrication to improve the bearing performance characteristics. The couple stress fluid takes into account the properties of lubricants with additives. Stokes [4] proposed a simplified couple stress fluid model based on micro-continuum theory. Lin [5] and Mokhiamer et al. [6] analyzed the effects of couple stresses on finite journal bearing characteristics. Li and Chu [7] presented the influence of porous media and couple stress models on thin film lubrication of journal bearing. Both surface adsorbent layer and couple stress fluid models improve the performance characteristics of journal bearing. The influence of surface adsorbent layer using the couple stress fluid model is of considerable interest in the analysis of hydrodynamic lubrication.

Corresponding author: tvvlnrao@gmail.com

Nanoparticle additive lubricants have higher viscosity and hence improve the load carrying capacity of fluid film bearings. Shenoy et al. [8] presented the effect of nanoparticle additive lubricants on the static performance characteristics of an externally adjustable fluid film bearing. Nair et al. [9] investigated the performance characteristics of hydrodynamic journal bearing with nanoparticle additive lubricants. Theoretical models used to approximate the viscosity of nanofluids have been derived from the work of Einstein based on spherical particle suspensions in viscous fluids. Einstein's formula is applicable to relatively low particle volume fraction. Brinkman [10] extended the Einstein equation to a more generalized form considering moderate particle volume fraction. Batchelor [11] considered the effect due to Brownian motion of spherical shape nanoparticle additives to predict the viscosity of nanofluids. Recent experimental [12] and empirical [13] models provide a comparison of earliest works [10,11] on viscosity of nanofluids.

The motivation of this theoretical study is to analyze improvements in the bearing performance characteristics (nondimensional load capacity and coefficient of friction). The purpose of this analysis is to investigate a three-layered film structure (adsorbent and core layers) of nanoparticle additive lubricants with couple stress fluid properties. The objective is to present one-dimensional (infinitely long bearing approximation [1]) analysis of nanoparticle additive couple stress fluids in three-layered journal bearing. A modified form of Reynolds equation is derived for couple stress fluids. Reynolds boundary conditions are used to solve the pressure distribution. The nondimensional pressure and shear stress expressions obtained. The parameters used to analyze the nondimensional load capacity and coefficient of friction are: (i) surface adsorbent fluid film layer thickness (Δ); (ii) dynamic viscosity ratio of surface to core (β_s); (iii) nanoparticle volume fraction (ψ); and (iv) couple stress parameter (λ). The proposed model is applicable for thin film lubrication in which the film is of the order of several molecular lengths thickness [2].

2. Analysis

The schematic of three-layered journal bearing is shown in Figure 1. Journal bearing is lubricated with viscous fluid which forms a film that separates the load-bearing surfaces [1]. The three-layered film in journal bearing is described through adsorbent layer at surfaces of higher viscosity than core layer. The adsorbent layers (film region I: $0 \leq y \leq \delta$ and film region III: $h - \delta \leq y \leq h$) at surfaces of higher viscosity than core layer ($\beta_s = \frac{\mu_s}{\mu_c}$, $\mu_c = \mu_n$ and $\beta_n = \frac{\mu_n}{\mu_f}$) and the core layer (film region II: $\delta \leq y \leq h - \delta$) are modeled using couple stress fluids.

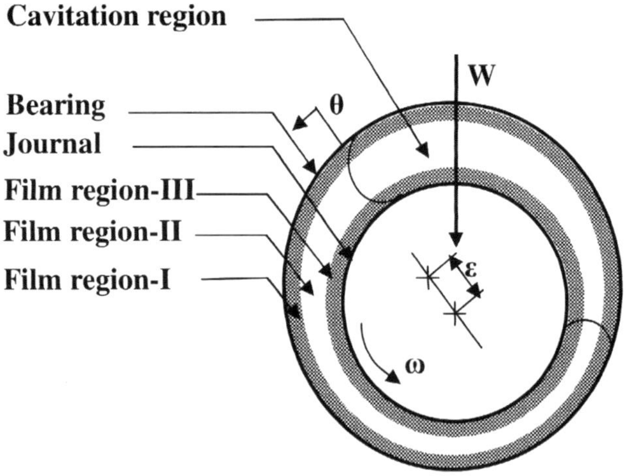

Figure 1. Geometry of three-layered journal bearing

In the present one-dimensional (infinitely long bearing approximation [1]) analysis, the variation of pressure across the fluid film is assumed to be negligible and pressure in the three-layered journal bearing is a function of sliding direction. Pressure distribution is obtained by using Reynolds boundary conditions.

The simplified momentum equation for couple stress fluid is ($0 \le y \le \delta$ for $j = 1$, $\delta \le y \le h-\delta$ for $j = 2$ and $h-\delta \le y \le h$ for $j = 3$)

$$\frac{1}{\eta}\frac{dp}{dx} = \frac{\mu_j}{\eta}\frac{d^2u_j}{dy^2} - \frac{d^4u_j}{dy^4} \tag{1}$$

The boundary conditions for velocity at the bearing surface, at the interface of bearing adsorbent and core layer, at the interface of core and journal adsorbent layer, and at the journal surface respectively are

$$y = 0:\ u_1 = 0 \text{ and } \frac{d^2u_1}{dy^2} = 0 \tag{2}$$

$$y = \delta: u_1 = u_2 = u_{12}, \mu_1\frac{du_1}{dy} = \mu_2\frac{du_2}{dy}, \frac{d^2u_1}{dy^2} = 0 \text{ and } \frac{d^2u_2}{dy^2} = 0 \tag{3}$$

$$y = h-\delta: u_2 = u_3 = u_{23} \text{ and } \mu_2\frac{du_2}{dy} = \mu\frac{du_3}{dy}, \frac{d^2u_2}{dy^2} = 0 \text{ and } \frac{d^2u_3}{dy^2} = 0 \tag{4}$$

$$y = h: u_3 = u_j \text{ and } \frac{d^2u_3}{dy^2} = 0 \tag{5}$$

The nanoparticle additive fluid dynamic viscosity increases with the particle volume fraction. Based on Batchelor [11] model, the dynamic viscosity ratio of nano to base fluid is

$$\beta_n = \frac{\mu_n}{\mu_f} = (1 + 2.5\psi + 6.5\psi^2) \tag{6}$$

Integrating (1) using the boundary conditions in (2)-(5), the non-dimensionl velocity distribution in bearing adsorbent layer ($0 \le Y \le \Delta$), in core layer ($\Delta \le Y \le H-\Delta$), and in journal adsorbent layer ($H-\Delta \le Y \le H$) respectively are expressed as

$$0 \le Y \le \Delta: U_1 = U_{12}\frac{Y}{\Delta} + \frac{1}{2\beta_s\beta_n}\frac{dP}{d\theta}Y(Y-\Delta) + \lambda_s^2\frac{1}{\beta_s\beta_n}\frac{dP}{d\theta}C_1 \tag{7}$$

$$\Delta \le Y \le H-\Delta: U_2 = U_{12} + (U_{23} - U_{12})\left(\frac{Y-\Delta}{H-2\Delta}\right) + \lambda_n^2\frac{1}{\beta_n}\frac{dP}{d\theta}C_2 + \frac{1}{2\beta_n}\frac{dP}{d\theta}(Y-\Delta)(Y-H+\Delta) \tag{8}$$

$$H-\Delta \le Y \le H: U_3 = U_{23} + (1-U_{23})\left(\frac{Y-H+\Delta}{\Delta}\right) + \frac{1}{2\beta_s\beta_n}\frac{dP}{d\theta}(Y-H)(Y-H+\Delta) + \lambda_s^2\frac{1}{\beta_s\beta_n}\frac{dP}{d\theta}C_3 \tag{9}$$

where

$$C_1 = \left[1 + \frac{\sinh\left(\frac{Y-\Delta}{\lambda_s}\right) - \sinh\left(\frac{Y}{\lambda_s}\right)}{\sinh\left(\frac{\Delta}{\lambda_s}\right)}\right], \quad C_2 = \left[1 + \frac{\sinh\left(\frac{Y-H+\Delta}{\lambda_n}\right) - \sinh\left(\frac{Y-\Delta}{\lambda_n}\right)}{\sinh\left(\frac{H-2\Delta}{\lambda_n}\right)}\right], C_3 = \left[1 + \frac{\sinh\left(\frac{Y-H}{\lambda_s}\right) - \sinh\left(\frac{Y-H+\Delta}{\lambda_s}\right)}{\sinh\left(\frac{\Delta}{\lambda_s}\right)}\right] \tag{10}$$

$$U_{12} = F_1 - \frac{1}{\beta_n}\frac{dP}{d\theta}F_2, \quad U_{23} = F_3 - \frac{1}{\beta_n}\frac{dP}{d\theta}F_4 \tag{11}$$

$$F_1 = \frac{-E_{12}E_{231}}{E_{11}E_{22}-E_{12}E_{21}}, \quad F_2 = \frac{E_{22}E_{13}-E_{12}E_{232}}{E_{11}E_{22}-E_{12}E_{21}},$$
$$F_3 = \frac{E_{11}E_{231}}{E_{11}E_{22}-E_{12}E_{21}}, \quad F_4 = \frac{-E_{21}E_{13}+E_{11}E_{232}}{E_{11}E_{22}-E_{12}E_{21}} \tag{12}$$

$$E_{11} = E_{22} = \frac{\beta_s}{\Delta} + \frac{1}{(H-2\Delta)}, \quad E_{12} = E_{21} = -\frac{1}{(H-2\Delta)},$$
$$E_{13} = -\lambda_s H_1^* - \lambda_n H_2^* + \frac{1}{2}(H-\Delta), \quad E_{231} = \frac{\beta_s}{\Delta}, \quad E_{232} = E_{13} \tag{13}$$

$$H_1^* = \left[\coth\left(\frac{\Delta}{\lambda_s}\right) - \operatorname{csch}\left(\frac{\Delta}{\lambda_s}\right)\right], \quad H_2^* = \left[\coth\left(\frac{H-2\Delta}{\lambda_n}\right) - \operatorname{csch}\left(\frac{H-2\Delta}{\lambda_n}\right)\right] \tag{14}$$

$$\lambda_s = \frac{\lambda}{\sqrt{\beta_s\beta_n}}, \quad \lambda_n = \frac{\lambda}{\sqrt{\beta_n}} \tag{15}$$

$$H = 1 + \varepsilon\cos\theta \tag{16}$$

The equation of continuity across the film is

$$Q = \int_0^{\Delta} U_1 dY + \int_{\Delta}^{H-\Delta} U_2 dY + \int_{H-\Delta}^{H} U_3 dY \tag{17}$$

Simplifying the equation of continuity across the film, yields

$$\frac{dP}{d\theta} = \beta_n\left(\frac{G_1 - Q}{G_2}\right) \tag{18}$$

where

$$G_1 = \frac{1}{2}\Delta + \frac{1}{2}(F_1 + F_3)(H - \Delta) = \frac{1}{2}H \tag{19}$$

$$G_2 = \frac{1}{2}(F_2 + F_4)(H - \Delta) + \frac{1}{12\beta_s}\left[2\Delta^3 + \beta_s(H - 2\Delta)^3\right] - \frac{1}{\beta_s}\left[2\lambda_s^2\Delta + \beta_s\lambda_n^2(H - 2\Delta)\right] + \frac{2}{\beta_s}\left[2\lambda_s^3 H_1^* + \beta_s\lambda_n^3 H_2^*\right] \tag{20}$$

For $\Delta = 0$, G_2 in (20) reduces to

$$G_2 = \frac{1}{12}H^3 - \lambda_n^2 H + 2\lambda_n^3\tanh\left(\frac{H}{2\lambda_n}\right) \tag{21}$$

For $\lambda = 0$, G_2 in (20) reduces to

$$G_2 = \frac{1}{12\beta_s}\left[2\Delta^3 + \beta_s(H - 2\Delta)^3 + 6\Delta(H - \Delta)^2\right] \tag{22}$$

The Reynolds boundary conditions are

$$P|_{\theta=0} = 0, \qquad P|_{\theta=\theta_r} = 0 \quad \text{and} \quad \left.\frac{dP}{d\theta}\right|_{\theta=\theta_r} = 0 \tag{23}$$

Integrating (18) and substituting the first boundary condition given in (23), yields the nondimensional pressure profile as

$$P = \beta_n\left(\int_0^{\theta}\frac{G_1}{G_2}d\theta - Q\int_0^{\theta}\frac{1}{G_2}d\theta\right) \tag{24}$$

Substitution of the Reynolds boundary conditions for nondimensional pressure at film rupture in (23) and simplifying results in Q as

$$Q = \frac{\int_0^{\theta_r}\frac{G_1}{G_2}d\theta}{\int_{\theta_g}^{\theta_r}\frac{1}{G_2}d\theta} \tag{25}$$

Substituting the pressure gradient boundary condition given in (23) in the expression for nondimensional pressure gradient in (18), results in

$$Q = G_1|_{\theta=\theta_r} \tag{26}$$

The Newton-Raphson iterative procedure is used to solve simultaneously both θ_r and Q using (25) and (26).

The nondimensional load capacity is

$$W = \sqrt{W_\varepsilon^2 + W_\phi^2} \tag{27}$$

where $W_\varepsilon = -\int_0^{\theta_r} P\cos\theta\, d\theta$ and $W_\phi = \int_0^{\theta_r} P\sin\theta\, d\theta$

The nondimensional shear stress in the journal bearing at $Y = H$ is obtained as

$$\Pi|_{Y=H} = \beta_s\beta_n\left.\frac{dU_3}{dY}\right|_{Y=H} = \beta_n\left[(1 - F_3)\left(\frac{\beta_s}{\Delta}\right) + \left(\frac{G_1 - Q}{G_2}\right)\left(F_4\frac{\beta_s}{\Delta} + \frac{\Delta}{2} - \lambda_s H_1^*\right)\right] \tag{28}$$

For $\Delta = 0$, $\Pi|_{Y=H}$ in (28) reduce to

$$\Pi|_{Y=H} = \beta_n\left[\frac{1}{H} + \left(\frac{G_1-Q}{G_2}\right)\left(-\lambda_n \tanh\left(\frac{H}{2\lambda_n}\right) + \frac{1}{2}H\right)\right] \tag{29}$$

For $\lambda = 0$, $\Pi|_{Y=H}$ in (28) reduce to

$$\Pi|_{Y=H} = \beta_n\left[\frac{\beta_s}{2\Delta+\beta_s(H-2\Delta)} + \left(\frac{G_1-Q}{G_2}\right)\frac{H}{2}\right] \tag{30}$$

The nondimensional friction force on the journal surface is obtained by integrating the shear stress along the journal surface as

$$F = \int_0^{\theta_r} \Pi \, d\theta \tag{31}$$

The nondimensional friction coefficient is calculated as $C_f = \left(\frac{R}{C}\right)\frac{f}{w} = \frac{F}{W}$.

3. Results and Discussion

A nanoparticle additive couple stress fluid film lubricated three layered journal bearing is considered in the analysis. The parameters used in the analysis are: journal eccentricity ratio (ε)=0.2; dynamic viscosity ratio of surface to core layer (β_s=10); surface (journal and bearing) adsorbent fluid film layer thickness (Δ=0.001-0.1); couple stress parameter (λ=0.0, 0.01, 0.1) and nanoparticle volume fraction (ψ=0.0-0.04). The influence of surface (journal and bearing) adsorbent fluid film layer thickness (Δ) and nanoparticle volume fraction (ψ) on the nondimensional load capacity and coefficient of friction are analyzed. Based on the analysis presented in this theoretical study, results of nondimensional load capacity and coefficient of friction for a three-layered film are compared with a conventional film ($\Delta = 0$).

Figures 2a-2b show the non-dimensional load capacity (W) of a three-layered journal bearing lubricated with nanoparticle additive couple stress fluids. The variation of non-dimensional load capacity (W) with surface adsorbent (journal and bearing) fluid film layer thickness (Δ) and nanoparticle volume fraction (ψ) are presented. As shown in Figure 2a, the non-dimensional load capacity is significantly increases with increase in surface adsorbent fluid film layer thickness (Δ) and dynamic viscosity ratio of surface to core layer (β_s). The non-dimensional load capacity (W) also increases with increase in (i) increase in couple stress parameter (λ) and (ii) nanoparticle volume fraction (ψ). However, the configuration with lower surface adsorbent fluid film layer thickness (Δ<0.01) does not generate enhancement in the non-dimensional load capacity (W). As shown in Figure 2b, the non-dimensional load capacity (W) increases with nanoparticle volume fraction (ψ=0.0-0.04). The influence of increase in surface adsorbent fluid film layer thickness (Δ=0.0-0.1) on the enhancement in the non-dimensional load capacity is higher compared to couple stress parameter (λ=0.01-0.1).

Figures 3a-3b show the coefficient of friction (C_f) for a three-layered journal bearing lubricated with nanoparticle additive couple stress fluids. For the parameters considered in the study as shown in Figure 3a, minimum coefficient of friction (C_f) is obtained for higher values of both surface adsorbent fluid film layer thickness (Δ) and dynamic viscosity ratio of surface to core layer (β_s). The influence of surface adsorbent fluid film layer thickness (Δ) on reduction in coefficient of friction (C_f) is more than the influence of couple stress parameters (λ). The reduction in coefficient of friction (C_f) is significant for surface adsorbent fluid film layer thickness (Δ=0.1), with increase in both dynamic viscosity ratio of surface to core layer (β_s) and couple stress parameters (λ). As shown in Figure 3b, the coefficient of friction (C_f) is not influenced by the increase in nanoparticle volume fraction (ψ=0.0-0.04). For λ=0.0, the coefficient of friction (C_f) is not influenced by nanoparticle volume fraction (ψ), as both nondimensional pressure and nondimensional shear stress increase by the magnitude of β_n as given in (24) and (30).

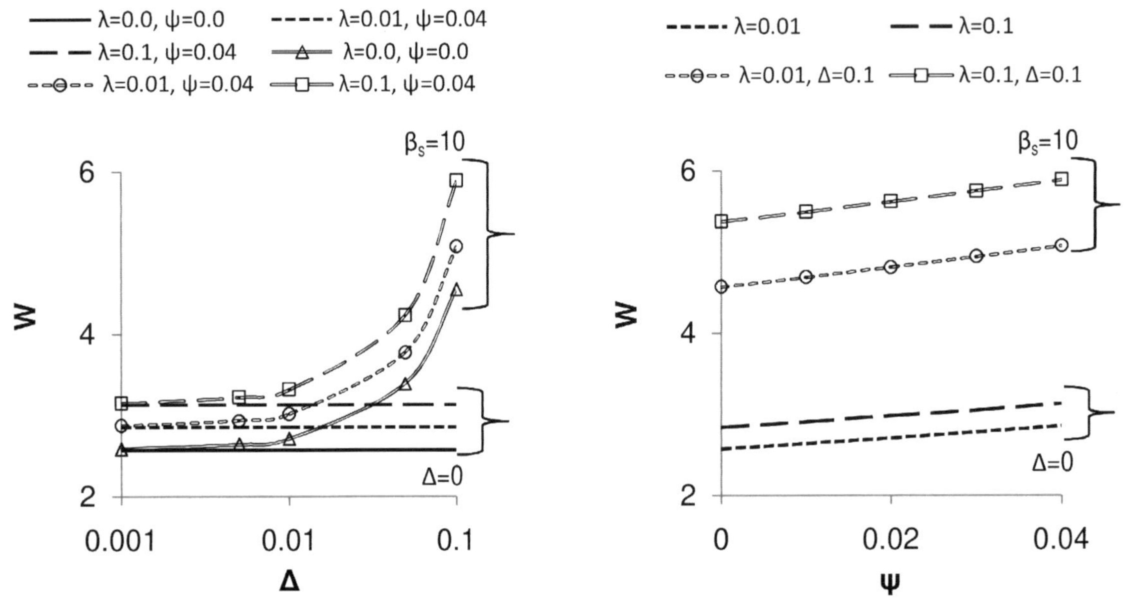

(a) (b)

Figure 2. Non-dimensional load capacity (ε=0.2)

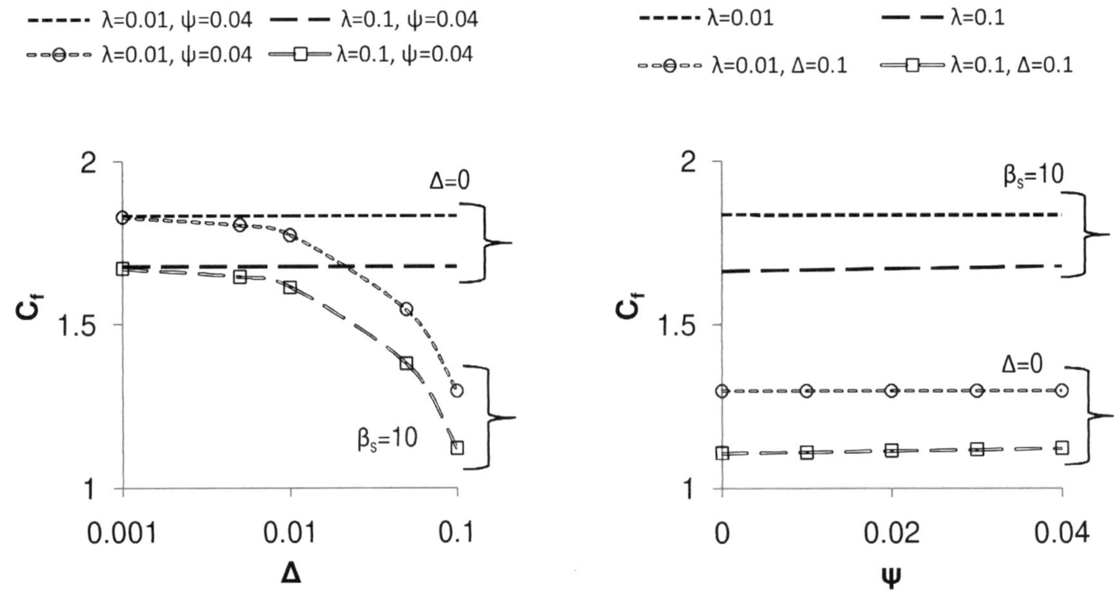

(a) (b)

Figure 3. Coefficient of friction (ε=0.2)

4. Conclusion

The present study evaluates improvement in load capacity and reduction in coefficient of friction for a nanoparticle additive couple stress fluid lubricated three-layered journal bearing. The nondimensional pressure and shear stress expressions are obtained using the Reynolds boundary conditions. The non-dimensional load capacity (*W*) increases significantly with increase in both surface adsorbent fluid film layer thickness (Δ) and dynamic viscosity ratio of surface to core layer (β_s). The influence of couple stress parameter (λ) is greater for higher surface adsorbent fluid film layer thickness (Δ). The non-dimensional load capacity (*W*) also increases with increase in nanoparticle volume fraction (ψ). The coefficient of friction (C_f) decreases significantly with increase in surface adsorbent fluid film layer thickness (Δ) and dynamic viscosity ratio of surface to core layer (β_s). The increase in nanoparticle volume fraction (ψ) does not yield improvement in the coefficient of friction (C_f). A three-layered journal bearing lubricated with nanoparticle additive couple stress fluids has a potential to increase the load capacity and reduce the friction coefficient.

Acknowledgement

The authors greatly appreciate the support provided by Universiti Teknologi PETRONAS for this research.

Nomenclature

C	Radial clearance, m
f	Friction force, N; $F = fC/\mu_f URL$
h, H	Film thickness, m; $H = h/C$
L	Length, m
p	Pressure distribution, N/m^2; $P = pC^2/\mu_f UR$
q	Volume flow rate per unit length along film thickness, m^2/s; $Q = q/UC$
R	Journal radius, m
u	Velocity component along circumferential direction
U	Shaft speed, m/s; $U = \omega R$
w	Static load, N; $W = wC^2/\mu_f UR^2L$
x	Coordinate along circumferential direction, m; $\theta = x/R$
y	Coordinate along radial direction, m; $Y = y/C$
β_n, β_s	Dynamic viscosity ratios of nano to base fluid, surface to core layer respectively
Δ	Surface (journal or bearing) adsorbent fluid film layer thickness; $\Delta = \frac{\delta}{C}$
ε	Journal bearing eccentricity ratio
μ	Fluid viscosity, Ns/m^2
η	Material constant for couple stress, kgm/s
λ	Couple stress parameter; $\lambda = \left(\sqrt{\eta/\mu}\right)/C$
ψ	Particle volume fraction
θ	Angular coordinate measured from the direction of maximum film thickness in journal bearing
θ_r	Angular extent of film rupture for journal bearing
τ	Shear stress component, N/m^2; $\Pi = \tau C/\mu_f U$
ω	Angular velocity of journal bearing, rad/s
Subscripts	
c, s	Core and adsorbent fluid film layer
f, n	Base and nanoparticle additive fluid
r	Extent of outlet film in journal bearing
1,2,3	Bearing adsorbent, core and journal adsorbent fluid layers
12	Fluid film interface of the bearing adsorbent and core layer

23	Fluid film interface between the core and journal adsorbent layer
ε	Along the radial direction
ϕ	Along the tangential direction

References

[1] Szeri A Z 2010 *Int. J. Eng. Sci.* 48: 1622-32.
[2] Tichy J A 1995 *Tribol. T.* 38: 577-82.
[3] Meurisse M-H and Espejel G M 2008 *P I Mech Eng J-J Eng.* 222: 369-80.
[4] Stokes V K 1966 *Phys. Fluids* 9: 1709-15.
[5] Lin J-R 1997 *Wear* 206: 171-8.
[6] Mokhiamer U M, Crosby W A and El-Gamal H A A 1999 *Wear* 224: 194-201.
[7] Li W-L and Chu H-M 2004 *Acta Mech.* 171: 189-202.
[8] Shenoy B S, Binu K G, Pai R, Rao D S and Pai R S 2012 *Tribol. Int.* 45: 38-42.
[9] Nair K P, Ahmed M S and Al-qahtani S T 2009 *Int. J. Nanoparticles* 251-62.
[10] Brinkman H C 1952 *J. Chem. Phys.* 20: 571-81.
[11] Batchelor G K 1977 *J. Fluid Mech.* 83: 97-117.
[12] Duangthongsuk W and Wongwises S 2009 *Exp. Therm. Fluid. Sci.* 33: 706-14.
[13] Hosseini S M, Moghadassi A R and Henneke D E 2010 *J. Therm. Anal. Calorim.* 100: 873-7.

Bluff Body Fluid Interactions Modelling for Micro Energy Harvesting Application

M S Bhuyan[1], B Y Majlis[1], M Othman[2], Sawal H Md Ali[3], C Kalaivani[4], and S. Islam[1]

[1]Institute of Microengineering and Nanoelectronics (IMEN), UniversitiKebangsaan Malaysia (UKM), 43600 UKM Bangi, Selangor, Malaysia.
[2] Ministry of Science, Technology and Innovation (MOSTI), Federal Government Administrative Centre, 62662, Putrajaya, Malaysia.
[3] Department of Electrical, Electronic and Systems Engineering, Faculty of Engineering and Built Environment, 43600 UKM Bangi, Selangor, Malaysia.
[4] Institute of Space Science, Faculty of Engineering and Built Environment, UniversitiKebangsaan Malaysia, 43600 UKM Bangi, Selangor, Malaysia.

E-mail: shakir_dhaka@yahoo.com, shabiul@ukm.my

Abstract. In this paper, we have presented a MEMS-based piezoelectric fluid-flow based micro energy harvester. The design and modelling of the energy harvester structure was based on a piezoelectric cantilever affixed to a bluff-body. In a cross fluid flow, pressure in the flow channel, in the wake of the bluff body, fluctuates with the same frequency as the pressure variation caused by the Kármán Vortex Street. This fluctuation of pressure in the flow channel causes the piezoelectric cantilever, trailing the bluff-body, to vibrate in a direction normal to the fluid flow direction. COMSOL finite element analysis software are used for the evaluation of various mechanical analysis of the micro energy harvester structure like, physical the Stress and Strain state in the cantilever structures, Eigen frequency Analysis, Transient analysis to demonstrate the feasibility of the design. Detailed steps of modelling and simulation results of the uniform cantilever were explained. The results confirm the probability of the fluid flow based MEMS energy harvester.

1. Introduction

Energy Harvesting (EH) technology has been explored for more than a decade as an enabling technology to power autonomous systems such as Wireless Sensor Networks (WSNs). WSN are made of several sensor nodes capable of collecting surrounding environmental information (for example, temperature, vibrations, light, etc.), to turn it into numeric data and to send to a base station. Recent widespread development of WSNs allowing real-time monitoring of many application areas like automotive, bio-medical, industrial process, environment etc [1]. All application area demand an economical energy source at micro or nano scale, for continuous operation of the low-powered WSN node. More important demand for the WSN nodes is to avoid the requirement of replacement of finite power stores like primary batteries for autonomous operation. Therefore, considerable academic and industrial research effort is focused on harvesting electrical energy from natural resources such as flowing water, rain, tides, wind, sunlight, geothermal heat and biomass. Global renewable energy consumption at micor/nano scale is growing very rapidly [2].Many application areas, such as automotive, transportation, industry and aeronautics, have a strong interest in developing and using WSN to increase their productivity (real-time monitoring), reduce their costs or limit machine downtimes (predictive maintenance). Figure 1 shows the basic modules of an autonomous sensor node with EH module to produce a power range of 10 μW to 100μW, the general power requirement standard for MEMS or NEMS based sensors [3].

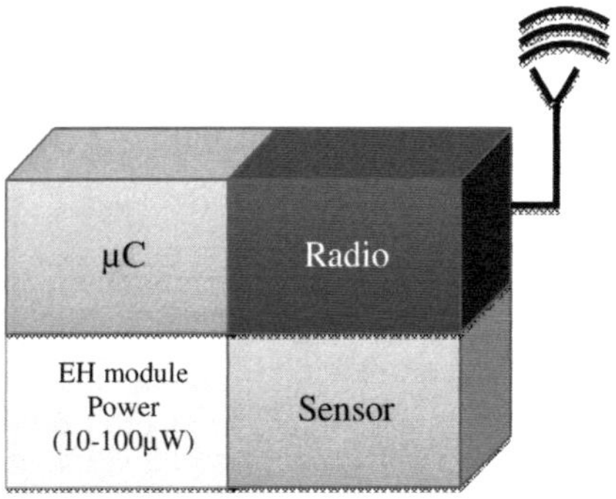

Figure1. Basic power requirement for an autonomous node and other modules.

When a bluff-body is placed in a fluid flow channel whose direction is perpendicular to the axis of the bluff- body, i.e., in a cross-flow, as shown in figure 2, the bluff-body structure will try to vibrate in a direction normal to the flow direction [4]. This flow induced vibration could be caused by the turbulence generated by the flow around and in the wake of the cylinder. The turbulence represents the random fluctuations of the flow and is not periodic. The turbulence excited vibration can occur at any flow velocity with increasing amplitude, with the increase of the flow velocity and a frequency around the natural frequency of the bluff-body structure. The flow induced vibration could also be caused by the periodic vortex shedding from the bluff-body edge surface. When a vortex is shed from one side of the bluff-body, a pressure difference forms between this side and the other side of the bluff-body causing a net force exerted on the bluff-body in the direction perpendicular to the flow. A vortex shed from the other side causes a force in the opposite direction. Thus the bluff-body is set to vibrate due to the alternate shedding of vortices named as Vortex-Induced Vibration (VIV). The pattern of periodic, alternating vortex shedding that occurs in the flow behind the body is referred as von Karman's vortex street [5] as shown in figure 2.

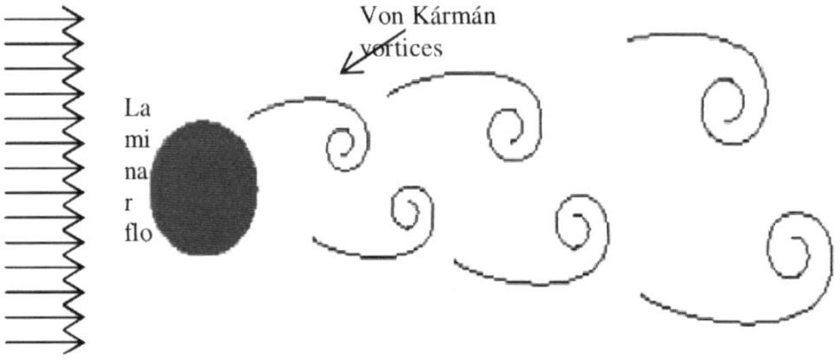

Figure 2. Von Kármán vortex streets forming in the wake of a Bluff-body.

In recent years, harvesting energy from fluid flow has gained increasing interest in the research community. Researchers have developed energy harvesting devices based on fluid induced vibration. Two types of device have been developed; one relies on the conversion of air motion to air steady oscillation, the other on the drag force of the airflow. In the first type of energy harvester's researchers uses Helmholtz resonators to convert airflow motion to steady air oscillation [6]. In the second type of energy harvester, the lift force of the fluid flow causes vibration on a piezoelectric cantilever or magneto-electric component which, in turn, converts kinetic energy of motion into electricity [6]. For vibration based energy harvesting device, it is important to predict the frequency of vibrations at various fluid speeds and thereby identify the desirable resonances between the bluff-body structure and the vortex shedding.

In this paper, we have presented various mechanical analyses for a MEMS-based piezoelectric fluid-flow micro energy harvester. COMSOL finite element analysis software are used for the evaluation of various mechanical analysis of the micro energy harvester structure like, physical the Stress and Strain state in the cantilever structures, Eigenfrequency Analysis, Transient analysis to demonstrate the feasibility of the design. A detailed step of modeling and simulation results of the uniform cantilever is explained. The results confirm the probability of the fluidflow based MEMS energy harvester.

2. Geometrical modelling of the Energy Harvester structure

This section explains the geometric modeling of the energy harvester cantilever structure capable of harness energy from fluid flow. As shown in the figure 3, A piezoelectric film is placed on top of a flexible rectangle, which is located in the wake of a D-shaped bluff body. The COMSOL-multiphysics CAD tools were used to create the 3D composite solid structure, using Boolean operations like union, intersection, and difference of the bluff-bodies. Geometric property range and values are provided in Table 1.

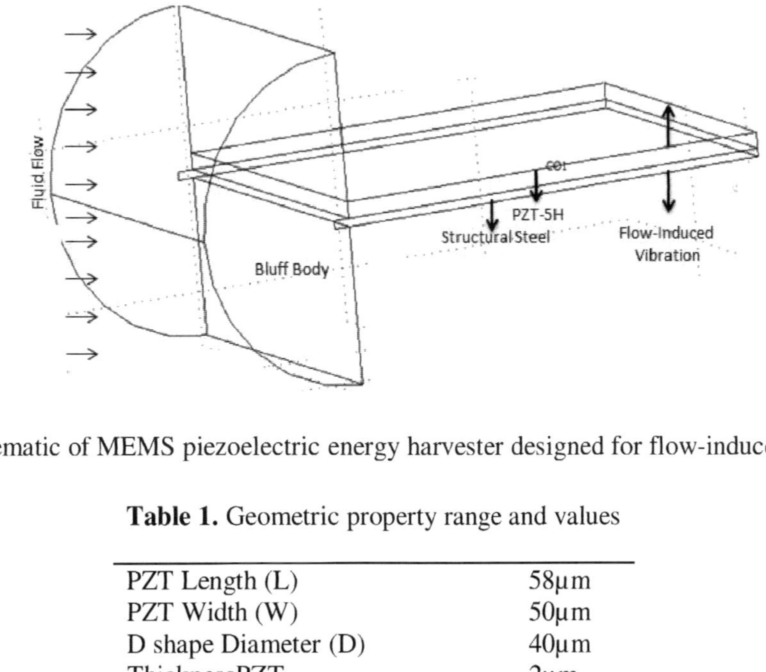

Figure 3. Schematic of MEMS piezoelectric energy harvester designed for flow-induced vibration.

Table 1. Geometric property range and values

PZT Length (L)	58μm
PZT Width (W)	50μm
D shape Diameter (D)	40μm
ThicknessPZT	2μm
Thickness Stainless Steel	1μm

3. Mechanical analysis of the Energy Harvester

The EH device depicted in the above figure 3 is part of a support mechanism and is subjected to both mechanical and electric loads. In this section, we have presented different analysis steps and results, to carry out a detailed mechanical analysis of the EH structure using MEMS Structural Mechanics Module. To specify the EH model's physics, we have defined material properties in 'Subdomain

Setting' and 'Boundary Setting' dialog box's, such as constraints, external forces and effects (such as mechanical constraints or loads), voltages, and temperatures etc.

3.1. Subdomain Setting conditions of the EH structure

In the Subdomain Setting the structural steel domain is an isotropic material model (the bluff-body and the lower part of the cantilever structure), and the PZT structure which has a length of 58μm, width 50 μm and thickness of 2 μm is Lead ZirconateTitanate (PZT-5H). Both properties with values are provided in Table 2.

Table 2. Material property range and values.

Lead ZirconateTitanate (PZT-5H)				Structural Steel			
Description	Quantity	Value/ Expression	Unit	Description	Quantity	Value/ Expression	Unit
Young's Module	E	2.0e11	Pa	Young's Module	E	200e9	Pa
Poisson's Ratio	v	0.33	-	Poisson's Ratio	v	0.33	-
Density	ρ	7500	Kg/m 3	Density	ρ	7850	Kg/m 3
Thermal expansion coeff.	a	1.2e-5	1/K	Thermal expansion coeff.	a	12.3e-6	1/K

3.2. Boundary Setting conditions of the EH structure

The boundary conditions resulted from the working conditions of the energy harvester model shown in figure 4. For the mechanical part a constraint of zero movement on the bluff-body has been considered. The cantilever part of the EH structure with the PZT substrate module (length of 550 mm, width 50 mm and thickness 5 mm) remain free to vibrate in response to a applied load on the free end of the cantilever beam. A distributed load is applied of 3.106 N/m^2in the z-direction on the PZT upper face.

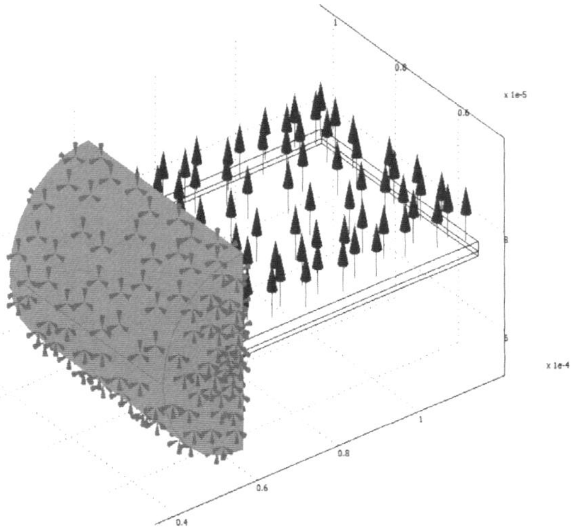

Figure 4. The MEMS EH geometry together with fixed constraints symbols (red arrows) and load symbols on the cantilever (blue arrows).

3.3. Static Analysis

A static analysis has no explicit or implicit time dependencies. It corresponds to the steady state of a transient analysis with constant (in time) boundary conditions and material properties. The purpose of this analysis is to find the magnitudes and locations of the maximum strain, stress, and electrical potential on the cantilever beam when an external static load is applied to the beam's free end, compare it with the material's yield strength, and to check that the deformation of the cantilever is within the limits of the design criteria. Figure 5 shows the maximum stress calculated with Von Mises criteria determined in theleft side of the beam (in the vicinity of the clamping side with the D shaped Bluff Body)andthe minimum value in the right side of the beam. Table shows the Von Mises effective stress calculation for different load, and using the Point Evaluation feature, obtained the deformation of a point on the geometry. The largest displacement occurs at the tip where the load is applied.

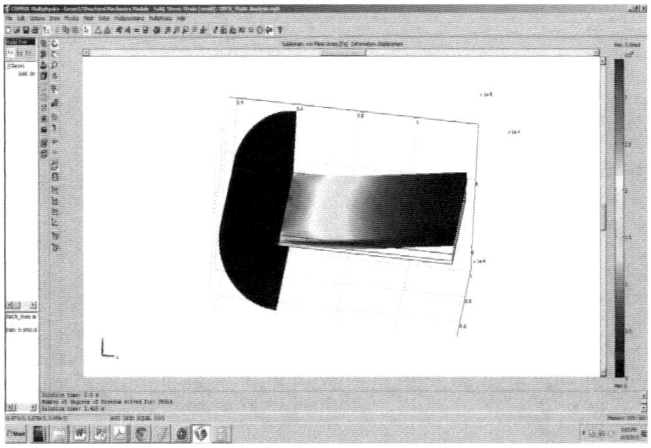

Figure 5.Maximum stress with Von Mises.

Table 3. Different Load applied and corresponding tip displacement

Fz	Von Mises Stress [Pa]	Z- displacement [μm]
3e6	$3.8185e^{9}$	8.765783
2e6	$2.545667e^{9}$	5.843855
1e6	$1.272833e^{9}$	2.921928
0.5e6	$6.364167e^{8}$	1.460964

3.4. Eigenfrequency Analysis

An Eigenfrequency analysis finds the undampedeigenfrequencies and mode shapes of a model sometimes referred to as the free vibration of a structure. The eigenfrequency analysis is carried out to determine the natural frequency of the EH structure, and therefore, setting up the excitation frequency of the dynamic load in the time-dependent analysis.The purpose of the eigenfrequency analysis is to find the first 6 modes of frequencyof the cantilever beam and their corresponding shape of deformation. These modes offrequency values are used later in finding the beam's damping factor, and in settingthe excitation frequency that makes the beam vibrates near its resonance and,therefore, gives maximum electrical potential.The values of the modes of frequency are illustrated in the following table. Figure 6 shows the corresponding deformation shapes for the highest and the lowest eignfrequencies of the EH model.

Table 4. Six lowest Eigen frequencies

Mode	Eigen frequency (Hz)
1	7.629872e5
2	2.030683e6
3	4.65233e6
4	7.111031e6
5	7.459217e6
6	8.491459e6

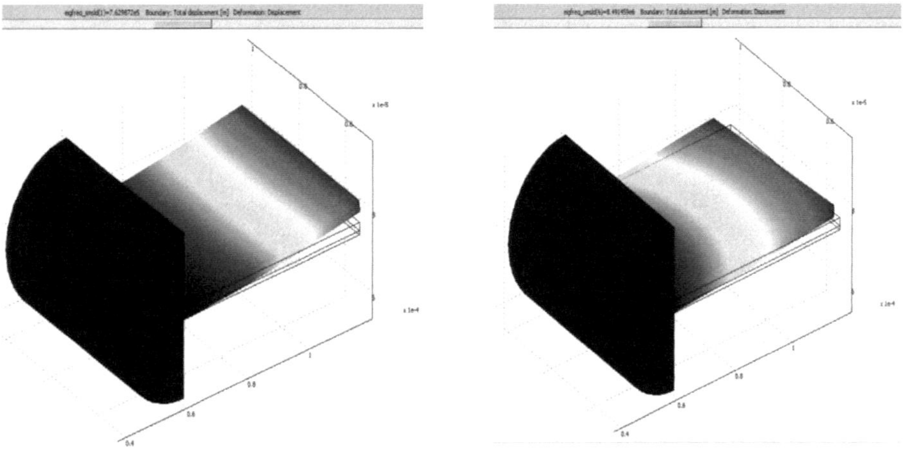

Figure 6. The highest and the lowest eigen mode of the EH model.

3.5. Transient Analysis

This type of analysis solves for the transient solution of displacements and velocities as functions of time [8].The purpose of this analysis is to find the transient response of the EH model from a harmonic load during the first five periods shown in figure 7. Figure 8 shows the displacement of the cantilever tip in the z-direction on the loaded face. The excitation frequency is set to $7.629872e^{5}$Hz, which is between the first and second eigenfrequency found in the eigenfrequency analysis.

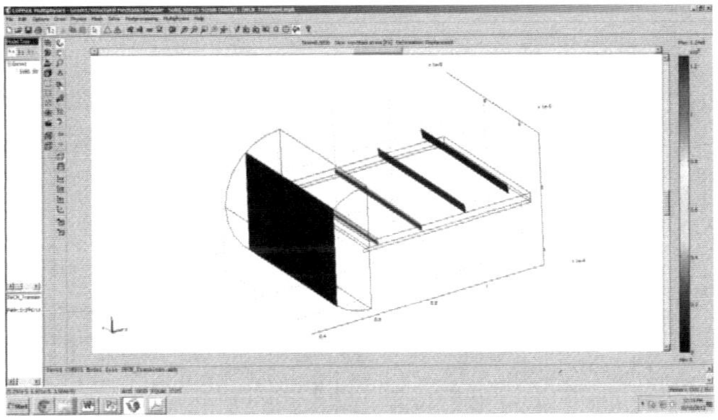

Figure 7. Slice plot of the von Misesstress shows the last time stepon the MEMS EH structure at 0.0036s.

Plot graph in figure 8 shows quantitative view of the time evolution of the cantilever displacement.

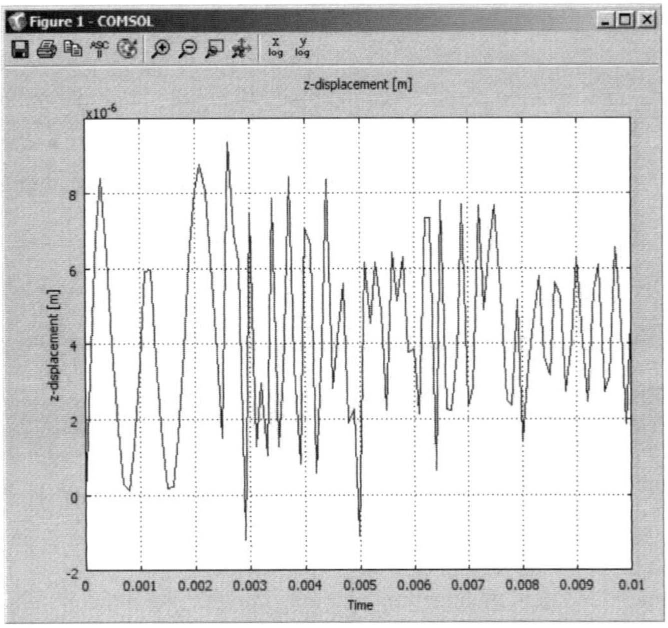

Figure 8. Displacement in the z-direction at a tip point on the loaded face.

4. Conclusion

In summary, we reported various mechanical analysis results of a MEMS micro energy harvester to extract energy from the Kármán Vortex Street behind a bluff body using COMSL FEA software [7]. These results will be compared with the experimental results of the fluid-flow based EH, presently under development at our lab.The initial results confirm the probability of the fluidflow based MEMS based energy harvester and necessary optimization directions.

Journal of Physics: Conference Series **431** (2013) 012024 doi:10.1088/1742-6596/431/1/012024

References

[1] Horowitz B. St. "Design and characterization of compliant backplate Helmholtz resonators" MSc Thesis, University of Florida. 2001.
[2] Seong-Hyok Kim et al., "An electromagnetic energy scavenger from direct airflow", *J. Micromech. Microeng.* vol. **19**, 2009.
[3] Cuadras A. and Ovejas V. J., "Piezoelectricity from the wind", *Proc. Eurosensors XXII,* pp. 305–8, 2008.
[4] Akaydin H. D., Niell E,, and Yiannis A., "Energy harvesting from highly unsteady fluid flows using piezoelectric materials", *J. Intel. Mat. Sys. Struct.* vol.211263–78., 2010.
[5] http://en.wikipedia.org/wiki/K%C3%A1rm%C3%A1n_vortex_street.
[6] Ming Zhaoa and Liang Chengb, "Numerical simulation of vortex-induced vibration of four circular cylinders in a square configuration," *Journal of Fluids and Structures*, vol. 31, 2012.
[7] http://en.wikipedia.org/wiki/Navier%E2%80%93Stokes_equations.
[8] C.H.K. Williamson and G.L. Brown, "A Series in to represent the Strouhal-Reynolds number relationship of the cylinder wake," *J. Fluids Struct., v*ol. 12, 1998.

Designing a Battery-Less Piezoelectric based Energy Harvesting Interface Circuit with 300 mV Startup Voltage

M R Sarker[1], Sawal H Md Ali[2], M Othman[3], and Shabiul Islam[1]

[1]Institute of Microengineering and Nanoelectronics (IMEN), UniversitiKebangsaan Malaysia (UKM), 43600 UKM Bangi, Selangor, Malaysia.
[2] Department of Electrical, Electronic and Systems Engineering, Faculty of Engineering and Built Environment, 43600 UKM Bangi, Selangor, Malaysia.
[3] Ministry of Science, Technology and Innovation (MOSTI), Federal Government Administrative Centre, 62662, Putrajaya, Malaysia.

Email: sklobanrahman@yahoo.com; shabiul@ukm.my

Abstract. This paper presents a designing a battery-less piezoelectric based energy harvesting interface circuits with 300mV step-up voltage. A technique (i.e., DC-DC Step-Up converter) has chosen for designing the startup voltage with low voltage energy (i.e., 300mV). The proposed method consumes very little power, and is especially suitable for the ambient environmental source, where energy harvested power is very low. The energy harvesting interface circuit consists of MOSFET bridge ac-dc rectifier, voltage regulator, dc-dc step-up converter and an energy storage device with capacitor at the output terminal, replacing this by external battery. This paper will study results these important issues regarding the efficiencies of the energy harvesting power conversion interface circuits considering the storage device low voltage. The achievement of our development circuit is able to boost up minimum 1.67 V for input DC voltage of 300mV. The overall circuit efficiency is greater than 80 % following the simulation results. This research has focused on the application of Wireless Sensor Network (WSN) and bio-medical device can be operated without battery.

1. Introduction

Using of piezoelectric elements from ambient vibrations to harvest energy of great interest over the past few years. One of the most widely used power harvesting techniques for micro-power applications uses piezoelectric materials to convert any type of energy to electrical energy. Micro Energy Harvesters are small electromechanical devices, which harvest ambient energy and then convert it into electrical energy, as shown in Table-1 [1]. Table [2] shows a block diagram of ambient energy harvesting system. The choice of the source depends primarily on the specification of the power requirement for an application and design feasibility [2, 3]. The process of extracting unused energy from the ambient environment and converting it into a usable form of electrical energy is known as Energy Harvesting. With recent growth in the development of low-power electronic devices such as microelectronics and wireless sensor nodes, as well as the global interest in the concept of "piezoelectric" the topic of energy harvesting has received much attention in the past decade. Piezoelectric materials have received the most attention for obtaining electric energy from the surrounding environment for their ability to directly convert vibrations into electrical energy.

Piezoelectric materials can be used as a means of transforming ambient vibrations into electrical energy that can be stored and used to power other devices. Usually, energy harvesting systems based on piezoelectric devices can be summarized to have three core components: piezoelectric devices, electric energy storage, and power electronic interface. Different power harvesting methods, one example is Wireless Sensor Network (WSN), which has potential to be used in many areas health monitoring [4, 5, 6], piezoelectric conversion systems [7, 8]. Traditional electrochemical batteries are usually used in powering the WSNs. But replacing batteries, due to their limited lifetime, is not very convenient. The labor and cost associated with changing hundreds or thousands of batteries would be troublesome and expensive in maintaining the networks. Moreover, in many cases, the WSNs are installed in remote areas; therefore, it is difficult to retrieve for battery replacement. Without replacing batteries, the WSNs will stop working as soon as the power runs out. Therefore, it is desired that the WSNs can acquire energy from the ambient environment to have a continuous power supply [9, 10]. A complete WSN node in signal processing and communication area will normally have a power budget of 100 μW. Achieving this requires good application knowledge and well designed energy harvester to enable an optimal balance between power and resolution. The ultimate goal of this research work is to design a micro-power module (i.e., Low voltage circuit) that can be integrated into bio-medical device. Therefore, in this research work, we have a way to develop a low voltage energy harvester system for Ultra Low Power (ULP) Bio-medical application that energy harvester to integrate with WSN node. In this paper, we present the development of a low voltage energy harvesting circuit (based on simulation) using piezoelectric elements. The paper is also organized as follows: Section II describes the modeling of piezoelectric based Energy Harvesting system, such as piezoelectric element, AC-DC rectifier, temporary storage devices, DC-DC Converter (Step-Up). Section III describes the simulation results in PSPICE. Finally, conclusion is given at section IV in the paper.

Table 1. Energy Harvesting Estimates [1]

Energy Source	**Harvested Power**
Vibration/ Motion	
Human	4 μW/cm2
Industry	100 μW/cm2
Temperature Difference	
Human	25 μW/cm2
Industry	1–10 mW/cm2
Light	
Indoor	10 μW/cm2
Outdoor	10 mW/cm2
RF	
GSM	0.1 μW/cm2
WiFi	0.001 μW/cm2

Table 2. Ambient Energy system [9]

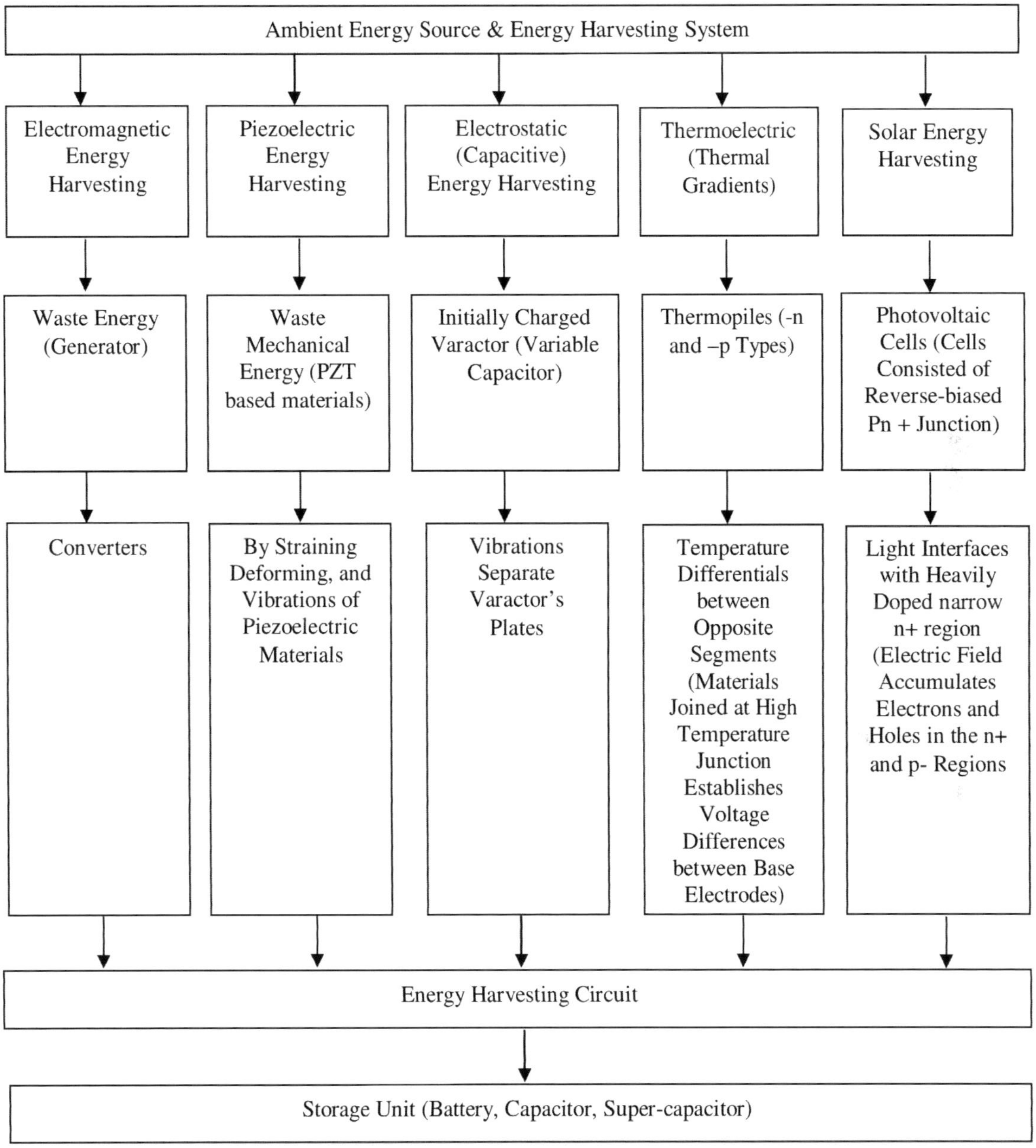

2. Development of piezoelectric based low voltage Energy Harvesting interface circuit

This section provides the modeling on piezoelectric based energy harvesting system considering of piezoelectric element, AC-DC rectifier, voltage regulator, DC-DC converter (step-up) and Temporary storage device. Here we mention all these characteristics are illustrated through selective block diagram. A simple produce is step-by-step presented to design the modeling of piezoelectric based energy harvesting system, as shown in Figure 2. To do the development of energy harvesting system, the PSPICE software has been performed on the proposed system. With the recent flow of micro scale devices, piezoelectric power generation can provide a convenient alternative to traditional power sources used to operate certain types of sensors/actuators, Piezoelectric materials can be used as a means of transforming ambient vibrations into electrical that can be stored and used to power other devices.

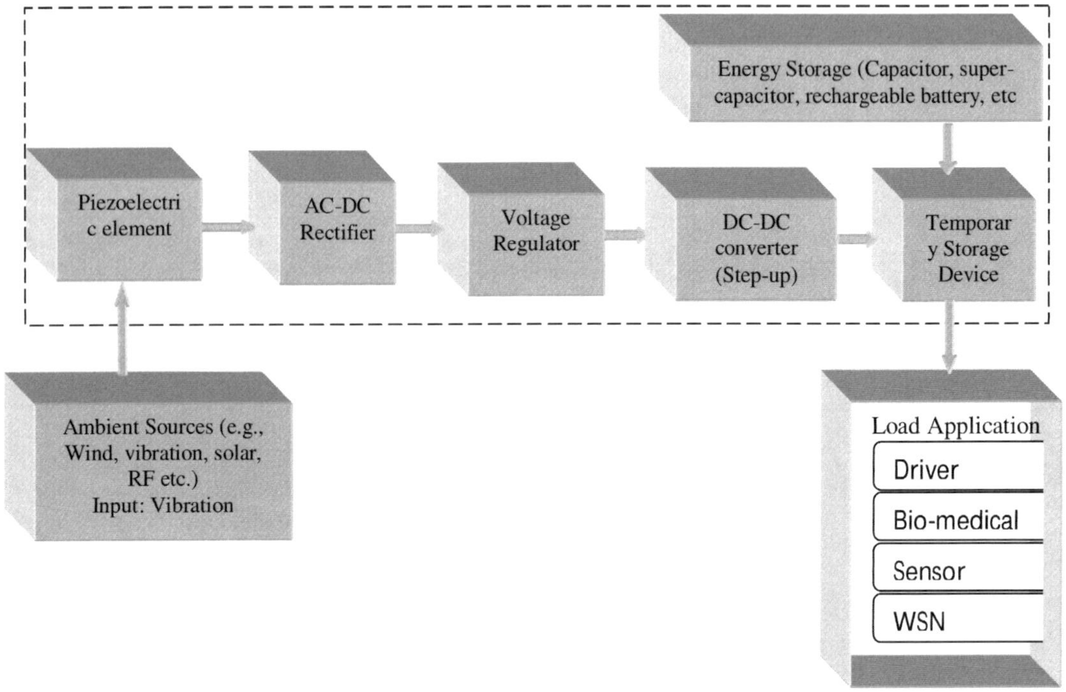

Figure 2. Propose Energy harvesting modules (Experimental sub-blocks).

2.1 Piezoelectric element
Piezo Element is a Piezo Buzzer and also called Piezo Diaphragms and Piezo Benders. Many mechanical vibration-based energy harvesting systems use a piezoelectric transducer as an AC power source, whose output voltage must first be rectified before being delivered to a load. This piezoelectric vibration based transducer very low voltage (300mV) which can be applied to the full-wave rectifier circuit that consist of DC voltage without any voltage lose.

2.2 AC-DC rectifier circuit
The first stage of our developed ULP energy harvesting circuit is AC-DC rectifier. The rectifier circuit posed the problem of having very low input at which most devices (diodes) do not work for our case i.e., the forward voltage potential is higher than the input. To eliminate this problem, MOSFET or thyristor needs to be used. In our research, we used MOSFET to design the AC-DC full-wave rectifier circuit to get suitable DC voltage without loss voltage. Figure 3 shown is the proposed block diagram of the AC-DC rectifier.

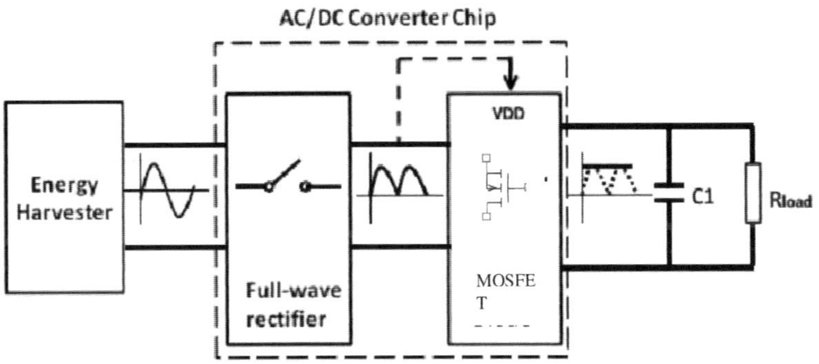

Figure 3. Block diagram of the development AC-DC Rectifier.

2.3 Voltage Regulator

Voltage regulator is another part in power conversion sequence. As the rectified voltage of the AC-DC converter (on the load resistor) has strong dependency on the distance from the source and amplitude of the input signal, therefore the need of a voltage regulator that converts an input voltage to a fixed output voltage level. It also minimizes any variation by fluctuations in the input voltage or the load voltage is essential. However, due to limitation on available power, the regulator must be efficient and work with low voltage level.

2.4 DC-DC step-up converter

The third block of the energy harvesting interface circuit is the DC-DC step-up converter and its schematic diagram shown in Figure 4. The corresponding block diagram is shown in Figure 2 and results shown in section 3. The development of DC-DC converter provides a boost-up output while input voltage is too low. The minimum start-up voltage is 1.67V of the development of DC-DC converter outputs while the input voltage range approximately 300 mV DC after rectified. The development of DC-DC converter used the DC input (V_{in}) = 0.3V. When switch control (M1) MOSFET is closed for time t_1, the inductor (L1) current rises and energy is stored in the inductor and if the switch is opened for time t_2, the energy stored in the inductor is transferred to load through thyristor (X1) and the inductor current falls. We had faced similar problems during to design DC-DC converter circuit were found, the (low input DC voltage 200mV). We used thyristor in the place of a diode to design this circuit. This is our new modified in our development DC-DC step up converter. The advantage of our modified new development DC-DC boost converter that it is able to increase the voltage using very low input source. We used MOSFET (M1) transistor and pulse (V5) is V1=0V, TD=1m, TR= 0u, TF=0u, PW=7ms, PER=10ms as a switching purpose for open and close. If a large capacitor (C1) is connected across the load as the output voltage is continuous and V_{out} becomes the average value that the voltage across the load can be stepped up by varying the duty cycle and the minimum output voltage is V_{in} when duty cycle = 0. We used inductor (L1) is 4.7uH for storage the current, Resistor (R1) is 10kΩ used for the better output, we used capacitor (C1) is 100uF for store voltage and filtering of the output voltage to reduce ripple. Finally we are able to achieved the output voltage V_{out} (1.67V) after using the DC-DC converter (Step-up).

For AC-DC rectifier circuit we used four MOSFET bridge rectifier and we got output (300mV) from this circuit without loss any voltage. The main focus is the paper by using DC-DC converter

(Step-Up) boost up the rectified voltage. Finally we got output (1.67V) after using the DC-DC converter (Step-Up).

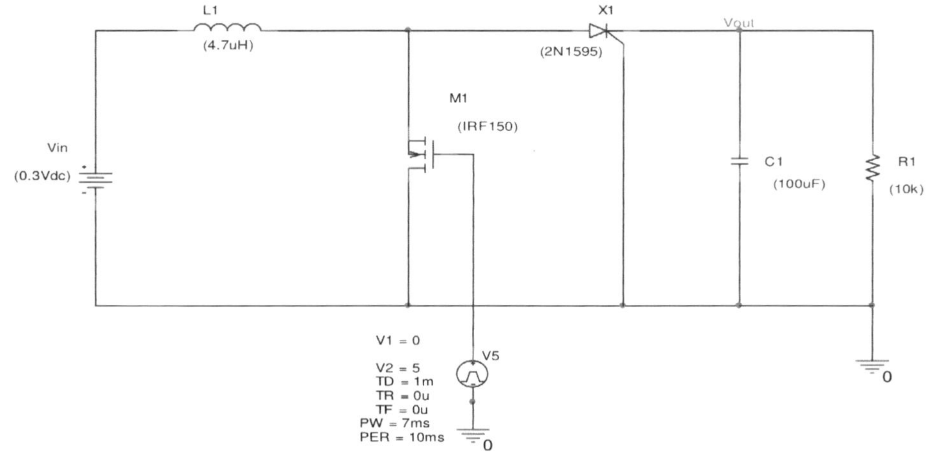

Figure 4. Development of DC-DC converter (Step-up) circuits.

3. Simulation Results in Pspice software

The piezoelectric element used sinusoidal (V_{in}) voltage source as an input source. We used voltage source and internal series capacitor, and for voltage source we use capacitor as a series, output as shown in Figure 5 (a). For the AC-DC rectifier circuit we used four MOSFET-bridge rectifiers. We got (300mV) after rectified the piezoelectric voltage output as shown in Figure 5(b). The voltage (rectifier voltage) can be modified by controlling the DC-DC converter (Step-Up) between the two energy voltage the AC-DC rectified voltage and after using the DC-DC converter (Step-Up) circuit output. Using the AC-DC rectifier, and DC-DC converter (Step-Up), the energy storage device with the rectifier voltage output as shown in Figure 5(c) and 5(d).

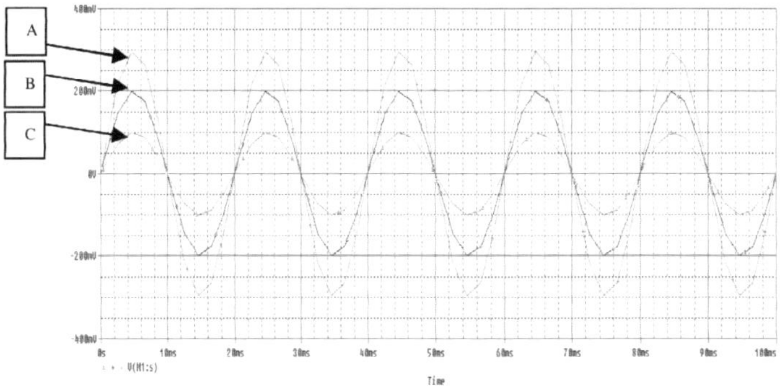

Figure 5(a). Input voltage of the piezoelectric transducer.

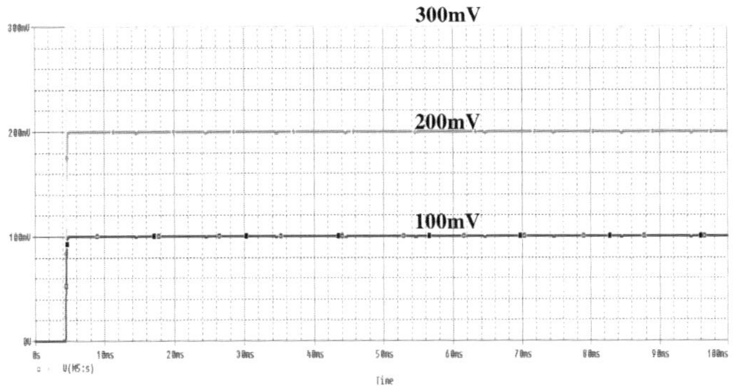

Figure 5(b). Rectified output voltage.

The curve on the Figure 5 (a) shows output of the piezoelectric circuit. The curve [A] observe is the input voltage of the piezoelectric transducer maximum 300mV AC voltage, curve [B] observe 200mV AC voltage and curve [C] observe 100mV AC voltage.

The Figure 5 (b) is the output of the AC-DC MOSFET-bridge rectifier circuit. Finally, after analysis we are able to get full wave AC-DC rectifier voltage (without loss any voltage and without any ripple). In Figure 4.9 (e) shows the final output DC voltage at different inputs AC single (between 100mV to 300mV).

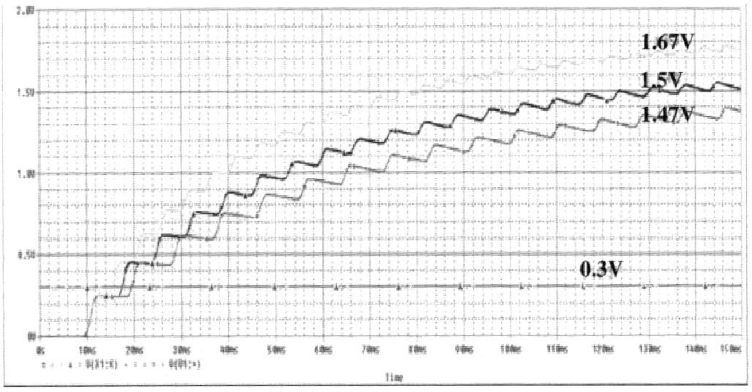

Figure 5(c). Voltage waveform of switch with different PER.

We consider PER values duration between 1m to 9m and observed voltage and current. After analysis we find out suitable value of target voltage. Specify its start value, end value and increment as shown in Figure 5(c). From this Figure it can be observed that the voltage increase and decrease with values of switch. However, beyond 5m switch parameter value the voltage exceed our goal. Therefore, values for 5ms switch are match to achieve our expected results 1.67V.

The Figure 5(d) shown is the final output of the DC-DC converter circuit. There are two curves in the Figure. Here the curve [D] mention is the input of piezoelectric vibration sensor after AC-DC rectifier circuit full DC voltage 300mV, and the curve [E] denoted is the final output of the Step-Up DC-DC converter circuit is (1.67V). From this Figure, it can be observed that the curve is become bending, initially it takes some time to reach the voltage after 30ms it becomes constant. This phenomenon is also clearer from these Figures.

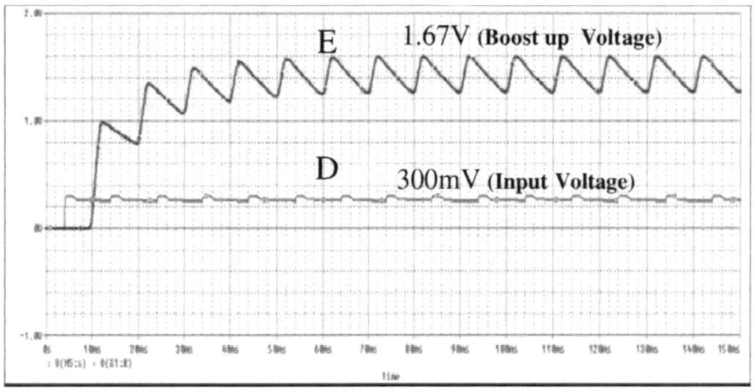

Figure 5(d). Voltage Output of the DC-DC Converter (Step-UP).

4. Conclusion

The energy harvesting circuits consisting of piezoelectric element, AC-DC rectifier, voltage regulator, DC-DC converter (Step-Up) and temporary storage device have developed by PSPICE simulation. The main focus of the research work is the development of the DC-DC Converter (Step-Up) of 1.67V. The overall circuit efficiency is about 80%, following the simulation results. Research work have focused on the markets for ultra-low power energy harvesting of vibration to electricity generation technology covering several primary applications – WSN, building automation wireless, battery-less, low-power switches, automotives, medical uses such as body area networks; precision agriculture; and consumer electronics. Future work will be focused on the design of the hardware implementation to verify the simulation result etc.

References

[1] M. R. Sarker, M.R., S.H.M. Ali, M. Othman and S. Islam, 2011. Designing a low voltage energy harvesting circuits for rectified storage voltage using vibrating piezoelectric. Aust. J. Basic Applied Sci., 5: 341-345.

[2] K.Finkenzeller, RFID Handbook. "Fundamentals and Applications in Contactless Smart Cards and Identification", John Wiley, Chicester, Sussex, Second edition, 2003.

[3] A-L. Annala, I. Oy, and U. Friedrich, "Passive Long Distance Multiple Accesses UHF RFID System", Palomar Project, European Commission, Public Report, Project Number 1ST 1999-10339, Nov. 2002.

[4] L. Chao, T. Chi-Ying, and Wing-Hung ki, " A Batteryless vibration-based Energy Harvesting System for Ultra Low Power Ubiquitous Application", pp.1349-1352, june 2007.

[5] J.Kymissis, C. Kendall, J. Paradiso, and N. Gershenfeld, "Parasitic power harvesting in shoes," in *Second International Symposium on Wearable Computers,* 1998, pp. 132-139.

[6] F.J. DiSalvo, "Thermoelectric cooling and power generation," *Science*, vol. 285, no. 5428, pp. 703-706, July 1999More references.

[7] K. Finkenzeller, RFID Handbook. "Fundamentals and Applications in Contactless Smart Cards and Identification", John Wiley, Chicester, Sussex, Second edition, 2003.

[8] A.L, Annala, I. Oy, and U. Friedrich, "Passive Long Distance Multiple Accesses UHF RFID System", Palomar Project, European Commission, Public Report, Project Number 1ST 1999-10339, Nov. 2002.

[9] M.,Renaud, T. Sterken, A. Schmitz, P. Fiorini, C. Van Hoof, R. Puers, "Piezoelectric Harvesters and MEMS Technology, Fabrication, Modeling and Measurements", International Solid-State Sensors, Actuators and Microsystems Conference, pp. 891-894, June 2007.

[10] Geffrey, K. Ottman," Adaptive Piezoelectric Energy Harvesting Circuit for wireless remote Power Supply", pp.669-676,Sep 2007.

Design and Optimization of 22 nm Gate Length High-k/Metal gate NMOS Transistor

Afifah Maheran A H[1], Menon P S[1], I Ahmad, S Shaari[1], H A Elgomati[1], F Salehuddin[2]

[1] Institute of Microengineering and Nanoelectronics (IMEN), Universiti Kebangsaan Malaysia (UKM), 43600 UKM Bangi, Selangor, Malaysia
[2] Centre for Micro and Nano Engineering (CeMNE), College of Engineering, Universiti Tenaga Nasional (UNITEN), 43009 Kajang, Selangor, MALAYSIA

e-mail: susi@eng.ukm.my

Abstract. In this paper, we invented the optimization experiment design of a 22 nm gate length NMOS device which uses a combination of high-k material and metal as the gate which was numerically developed using an industrial-based simulator. The high-k material is Titanium dioxide (TiO_2), while the metal gate is Tungsten Silicide (WSi_x). The design is optimized using the L9 Taguchi method to get the optimum parameter design. There are four process parameters and two noise parameters which were varied for analyzing the effect on the threshold voltage (V_{th}). The objective of this experiment is to minimize the variance of V_{th} where Taguchi's nominal-the-best signal-to-noise ratio (S/N Ratio) was used. The best settings of the process parameters were determined using Analysis of Mean (ANOM) and analysis of variance (ANOVA) to reduce the variability of V_{th}. The results show that the V_{th} values have least variance and the mean value can be adjusted to 0.306V ±0.027 for the NMOS device which is in line with projections by the ITRS specifications.

1. Introduction

The replacement of Silicon Dioxide (SiO_2) as the gate dielectric with a high dielectric (high-k) material in future complementary metal-oxide-semiconductor (CMOS) technology is being widely studied nowadays. A number of different high-k materials have been proposed and analyzed to achieve the future trends in semiconductor where the decrease in the size of devices causes the decrease in the size and the thickness of the dielectric gate. As the physical thickness of SiO_2 - based gate oxides approaches 2 nm, a number of fundamental problems occur such as the increment of gate leakage current, oxide breakdown, boron penetration from the polysilicon (poly-Si) gate electrode and channel mobility problem [1, 2]. Each one of these phenomenon is important for device operation. In other words, the conventional device-scaling scenario involving scaling down of SiO_2 - based dielectrics up to below 1 nm becomes impractical now[1].

During early high-k dielectric development, material incompatibility between high-k and the poly gate electrode was discovered [3, 4]. One of the issues was the high defect rates at the interface of high-k and poly-Si causing the device's lower electrical mobility. The latter was attributed to charge scattering, which is intrinsic to the physics of an integrated poly and high-k device [5]. Therefore, self-aligned silicide (SALICIDE) technology has been widely used to reduce the sheet resistance (R_s) of poly-Si gates [6].

Then there is other solution by replacing poly-Si with a metal gate electrode. Replacing poly-Si gate materials with metal is the answer of the compatibility issues between the high-k dielectric and poly electrode. Significant effort and research was focused on finding metal electrodes with the correct work function [7]. However, the metal work functions are needed to adjust the NMOS and PMOS devices' threshold voltage (V_{th}).

Current research efforts are focused on metal oxide materials with high-k values (e.g., ZrO_2, HfO_2, Al_2O_3, Y_2O_3, La_2O_3 and TiO_2) since their introduction produce a physical thicker oxide, causes reduction in gate leakage current and in the same time maintaining a large gate capacitance [8].From previous research, the scaled down design of 22 nm gate length NMOS device from 32 nm gate length was a success [9]. This paper is the continuation from the previous work where the device design is optimized using Taguchi's L9 orthogonal array.The results show that a 22 nm gate length NMOS transistor with V_{th} value of 0.306V that is well within ITRS 2011 prediction can be produced numerically [10].

2. Materials & methods

The fabrication process steps are as follows. The substrate used for the experiment was a p- type silicon with a <100> orientation. It is then followed by growing an oxide layer, at the top of the bulk silicon using dry oxygen. P-well implantation process was done, using this oxide layer as a mask. This was done using Boron as dopant with a dose of $3.75x10^{12}$ ions/cm^2. The silicon wafer then underwent the annealing process at 900 °C in a Nitrogen environment, and followed by dry oxygen in order to ensure that boron atoms are being spread properly in the wafer. The next step was to produce a Shallow Trench Isolator (STI) of 130 Å thickness [11]. In order to form the STI layer, the wafer was oxidized in dry oxygen for 25 minutes. Then, a 1000 Å nitride layer was deposited on top of the oxide layer by applying low pressure chemical vapour deposition process (LPCVD), followed by a photo resist deposition with a thickness of 1.0µm. The trench depth of 3200 Å was achieved. Thereafter, a sacrificial oxide layer was grown and then etched followed by a sacrificial nitride layer whereby the trench is then completed.

Then the high-k material, TiO_2 was deposited for thickness of 2 nm [12] followed by etching to get the desired thickness and was adjusted to produce a 22 nm gate length. The next step was to implant the N well active area, in order to adjust the threshold voltage, V_{th} value. The dosage for boron was $6.98x10^{12}$ atom/cm^2. Tungsten silicide (WSi_x) will then be deposited on the top of the bulk device with thickness of 8nm and etched accordingly to produce the gate contact point as desired [13]. Later on, Halo implantation was performed in order to get an optimum performance for the NMOS device where Indium was implanted with the dose of $12.76x10^{12}$ ions/cm^2. The dosage was varied in order to get the optimum value [6, 14]. Then spacers were formed at each of the polysilicon sides, namely the source and drain regions respectively. Side wall spacers were used as a mask for source and drain implantation [15]. Then, there are source-drain implantations where Arsenic was firstly implanted with a dose of $5.1x10^{13}$ ions/cm^2 followed by phosphorous with a dose of $1.5x10^{12}$ ions/cm^2, to ensure the smooth current flow in the device. The next process was the development of 0.5 µm Borophosphosilicate Glass (BPSG) layer [16]. This layer acts as a pre metal dielectric (PMD). After Borophosphosilicate Glass (BPSG) deposition, the wafer undergoes annealing process at a temperature of 950 °C [16].

The next process was compensation implantation using phosphorous, with a dose of $3.71x10^{13}$ ions/cm^2 [17]. Then, aluminium layer was deposited on top of the structure and then it was etched accordingly to form the metal contact for the source and drain. At this stage the transistor design is completed. Figure 1(a) and Figure 1(b) shows the overall device structure and the enlarged figure of the 22 nm gate length of the high k/metal gate NMOS transistor respectively. Figure 2 shows the net doping profile of the device that was successfully designed. Then, the transistor undergoes electrical characteristic measurement in order to find the threshold voltage (V_{th}) in reference to ITRS 2011 [10].

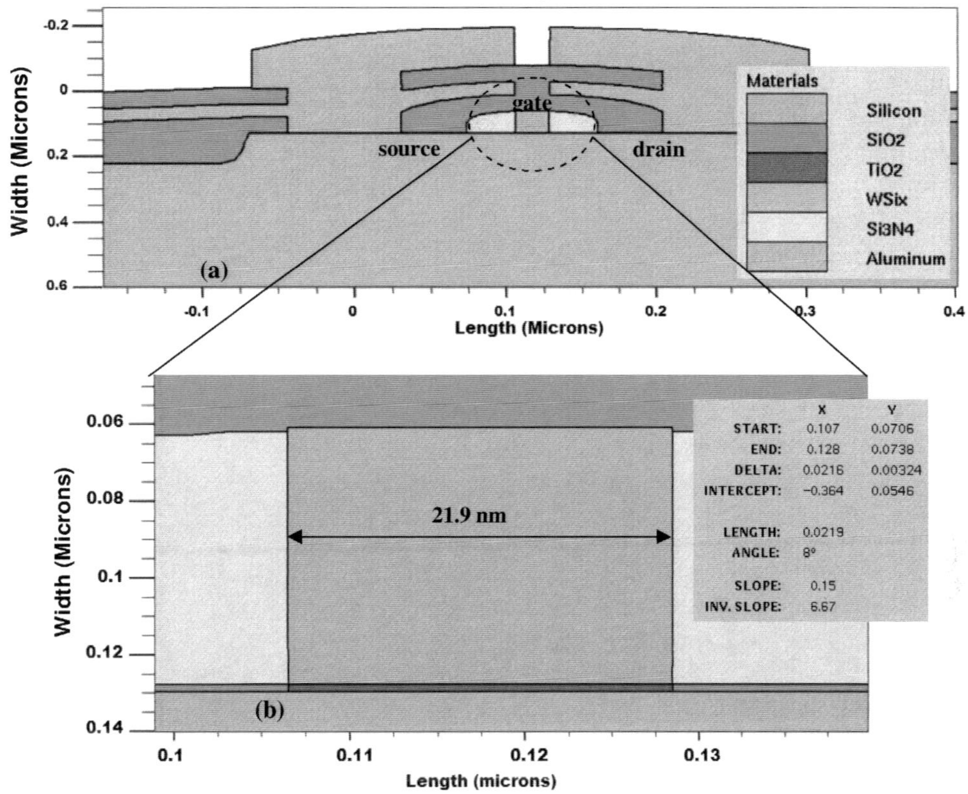

Figure 1(a). Completed NMOS transistor with 22 nm gate length; **(b).** Zoom up of a 22 nm gate length NMOS technology

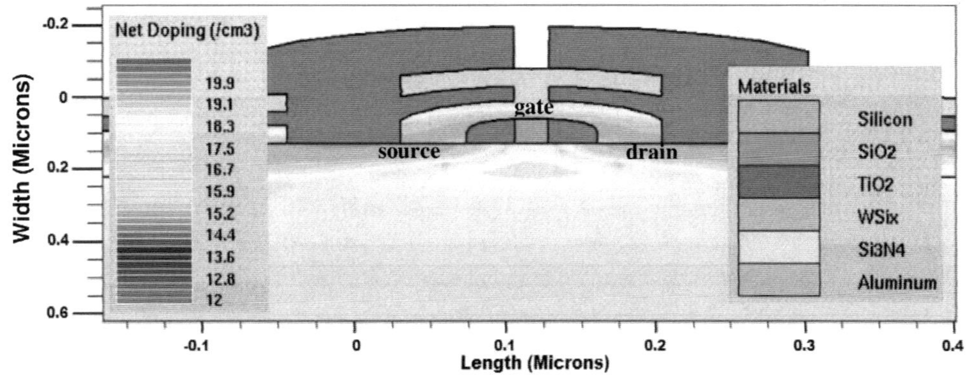

Figure 2. The doping profile of the NMOS transistor

2.1. Taguchi Orthogonal L9 Array Method

The Taguchi orthogonal L9 array method requires 4 process parameters and two noise factors were also chosen to complete the design of four sets of experiments consisting of 36 runs. The noise factors are included in order to get a more accurate design and to make the process parameters insensitive to variation. All these values of the process parameters of each levels and noise factors are listed in Table 1 and Table 2 respectively.

Table 1. Process parameters and their levels

Symbol	Process Parameter	Unit	Level 1	Level 2	Level 3
A	Halo Implant	atom/cm^3	1.270e^{13} (A1)	1.275 e^{13} (A2)	1.280 e^{13} (A3)
B	S/D Implant	atom/cm^3	5.100e^{13} (B1)	5.150 e^{13} (B2)	5.200 e^{13} (B3)
C	Compensation Implant	atom/cm^3	3.650 e^{13} (C1)	3.700 e^{13} (C2)	3.750 e^{13} (C3)
D	V_{th} Adjust Implantation	atom/cm^3	6.940 e^{12} (D1)	6.960 e^{12} (D2)	6.980 e^{12} (D3)

Table 2. Noise factors and their levels

Symbol	Noise Factor	Unit	Level 1	Level 2
X	Sacrificial Oxide layer	°C	900 (X1)	902 (X2)
Y	P-well Implantation Temperature	°C	850 (Y1)	852 (Y2)

3. Result and discussion

The optimized results from Taguchi Method were used to verify the predicted optimal design. The results of V_{th} were analyzed and processed in order to get the optimal design of the device.

3.1. Analysis for 22nm NMOS Device

The L9 Taguchi method analysis which consists of nine experiments is specified in the orthogonal array table. Four specimens were simulated for each of the parameter combinations. The completed response for V_{th} data is shown in Table 3

Table 3. V_{th} values for NMOS device

Exp. No	Threshold Voltage (Volts)			
	X_1Y_1	X_1Y_2	X_2Y_1	X_2Y_2
1	0.319202	0.317181	0.319114	0.317094
2	0.295836	0.295111	0.295804	0.295079
3	0.283015	0.282119	0.282976	0.282079
4	0.309481	0.307376	0.309391	0.307286
5	0.288823	0.287997	0.288787	0.287961
6	0.30077	0.300098	0.300741	0.300069
7	0.31957	0.317453	0.319484	0.317367
8	0.349681	0.347745	0.349603	0.347667
9	0.31119	0.308995	0.311102	0.308907

Based on the experiment results, the control factor that has the most effect on the device characteristics can be determined. The Signal-to-noise (S/N) ratio is used to discover the optimal process parameters when analyzing the experimental data. The V_{th} analysis in this experiment uses S/N of nominal-the-best [18, 19].

The utilization of S/N Ratio for each level of process parameters is to make the design more accurate. Regardless of the category of the performance characteristic, the larger the S/N Ratio resulted the better the performance characteristic. Therefore, the optimal level of the process parameters is the level with the highest S/N Ratio [20, 21].

The S/N Ratio is selected in order to obtain a threshold voltage value which is close or equal to the given target value (0.306V), which is also known as nominal value according to ITRS specification [22]. The S/N Ratio (Nominal-the-best), η can be expressed as [19]:

$$\eta = 10\text{Log}_{10}\left[\frac{\mu^2}{\sigma^2}\right] \quad (1)$$

Where:

$$\mu = \frac{Y_i + \ldots\ldots + Y_n}{n} \quad (2)$$

$$\sigma^2 = \frac{\sum_{i=1}^{n}(Y_i - \mu)^2}{n-1} \quad (3)$$

Where n is number of tests, Y_i is the experimental value of the threshold voltage, μ is mean and σ is variance. For nominal-the-best analysis, there are two types of factor of interest which are the dominant and the adjustment factors. By applying (2)-(4), the η for each device were calculated and given in Table 4 [19].

Table 4. Mean, variance and S/N ratios analysis

Exp No.	Mean	Variance	S/N Ratio (Mean)	S/N Ratio (Nominal-the-Best)
1	3.18E-01	1.36E-06	-9.95E+00	48.71
2	2.95E-01	1.76E-07	-1.06E+01	56.97
3	2.83E-01	2.68E-07	-1.10E+01	54.73
4	3.08E-01	1.48E-06	-1.02E+01	48.08
5	2.88E-01	2.28E-07	-1.08E+01	55.62
6	3.00E-01	1.51E-07	-1.04E+01	57.77
7	3.18E-01	1.50E-06	-9.94E+00	48.31
8	3.49E-01	1.25E-06	-9.15E+00	49.87
9	3.10E-01	1.61E-06	-1.02E+01	47.76

Referring to Table 4, rows 2, 3, 5 and 6 gives a S/N Ratio value of 56.97 dB, 54.73 dB, 55.62 dB and 57.77 dB respectively. This indicates that the process parameter combinations in these rows give the best insensitivity for the response characteristics. Since the experimental design is orthogonal, the effect of each process parameter on the S/N Ratio at different levels can be separated out.

The S/N Ratio for each level of the process parameters are summarized in Table 5. In addition, the total mean of the S/N Ratio for the experiments are also calculated and listed in Table 5.

Table 5. S/N ratio for the threshold voltage

Symbol	Process Parameter	S/N Ratio (Mean)			Total Mean S/N	Max - Min
		Level 1	Level 2	Level 3		
A	Halo Implantation	53.47	53.82	48.65	51.98	5.17
B	S/D Implantation	48.37	54.15	53.42		5.78
C	Compensation Implantation	52.12	50.94	52.89		1.95
D	V_{th} Adjust Implantation	50.70	54.35	50.90		3.65

The factor effect graph for the S/N Ratio of the experiments is shown in Figure 3. The dashed line in the graph represents the values of the overall- mean of the S/N Ratio. While the line of the graphs at the top and bottom represents the S/N Ratio (Nominal-the-best) and the S/N Ratio (Mean) respectively. In the S/N Ratio analysis, the larger the S/N Ratio resulted the better the quality characteristic for the threshold voltage. Referring to the graph, from the left, the slopes correspond to the Halo Implantation (Factor A), followed by S/D implantation (Factor B), Compensation Implantation (Factor C) and lastly V_{th} Adjust Implantation (factor D) respectively. The slopes show that the Halo implantation (Factor A) dose has found to be dominant factor because it has the maximum S/N ratio since the slope is the sharpest and highest compared to the others. While the Compensation Implantation (Factor C) was found to be the adjustment factor because it has small effect on the S/N Ratio (Nominal-the-best) and in the same time has large effect on S/N Ratio (Mean).

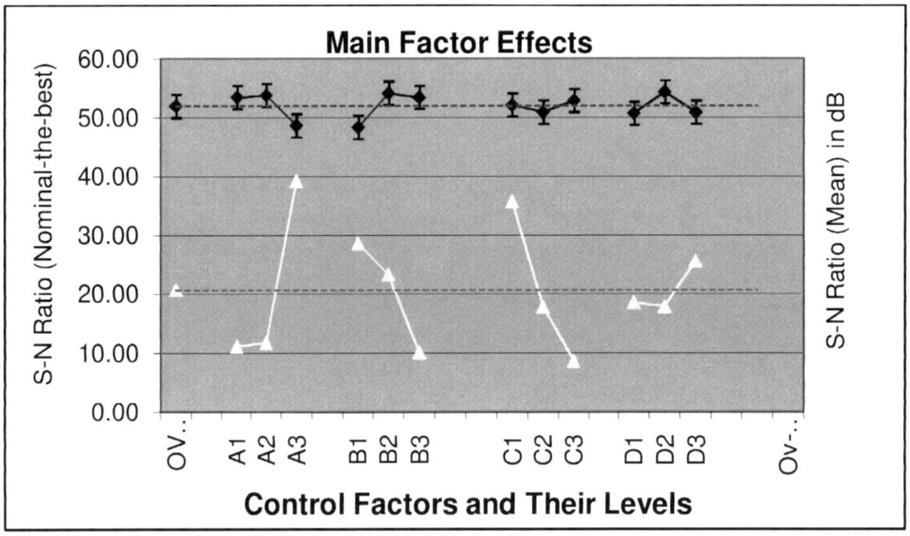

Figure 3. S/N graph for threshold voltage

3.2 Analysis of Variance (ANOVA)

The priority of the process parameters with respect to the V_{th} was investigated to determine the accuracy of the optimum combinations. The result of ANOVA for the NMOS device is presented in Table 6. The percent factor effect on S/N Ratio indicates the priority of a factor (process parameter) to reduce variation. The high percentage of a factor effect on mean contributes to the greatest influence on the stability of V_{th} with respect to the noise parameters.

Table 6. Result of ANOVA

Symbol	Process Parameter	Degree of Freedom	Sum of Square	Mean square	Factor Effect on S/N Ratio (%)	Factor Effect on Mean (%)
A	Halo Implantation	2	50	25	36	46.20
B	S/D Implantation	2	60	30	42	16.38
C	Compensation Implantation	2	6	3	4	34.06
D	V_{th} Adjust Implantation	2	25	13	18	3.36

Obviously the results show that the Halo implantation with a percentage of 46.20 % has the most dominant impact to the threshold voltage of the device, followed by compensation implantation at 34.06 % while 16.38 % and 3.36 % is owned by S/D implantation and V_{th} adjust implant respectively.

3.3 Confirmation of Optimum Run

Based on all the analyzed data, it can be deduced that, the compensation implantation is said to be as an adjustment factor because it has a small effect on the S/N Ratio (Nominal-the-best) and has a large effect on the mean. As a result the value of this factor can be adjusted. The adjustments are done to get the threshold voltage as close as possible to the nominal value or the target value. The values of the control factors are varied in compliance to Taguchi method until the best result is achieved. The best predicted setting by Taguchi method for the process parameters for the device is shown in Table 7.

Table 7. Best setting of the process parameters

Symbol	Process Parameter	Unit	Level	Best Value
A	Halo Implantation	atom/cm^3	2	1.275e^{13}
B	S/D Implantation	atom/cm^3	1	5.1e^{13}
C	Compensation implantation	atom/cm^3	-	Sweep (3.65 e^{13} to 3.75e^{13})
D	V_{th} Adjust Implantation	atom/cm^3	3	6.98e^{12}

Using the above parameters, the final simulation was performed to verify the accuracy of the prediction. By setting the Halo Implantation to Level 2, S/D Implantation to Level 2, V_{th} Adjust Implantation to Level 2, the value of compensation implantation was adjusted to fall within the range of 3.65×10^{13} to 3.75×10^{13} in order to obtain a V_{th} value that is closer to 0.306 V. The sweep that was done with the Compensation Implantation resulted in the best parameter setting of 3.67×10^{13} atom/cm^3. These final parameters were then simulated with the noise factors to get the optimal result as noted in Table 8.

Table 8. Results of Confirmation Experiment with Added Noises

V_{th1} (n1,n1)	V_{th2} (n1,n2)	V_{th3} (n2,n1)	V_{th4} (n2,n2)	SNR (Mean)	SNR (Nominal-the-best)
0.305901	0.303762	0.305811	0.303672	0.304787	47.8

Finally, the experiment has resulted in the achievement of mean V_{th} value of 0.304787 V with the S/N Ratio of 47.8 dB. The values are in line with projections by ITRS 2011.

4. Conclusion

As a conclusion, the Taguchi method is a reliable method in achieving the optimum solution in fabricating nanoscale CMOS devices. The V_{th} is the main response in determining the functionality of the device. The numerical fabrication of a 22 nm gate length high-k/metal gate NMOS transistor was achieved successfully. In this experiment, the Halo implantation dose was the strongest effect and identified as the dominant factor while the Compensation Implantation as an adjustment factor where the threshold voltage is the main response studied in this experiment. By adding noise factors to the design, a more robust device design can be attained.

Acknowledgment

The authors would like to thank the Institute of Microengineering and Nanoelectronic (IMEN), Universiti Kebangsaan Malaysia (UKM), the Centre for Micro and Nano Engineering (CeMNE), the College of Engineering Universiti Tenaga Nasional (UNITEN), the University Teknikal Malaysia Melaka (UTeM) and the Ministry of Higher Education for moral, operational and financial support throughout the project. We would also like to thank P.R. Apte from Indian Institute of Technology Bombay (IITB) for knowledge on Taguchi method. Authors would also like to acknowledge Universiti Kebangsaan Malaysia for supporting this project using grant no GUP-2012-012.

References

[1] Evgeni P Gusev, Vijay Narayanan and Martin M Frank 2006 Advanced high-k dielectric stacks with polySi and metal gates: Recent progress and current challenges. *IBM Journal of Research & Development (IBMRD)* vol.50 (4-5) pp 387-410.

[2] H.Iwai 2009 Roadmap for 22 nm and beyond *Microelectronic Engineering 86;* pp 1520 -1528.

[3] M V Fischetti, D A Neumayer, E A Cartier 2001 Effective Electron Mobility in Si Inversion Layers in Metal-Oxide-Semiconductor Systems with a High-k Insulator: The Role of Remote Phonon Scattering *Journal of Applied Physics* (Vol. 90, No. 9) pp 4587-4608.

[4] S Datta, G Dewey, M Doczy, B S Doyle, B Jin, J Kavalieros, R Kotlyar, M Metz, N Zelick and R Chau 2003 High Mobility Si/SiGe Strained Channel MOS Transistors with HfO_2/TiN Gate Stack *IEDM Technical Digest.*

[5] D J Frank, W Haensch, G Shahidi, and O H Dokumaci 2006 Optimizing CMOS Technology for Maximum Performance *IBM J. Res. & Dev. 50* No. 4.5 pp 419-431.

[6] F Salehuddin, I Ahmad, F A Hamid and A Zaharim 2010 Characterization and Optimizations of Silicide Thickness in 45nm pMOS Device *International Conference on Electronic Devices, Systems and Applications (ICEDSA)* pp 287-291

[7] R. Chau, S Datta, M Doczy, B Doyle, J Kavalieros, and M Metz 2004 High-k/Metal-Gate Stack and Its MOSFET Characteristics *IEEE Electron Device Letter* (Vol. 25, No. 6) pp 408-410.

[8] Giulio Ferrari, J R Watling, S Roy, J R Barker and A Asenov 2007 Beyond SiO_2 technology: Simulation of the Iimpact of High-k Dielectrics on Mobility *Journal of Non-crystalline Solids 353* pp 630-634.

[9] Afifah Maheran A H, Menon P S, I Ahmad, H A Elgomati, B Y Majlis and F Salehuddin 2012 Scaling Down the 32 nm Gate Length NMOS Transistor to 22 nm *Proceedings of the 2012 IEEE International Conference on Semiconductor Electronics (IEEE-ICSE2012)*, pp 191-194.

[10] ITRS 2011. *www.ITRS2011.net*

[11] J W Sleight, I Lauer, O Dokumaci, D M Fried, D Guo, B Haran, S Narasimha, C Sheraw, D Singh, M Steigerwalt, X Wang, P Oldiges, D Sadana, C Y Sung, W Haensch, M Khare 2006 Challenges and Opportunities for High Performance 32nm CMOS Technology *IEDM Tech Digital.*

[12] M K Bera and C K Maiti 2006 Electrical Properties of Sio_2/Tio_2 High-K Gate Dielectric Stack *Materials Science in Semiconductor Processing* pp 909–917.

[13] L Pereira, P Barquinha, E Fortunato and R Martins 2008 Low Temperature High K Dielectric on Poly-Si Tfts *Journal of Non-Crystalline Solids 354* pp 2534–2537.

[14] H A Elgomati, B Y Majlis, I Ahmad, F Salehuddin, F A Hamid, A Zaharim, T Ziad Mohamad and P R Apte 2011 Investigation Of The Effect for 32nm PMOS Transistor and Optimizing Using Taguchi Method *Asian Journal of Applied Sciences*.

[15] J Laeng, Zahid A Khan and S Y Khu 2006 Optimizing Flexible Behaviour of Bow Prototype Using Taguchi Approach *Journal of Applied Sciences 6 (3)* pp 622-630.

[16] Sarcona G T, M Stewart, and M K Hatalis 1999 Polysilicon Thin Film Transistor using Self_Aligned Cobalt and Nickel Silicide Source and Drain Contacts *IEEE Electron Device Letter* (vol.20).

[17] Hashim U 2009 Statistical Design of Ultra-Thin SiO2 for Nanodevices *Sains Malaysiana 38(4)* pp 553-557.

[18] H A Elgomati, B Y Majlis, I Ahmad, A M Abdul Hamid, P M Susthitha 2012 Modelling of Process Parameters for 32nm PMOS Transistor Using Taguchi Method *Asian Modelling Symposium (AMS)* pp 40-45

[19] Madhav S Phadke, 2008 Quality Engineering Using Robust Design *Pearson Education Inc. And Dorling Kindersley Publishing Inc.* India.

[20] Abdullah H, Jurait J, Lennie A, Nopiah Z M, Ahmad I 2009 Simulation of Fabrication Process VDMOSFET Transistor Using Silvaco Software *European Journal of Scientific Research;* ISSN 1450-216X vol 29 no 4 pp 461-470.

[21] Naidu N V R 2008 Mathematical Model for Quality Cost Optimization. *Robotics and Computer-integrated Manufacturing*

[22] G P Syros 2003 Die casting process optimization using Taguchi methods. *Journal of Materials Processing Technology 135* pp 68-74.

3rd ISESCO International Workshop and Conference On Nanotechnology 2012 (IWCN2012) IOP Publishing
Journal of Physics: Conference Series **431** (2013) 012027 doi:10.1088/1742-6596/431/1/012027

Guided Modes in Uniaxial Chiral Waveguide of Circular Cross-Section under PEC Boundary

M. A. Baqir and P.K. Choudhury

Institute Microengineering and Nanoelectronics (IMEN), Universiti Kebangsaan Malaysia 43600 UKM Bangi, Selangor, Malaysia

Email: pankaj@ukm.my

Abstract. Propagation of electromagnetic waves through a circular waveguide of uniaxial anisotropic chiral medium is studied with the emphasis on the energy flux patterns. The outer surface of the guide is assumed to be bounded by a perfect electric conductor (PEC) medium. The dispersion relation of guide is derived by applying suitable boundary conditions, and the allowed values of propagation constants are computed. Energy flux patterns corresponding to three different types of uniaxial anisotropic chiral metamaterials are taken into account. The propagation of negative energy flux is observed, and attributed to the presence of backward waves in the waveguide structure.

1. Introduction

Chiral objects possess the property that their mirror images are not superimposable by any set of translation or rotation. In terms of chemistry, such objects are also known as enantiomers; their molecules have similar molecular formula but different structural formula. Chiral mediums possess the remarkable property that, when a plane wave is made to incident on it, the wave decomposes into two circularly polarized plane waves, one is left circularly polarized (LCP,–) and the other is right circularly polarized (RCP,+) [1].

Chiral metamaterials possess the property of negative reflection and refraction, which made these to be of attractive field of interest among the researchers. It has been investigated that the phenomenon of negative refraction can be achieved from these materials by increasing their chirality (or the coupling) parameter enough [2–6]. In such a case, one of the eigenwaves, i.e. either LCP or RCP, acquires negative refraction. Materials having negative refractive index possess many fabulous technological applications. For instance, many novel devices can be fabricated by the use of such materials, such as perfect lens, cloaks, oscillators, etc. [5–7].

Within the context, circular waveguides of chiral metamaterials have attracted the interest of many researchers, and the investigation of the propagation behavior of electromagnetic waves through such guides has been reported in the literature [8–13]. In this stream, uniaxial chiral metamaterials are of special kind in which chirality appears only in one direction. It has been reported that these materials can easily be fabricated by using miniature spirals wires or conducting springs in ordinary dielectric host mediums [14,15]. It is to be noted in this connection that chiral waveguides with unconventional shapes have also been reported in the literature [16,17], and also, the property of chirality may be introduced by the use of helical clad (or twisted clad) optical fibers [18–21].

In the present communication, we make an attempt to present the investigation of the propagation behavior of energy flux patterns through a circular waveguide consisting of uniaxial anisotropic chiral metamaterials; the outer surface of the guide being bounded by PEC (perfect electric conductor) medium. The results reveal that the energy flux remains confined near the central region of the guide, and greatly depends on the type of metamaterial used. The existence of backward waves is also observed.

2. Analytical Treatment

We consider a circular waveguide composed of homogeneous anisotropic chiral medium having its radius as a, as shown in Fig. 1. The outer surface of the guide is assumed to be coated with a PEC medium. The constitutive relations corresponding to uniaxial anisotropic medium can be described as [22]

$$D = \left[\varepsilon_t \overline{\overline{I}}_t + \varepsilon_z u_z u_z\right] \cdot E - j\kappa\sqrt{\varepsilon_0 \mu_0}\, u_z u_z . H \tag{1a}$$

$$B = \left[\mu_t \overline{\overline{I}}_t + \mu_z u_z u_z\right] \cdot H + j\kappa\sqrt{\varepsilon_0 \mu_0}\, u_z u_z . E \tag{1b}$$

In these equations, u_z is the unit vector along the axis of the guide, and μ_z and ε_z are, respectively, permeability and permittivity of the waveguide medium along the optical axis. Also, μ_t and ε_t are those along the transverse directions, and κ is the chirality parameter. Further, $\overline{\overline{I}}_t$ is the unit dyadic define as

$$\overline{\overline{I}}_t = \hat{x}\hat{x} + \hat{y}\hat{y} .$$

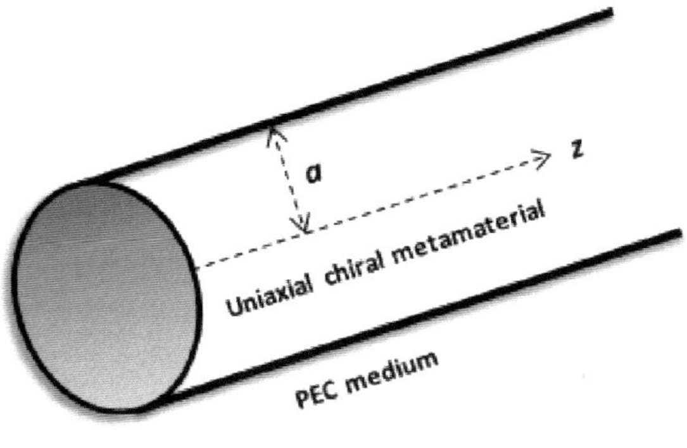

Figure 1. Circular waveguide made of anisotropic chiral metamaterial with a PEC boundary.

The cylindrical struture of the guide allows to consider the longitudinal component of the excited electromagnetic field as

$$E_z = J_n\left(kr\right)e^{jn\phi}$$

with $k = \sqrt{k_0^2 - \beta^2}$; k_0 and β, respectively, being the propagation constants in the free-space and the unbounded medium. As stated in the introductory part, the excited electromagnetic field in the chiral medium gets decomposed into two circularly polarized plane waves, viz. the LCP and the RCP [1], given as

$$E_{+z} = A_n J_n\left(k_+ r\right)e^{jn\phi} \tag{2a}$$

$$E_{-z} = B_n J_n \left(k_- r\right) e^{jn\phi} \tag{2b}$$

In the above Eqs. (2), A_n and B_n are unknown coefficients, and $J_n(\bullet)$ is Bessel function of the first kind with integer order n (i.e. $n = 0,1,2,...$). Now, the total electromagnetic field inside the chiral medium is the sum of the RCP and the LCP waves [1], and can be written as

$$E = E_+ + E_- \tag{3a}$$

$$H = \frac{j}{\eta}\left(E_+ - E_-\right) \tag{3b}$$

The electromagnetic field can be broken into transverse and longitudinal components as

$$E = \left(E_t + \hat{z}E_z\right)e^{-j\beta z} \tag{4a}$$

$$H = \left(H_t + \hat{z}H_z\right)e^{-j\beta z} \tag{4b}$$

with β as the propagation constant. The longitudinal and transverse field components are related to each other [13] as

$$E_t = -j\frac{\beta}{\lambda^2}\nabla_t E_z - j\frac{\omega\mu_t}{\lambda^2}\nabla_t H_z \times \hat{z} \tag{5a}$$

$$H_t = -j\frac{\beta}{\lambda^2}\nabla_t H_z - j\frac{\omega\varepsilon_t}{\lambda^2}\nabla_t E_z \times \hat{z} \tag{5b}$$

with $\nabla_t = \nabla - \hat{z}\frac{\partial}{\partial z}$, and $\lambda^2 = \omega^2\mu_t\varepsilon_t - \beta^2$.

Now, the longitudinal components must satisfy the wave equation [13], and it can be shown that the eigenvalues will be of the form

$$k_\pm^2 = \lambda^2\left[\frac{\varepsilon_z}{\varepsilon_t} + \frac{\mu_z}{\mu_t} \pm \sqrt{\left(\frac{\varepsilon_z}{\varepsilon_t} - \frac{\mu_z}{\mu_t}\right)^2 + 4\kappa^2\frac{\varepsilon_0\mu_0}{\varepsilon_t\mu_t}}\right] \tag{6}$$

Also, the corresponding eigenfunctions are given by [13]

$$\left(E_z, H_z\right) = \left(E_z, j\frac{\alpha}{\eta_t}E_z\right) \tag{7}$$

where $\alpha = \left(\frac{k^2}{\lambda^2} - \frac{\varepsilon_z}{\varepsilon_t}\right)\sqrt{\varepsilon_t\mu_t} / \left(k\sqrt{\varepsilon_0\mu_0}\right)$ and $\eta_t = \sqrt{\frac{\mu_t}{\varepsilon_t}}$.

The total longitudinal component of the electromagnetic field can be written as

$$E_{z1} = \left[A_n J_n(k_+ r) + B_n J_n(k_- r)\right] e^{jn\phi} e^{-j\beta z} \tag{8a}$$

$$H_{z1} = \frac{j}{\eta_t}\left[A_n \alpha_+ J_n(k_+ r) + B_n \alpha_- J_n(k_- r)\right] e^{jn\phi} e^{-j\beta z} \tag{8b}$$

and the transverse components can be derived from the longitudinal components by using Eq. (5). These transverse field components can be explicitly written as

$$E_{r1} = \left\{A_n\left[\frac{jnk_t}{\lambda^2 r}\alpha_+ J_n(k_+ r) - \frac{j\beta k_+}{\lambda^2} J_n'(k_+ r)\right] + B_n\left[\frac{jnk_t}{\lambda^2 r}\alpha_- J_n(k_- r) - \frac{j\beta k_-}{\lambda^2} J_n'(k_- r)\right]\right\} e^{jn\phi} e^{-j\beta z} \tag{9a}$$

$$E_{\phi 1} = \left\{A_n\left[\frac{n\beta}{\lambda^2 r} J_n(k_+ r) - \frac{k_t k_+ \alpha_+}{\lambda^2} J_n'(k_+ r)\right] + B_n\left[\frac{n\beta}{\lambda^2 r} J_n(k_- r) - \frac{k_t k_- \alpha_-}{\lambda^2} J_n'(k_- r)\right]\right\} e^{jn\phi} e^{-j\beta z} \tag{9b}$$

$$H_{r1} = \left\{A_n \frac{1}{\lambda^2 \eta_t}\left[-\frac{nk_t}{r} J_n(k_+ r) + \beta k_+ \alpha_+ J_n'(k_+ r)\right] + B_n \frac{1}{\lambda^2 \eta_t}\left[-\frac{nk_t}{r} J_n(k_- r) + \beta k_- \alpha_- J_n'(k_- r)\right]\right\} e^{jn\phi} e^{-j\beta z} \tag{9c}$$

$$H_{\phi 1} = \left\{A_n \frac{1}{\lambda^2 \eta_t}\left[\frac{j\beta n \alpha_+}{r} J_n(k_+ r) - k_+ k_t J_n'(k_+ r)\right] + B_n \frac{1}{\lambda^2 \eta_t}\left[\frac{j\beta n \alpha_-}{r} J_n(k_- r) - k_- k_t J_n'(k_- r)\right]\right\} e^{jn\phi} e^{-j\beta z} \tag{9d}$$

In the above Eq. (9), the prime represents the differentiation of Bessel function of the first kind with respect to the argument.

The outer surface of the guide is bounded with a PEC medium. The dispersion relation of the guide is derived by applying suitable boundary conditions at the interface of the uniaxial anisotropic and PEC mediums. According to the nature of PEC, the tangential components of the electric field vanish at the surface of the guide which, in our case, can be given as below:

$E_{z1} = 0$ at $r = a$

$E_{\phi 1} = 0$ at $r = a$

By employing the above boundary conditions, a set of two homogeneous equations can be obtained, and the determinent formed by the coefficients in those equations should be zero, in order to have a non-trivial solution to that set of equations. Thus, the characteristic eigenvalue equation for the structure can be written as equating the aforesaid determinant to zero. After some mathematical steps, the dispersion relation of the guide can finally be given as

$$\beta\alpha_- k_-^2 J_n(k_+ a)\left\{J_{n-1}(k_- a) - J_{n+1}(k_- a)\right\} - \beta\alpha_+ k_+^2 J_n(k_+ a)\left\{J_{n-1}(k_- a) - J_n(k_- a)\right\} = 0 \tag{10}$$

Further, the energy flux flowing through the guide can be derived by using Eq. (9). It can be shown that the equation corresponding to the energy flux finally assumes the form

$$
\begin{aligned}
S_z =& \frac{|A_n|^2}{2\lambda^2\eta_t}\left\{\frac{n^2\beta k_t}{\lambda^2 r^2}\alpha_+^2 J_n^2(k_+r) - \frac{n\beta^2 k_+^2}{2\lambda^2 r}\alpha_+ J_n(k_+r)\left(J_{n-1}(k_+r) - J_{n+1}(k_+r)\right)\right\} \\
&+ \frac{|B_n|^2}{2\lambda^2\eta_t}\left\{\frac{n^2\beta k_t}{\lambda^2 r^2}\alpha_-^2 J_n^2(k_-r) - \frac{n\beta^2 k_-^2}{2\lambda^2 r}\alpha_- J_n(k_-r)\left(J_{n-1}(k_-r) - J_{n+1}(k_-r)\right)\right\} \\
&+ \frac{A_n B_n}{2\lambda^4\eta_t r}\left\{n\beta\alpha_+ k_-(k_+r)\left(nk_t\alpha_- J_{n-1}(k_+r) - 0.25\times\beta k_-^2 J_{n-1}(k_-r) - J_{n+1}(k_+r)\right)\right. \\
&\left. -0.25\times n\beta^2 k_+^2\alpha_- J_n(k_-r)\left(J_{n-1}(k_+r) - J_{n+1}(k_+r)\right)\right\} \qquad (11)
\end{aligned}
$$

3. Results and Discussion

We are now in a position to investigate the features related to the propagation of energy flux through the guide of interest. As described before, the waveguide is loaded with a PEC medium. In order to perform computations of the sustained flux, we consider three different types of uniaxial anisotropic chiral mediums – Type I, Type II and Type III. These three types of material structures assume the parametric values as follows:

Type I: $\mu_z = \mu_t = \mu_0$, $\varepsilon_z = 2\times\varepsilon_0$ and $\varepsilon_t = 1.5\times\varepsilon_0$,
Type II: $\mu_z = \mu_t = \mu_0$, $\varepsilon_z = 4\times\varepsilon_0$ and $\varepsilon_t = -1.5\times\varepsilon_0$,
and Type III: $\mu_z = \mu_t = \mu_0$, $\varepsilon_z = -\varepsilon_0$ and $\varepsilon_t = 1.5\times\varepsilon_0$.

We first obtain the allowed values of propagation constants of the guide corresponding to the above stated three types of material mediums by the use of the dispersion relation, as given in Eq. (10). For this purpose, we assume the operating wavelength of the excited electromagnetic field as 1.5 μm. Also, the radius of the guide is taken to be 15 μm.

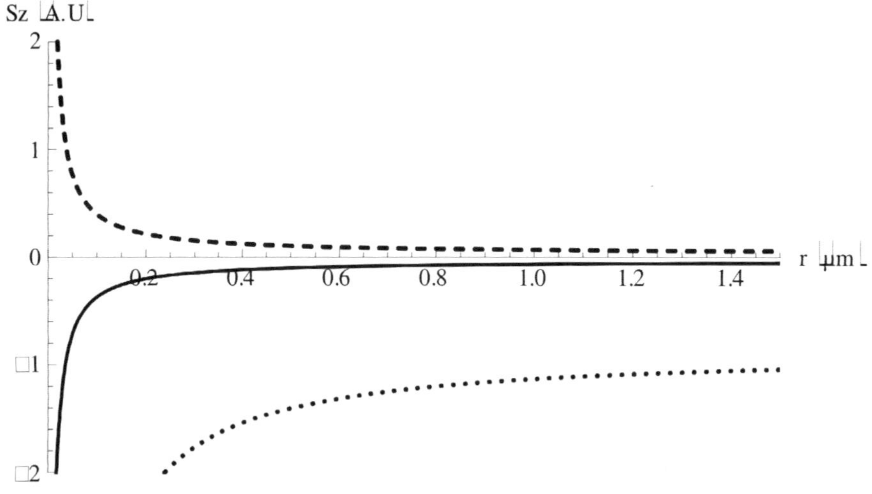

Figure 2. Energy flux in the guide with Type I material composition.

As the waveguide medium is assumed to be chiral in nature, the electric/magnetic fields are coupled with each other through the chirality parameter, resulting thereby the existence of hybrid modes in the guide with right- and left-circular polarizations. We make an attempt to investigate the propagation of energy flux, as depicted in Figs. 2, 3 and 4, corresponding to the three different low-order hybrid modes – viz. H_{01}, H_{11} and H_{-11} – considering the aforesaid material compositions. In Figs. 2, 3 and 4, the dotted lines correspond to the situations of the H_{01} mode, and the dashed and solid lines, respectively, represent those of the H_{11} and H_{-11} modes.

In the attempt to evaluate the allowed values of propagation constants, we found that when the Type I medium is in use, the propagation constant β corresponding to the H_{01} mode becomes 5.52×10^6 m^{-1}. Further, the H_{11} and H_{-11} modes exhibit the similar value of β, which is 5.82×10^6 m^{-1}. Figure 2 illustrates the energy flux characteristics under this situation. We observe that the flux is mostly bound near the central region of the guide, and it goes on decreasing with the increase of radial distance. Near the interface of the uniaxial chiral and PEC mediums, it almost disappears. Further, corresponding to the H_{01} and H_{-11} modes, the flux characteristics remain negative, though it is more negative for the hybrid H_{01} mode, which is more prominent near the central region of the guide, and gets attenuated as it reaches the waveguide interface. The negative flow of energy flux can be interpreted as the support of backward wave propagation in the guide of interest.

While using the Type II material for the waveguide, the propagation constant corresponding to the H_{01} mode is found to be 4.29×10^6 m^{-1}, and that for the H_{11} and H_{-11} modes as 4.43×10^6 m^{-1} (β-values are the same for both of these modes). We observe from Fig. 3 that, under the Type II kind of material composition, the H_{01} mode exhibits similar characteristics as seen while the Type I material was in use. The flux corresponding to this mode primarily remains negatively bounded near the central region of the guide, and goes on negatively decreasing while moving toward the waveguide interface. However, the H_{11} and H_{-11} modes show typical features that the flux remains independent of the radial variation in the waveguide dimension. As such, the distribution of energy flux due to these two modes becomes uniform throughout the guide. Interestingly, the features of flux corresponding to the H_{11} and H_{-11} modes become reversed (as compared to the situation of the waveguide with Type I material) in this case as the H_{11} mode exhibits negative flux whereas the H_{-11} mode positive.

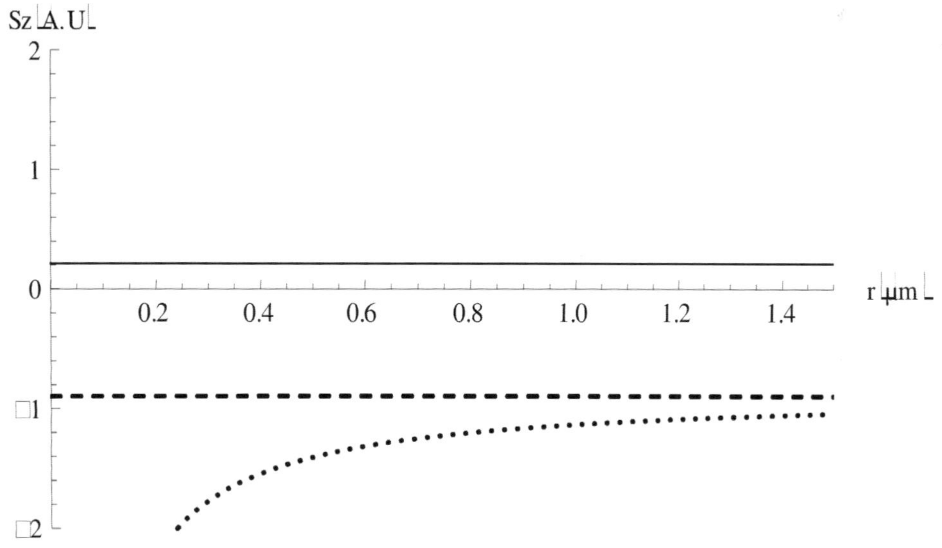

Figure 3. Energy flux in the guide with Type II material composition.

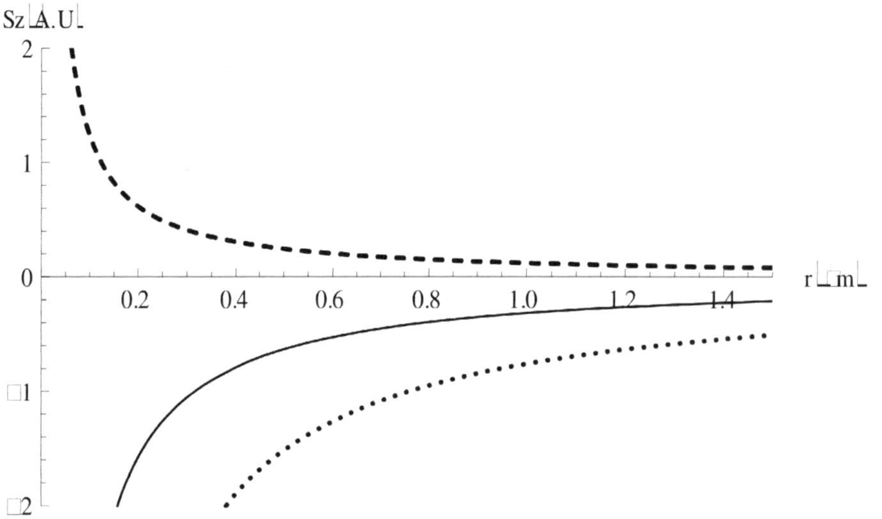

Figure 4. Energy flux in the guide with Type III material composition.

In the case of Type III medium, we evaluated the propagation constants corresponding to the H_{01} mode as 5.65×10^6 m^{-1}, and that for the H_{11} and H_{-11} modes as 5.73×10^6 m^{-1}. We find from Fig. 4 in this case that, for all the three low-order modes of interest, the flux characteristics are almost similar to that observed in the case of waveguide with Type I material composition in use. Thus, the flux remains more confined in the central region of the guide, and goes on decreasing with the increase in radial direction. The only difference from the situation of Fig. 2 is that the flux in the case of Type III material composition remains a little loosely bounded near the central region of the guide as compared to that under the usage of Type I material composition. As such, the flux characteristics of the guide remain greatly dependent on the material composition used for waveguide fabrication.

4. Conclusion

The paper presents numerical investigations of the energy flux patterns corresponding to the propagating modes considering three different types of uniaxial anisotropic chiral waveguides. It has been found that most of the energy flux remains confined near the central region of the guide, and it varies upon changing the type of medium through its constitutive parameters. Backward waves are also observed to be supported by the PEC loaded uniaxial anisotropic chiral medium, which is evident from the propagation of negative energy flux in the guide.

Acknowledgements

The authors are thankful to Prof. Burhanuddin Yeop Majlis, the Director of IMEN (Universiti Kebangsaan Malaysia), for constant encouragement and help. This work is partially supported by the Fundamental Research Grant Project (FRGS/1/2011/TK/UKM/01/16) by the Ministry of Higher Education (Malaysia); the authors are thankful to the Ministry. Also, the constructive criticisms by an anonymous reviewer are gratefully acknowledged.

References

[1] S. Bassiri, C.H. Papas, N. Engheta, Electromagnetic wave propagation through a dielectric slab interface and through a chiral slab, J. Opt. Soc. Am A. 5 (1998) 1450–1459.
[2] J.B. Pendry, A chiral route to negative refraction, Science 306 (2004) 1353–1355.
[3] S. Tretyakov, A. Sahvola, L. Jylha, Backward-wave regime and negative refraction in chiral composites, Phot. and Nanost. 3 (2005) 107–115.

[4] T.G. Mackay, A. Lakhtakia, Plane waves with negative phase velocity in Faraday chiral mediums, Phys. Rev. E 69 (2004) 026602-1–9.
[5] S. Zhang, Y.Park, J. Li, X. Lu, W.Zhang, X. Zhang, Negative refractive index in chiral metamaterials, Phy. Rev. Lett. 102 (2009) 023901.1–023901.4.
[6] J.B. Pendry, Negative refraction makes a perfect lens, Phy. Rev. Lett. 85 (2000) 3966–3969.
[7] D. Schuring, J.J. Mock, B.J. Justice, S.A. Cummer, J.B. Pendry, A.F. Starr, D.R. Smith, Metamaterial electromagnetic cloak at microwave frequencies, Science 314 (2006) 977–980.
[8] P.K. Choudhury, T. Yoshino, Characterization of optical power confinement in a simple chirofiber, Optik 13 (2002) 89–95.
[9] P.K. Choudhury, T. Yoshino, Dependence of optical power confinement on core/cladding chiralities in chirofibers, Microw. and Opt. Tech. Lett. 32 (2002) pp. 359–364.
[10] A. Nair, P.K. Choudhury, On the analysis of field patterns in chirofibers, J. Electromagn. Waves and Appl. 21 (2007) 2277–2286.
[11] M.A. Baqir, A.A. Syed, Q.A. Naqvi, Electromagnetic waves in a circular waveguide of chiral nihility metamaterial, Prog. in Electromagn. Res. M 16 (2011) 85–93.
[12] K.Y. Lim, P.K. Choudhury, Z. Yusoff, Chirofibers with helical windings – An analytical investigation, Optik 121 (2011) 980–987.
[13] J.F. Dong, J. Li, Characteristics of guided modes in uniaxial chiral circular waveguide, Prog. in Electromagn. Res. 124 (2012) 331–345.
[14] Q. Cheng, T.J. Cui, Negative refraction in uniaxially anisotropic chiral media, Phy. Rev. B 73 (2006) 113104-1–113104-4.
[15] Q. Cheng, T.J. Cui, Refection and refraction properties of plane waves on the interface of uniaxially anisotropic chiral media, J. Opt. Soc. Am. A 23 (2006) 3203–3207.
[16] P.K. Choudhury, On the propagation of electromagnetic waves through parabolic cylindrical chiroguides with small flare angles, Microw. and Opt. Tech. Lett. 33 (2002) 414–419.
[17] P.K. Choudhury, Partial electromagnetic wave guidance by a parabolic cylindrical chiroboundary with small flare angle, Optik 113 (2002) 177–180.
[18] D. Kumar, P.K. Choudhury, O.N. Singh II, Towards the dispersion relations for dielectric optical fibers with helical windings under slow- and fast-wave considerations – A comparative analysis, Prog. in Electromagn. Res. 80 (2008) 409–420.
[19] A.H.B.M. Safie, P.K. Choudhury, On the hybrid field patterns of helical clad dielectric optical fibers, Prog. in Electromagn. Res. 91 (2009) 69–84.
[20] P.K. Choudhury, D. Kumar, On the slow-wave helical clad elliptical fibers, J. Electromagn. Waves and Appl. 24 (2010) 1931–1942.
[21] P.K. Choudhury, Transmission through twisted clad liquid crystal optical fibers, Prog. in Electromagn. Res. 131 (2012) 169–184.
[22] I.V. Lindell, A.J. Viitanen, Plane wave propagation in uniaxial bianisotropic medium, Electron. Lett. 29 (1993) 150–152.

Angle Shifting in Surface Plasmon Resonance: Experimental and Theoretical Verification

W M Mukhtar[1], P Susthitha Menon[1], S Shaari[1], M Z A Malek[2] and A M Abdullah[2]
[1]Institute of Microengineering and Nanoelectronics, Universiti Kebangsaan Malaysia 43600 UKM Bangi, Malaysia
[2]Advanced Materials Research Centre, Lot 34, Jalan Hi-Tech 2/3, Kulim Hi-Tech Park, 09000 Kulim Kedah, Malaysia

E-mail:wmaisarah2101@yahoo.com, susi@eng.ukm.my, sahbudin@eng.ukm.my, zahid@sirim.my, makarimi@sirim.my

Abstract. This paper discussed a sensing property of the SPR sensor to detect changes in refractive index of the dielectric medium. The resonant angle shifting was observed as the dielectrics medium was changed from air to SiO_2 thin film. The effect of a free space wavelength to the excitation of SPP was investigated by employing laser sources with wavelengths transmission of 545nm and 632.8nm. In comparison with 545nm of wavelength transmission, we revealed that He-Ne laser source with wavelength of 632.8nm is suitable to produce a strong SPR signal due to its high *Q*-factor. To avoid misinterpretation between the destructive interference and SPR phenomenon, we proposed a simple theoretical approach using momentum conservation principle which concerning the wavevector values to validate the experimental results. By ignoring the imaginary part of the dielectric function of gold, we proved that this theoretical approach is still acceptable due to the excellent agreement between the theoretical and simulation approaches.

1. Introduction

Recently, researches in surface plasmon resonance (SPR) become a major interest in biomedical sensing application. The biosensing techniques are motivated by numerous applications in medical diagnostics and single molecule detection. SPR is an optic method to detect the small change of the refractive index on the metal surface [1, 2]. Many works have been conducted to improve the sensitivity and stability of the SPR sensor.

SPR phenomenon is represented by a sharp dip of curve which represents the minimum reflectance. A high *Q*-factor of SPR curve implies that the plasmons are strongly confined. The *Q*-factor is influenced by the thin film thicknesses, types of metal used, structures of nanoparticles,
thicknesses, types of metal used, structures of nanoparticles, deposition techniques, value of the free space wavelength, quality of substrates and the coupling methods [3-12]. Nonetheless, it is important to differentiate between the existences of SPR and destructive interference where both phenomena are interpreted as an absorption dip in the reflectance function. Consequently, with the support of a theoretical analysis by using plasmonic related equations such as Fresnel equation, Drude model analysis and Maxwell equations; the SPR phenomenon can be recognized [13-16].

In this paper, we seek to determine the SPR sensing characteristics using our custom made modified optical waveguiding assembly by monitoring the angle shifting phenomenon. We believe that this research output can be exploited gainfully in many sensing applications due to its sensitive response to the change of refractive index of the dielectric medium. A significant theoretical approach for the verification of the SPR phenomenon using a momentum conservation principle by matching the component of wavevector along the surface plasmon propagation interface, k_x with the wavevector of surface plasmon, k_{spp} has been discovered.

2. Theory

Surface plasmon is the localized charge density oscillations that occur in the interface between a metal and a dielectric medium. These oscillations propagate along the boundary between a thin metal film and a dielectric medium. The relationship of the surface plasmon wave vector (k_{spp}) between the metal and dielectric medium is given as

$$k_{spp} = k_o \sqrt{\frac{\varepsilon_{metal}\varepsilon_{dielectric}}{\varepsilon_{metal} + \varepsilon_{dielectric}}} \quad (1)$$

where k_o is the free space wave vector of the optical wave which can be calculated as $k_0 = \frac{2\pi}{\lambda_0}$, ε_{metal} and $\varepsilon_{dielectric}$ are respectively the dielectric functions of metal and the dielectric medium. During SPR, the energy of the incident light can be strongly coupled from photons onto the surface plasmon wave which decays via heat dissipation along the propagation path. The relationship between k_x and angle of incident light can be expressed as follows:

$$k_x = k_o n_{glass} \sin\theta_{inc} \quad (2)$$

where n_{glass} is the refractive index of the glass prism and θ_{inc} is the angle of incidence [17]. The conditions for the surface plasmon excitation at the interface between the metal and dielectric layers in a glass-metal-dielectric are given by

$$k_x = k_{spp} \quad (3)$$

Resonance is achieved by matching the projection of the wave vector of the incident light in the direction of the interface and wave vector of the surface plasmon oscillations as expressed in Eq. (3). Using a prism with dielectric constant value larger than 1 with the incident angle is larger than the critical angle, value of k_x can be determined. As the incident light satisfies the resonance condition, a strong absorption dip in the reflectance can be observed.

3. Methodology

Herein, we investigated the distinct resonance condition associated with the excitation of surface plasmons for different types of samples, namely single layer of Au (Au-air) and Au-SiO_2 (Au layer deposited prior SiO_2 layer). We fixed the thickness of SiO_2 and gold layer to 45nm and 50nm, respectively. Gold layer was sputtered on a borosilicate glass slide (RI: 1.503) using direct current (DC) sputtering technique; meanwhile SiO_2 layers was sputtered by radio frequency (RF) sputtering technique. The thin film thicknesses were determined using reflectometer (Brand: Filmetrics). We applied a curve fitting method with 96% of accuracy to obtain the thicknesses of the films. Prior to thin film sputtering, the substrate was cleaned by acetone in an ultrasonic bath for 15 min and rinsed with ethanol and deionized water, before dried with a nitrogen stream.

The SPR phenomenon was created using our custom made modified optical waveguiding assembly. He-Ne laser sources with different transmission wavelengths were employed, namely 545nm and 632.8nm. In SPR, it is sufficient to consider only *p*-polarized waves because no solutions exist for the case of *s*-polarization. Since the linearly polarized lasers were used in this experiment, the *p*-polarization state was obtained by rotating the position of laser into 90°. It is worth mentioning that the *p*-polarization state of the incident light can be achieved by observed the reflected light without the metal coated layer on the glass slide to obtain the critical angle and the Brewster angle

[18]. Once the TM polarization state was confirmed, the excitation of SPP was accomplished by illuminating the thin film surface normally with a focused laser beam using an angular interrogation technique. The focusing of linearly polarized beam was realized by coupled the light incident from monochoromatic *p*-polarized He-Ne laser (Brand: Thorlabs) using a convex lens (Brand: Thorlabs). The reflected light intensity from the metal-coated prism was measured by photodetector (Brand: Newport). The intensity will go through a minimum when the coupling with the plasmon wave occurs. To excite surface plasmon polaritons (SPP), we applied a prism coupling technique so that the phase matching between k_{spp} and k_x can be achieved. By employed Kretschmann configuration, the microscope glass slide and the half-cylinder prism were coupled using optical couplant gel which can reduce Fresnel reflection and the difference in the index of refraction between the glass slide and prism. Figure 1 depicts the schematic illustration of Kretschmann configuration for the SPP excitation.

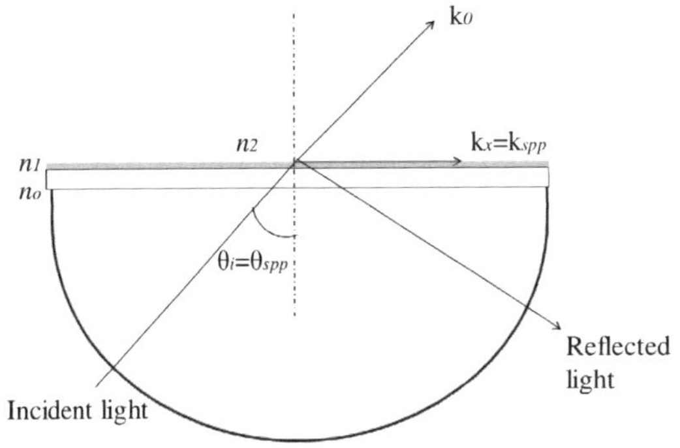

Figure 1. Illustration of SPR phenomenon which can be achieved by matching the projection of the wavevectors, k_x and k_{spp}

For the experimental results satisfaction, we ran a simulation using WINSPALL 3.02 simulator programme to investigate the accuracy of our experimental results with the simulated one. The theoretical study was performed to satisfy the principle of momentum conservation by considering two types of wavevectors, namely k_x and k_{spp}. The behaviour of λ_{spp} which is always less than the free space wavelength λ_0 was investigated. In this paper, we set the dielectric functions of gold as $\varepsilon_1 = -11.6 + 1.2i$ and $\varepsilon_1 = -6 + 2.2i$ for the transmission wavelength of 632.8nm and 545nm, respectively based on a plot of the dielectric constant from the paper of Johnson and Christy which is an extended work from Drude model [19]. For simplicity, we assume that the imaginary parts of the dielectric functions are small as compare with the real parts, so that they may be neglected [19]. By considering only the real parts, we investigated the SPP waves that propagate along the interface, which is between the metal thin film and the dielectric (i.e air and SiO_2). The losses in the plasmon's propagation which associated with the electron scattering were omitted due to the ignorance of the imaginary parts of the dielectric function.

4. Results

Prior observation of the SPR phenomenon, it is important to confirm the characteristic of light propagation in p-polarization state. The light reflectance behaviour for the combination of bare microscope glass slide and the prism using p-polarized laser results an excellent agreement with the theoretical as illustrated in figure 2 [20]. The critical angles for laser sources with wavelengths of 632.8nm and 545nm are 38.5° and 43°, respectively. Since the excitations of SPP occur above the critical angles, we expected that the location of SPR dips should be beyond these angles.

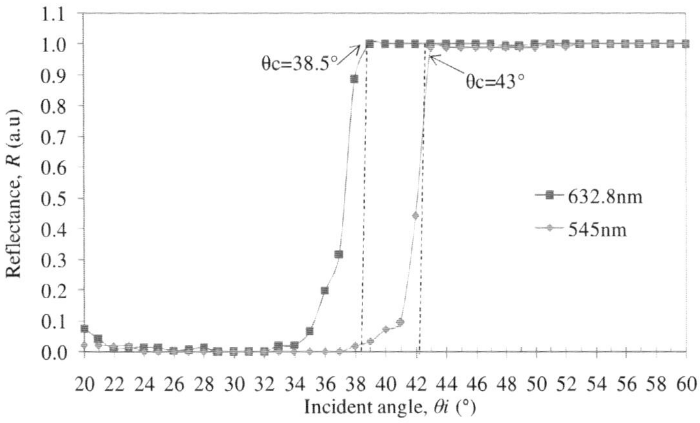

Figure 2. Behaviours of reflected light under TM-polarization state using lights sources with different wavelength

4.1Investigation of the angle shifting phenomena with the increment of refractive indices of the dielectrics medium

By using our custom made modified optical waveguiding assembly, the shifts of incident angles were investigated as the dielectrics medium were changed from air to SiO_2. In this experiment, we set the refractive indices values of the dielectrics medium namely air and SiO_2 as 1.00 and 1.47 respectively. Figure 3(a) illustrates the characteristics of SPP with wavelength transmission of 545 nm. A strongly shifted of SPR dips from 45.0° to 55.0° with the increment of minimum reflectance from 0.03 to 0.15 are clearly seen as the dielectric was replaced with SiO_2. Evidently, the Q-factor is increased with the increment of light transmission wavelength to 632.8nm as depicted in figure 3(b). With the additional layer of SiO_2 on top of the gold layer, the resonant angles shifted from 43.0° to 56.4°, as well as the minimum reflectances which increased from 0.13 to 0.28.

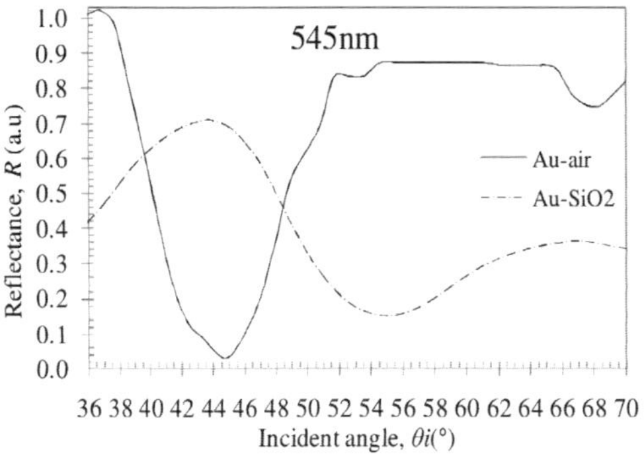

Figure 3(a). Angle shifting as the dielectric changed from air to SiO_2 using 545nm wavelength transmission (experimental)

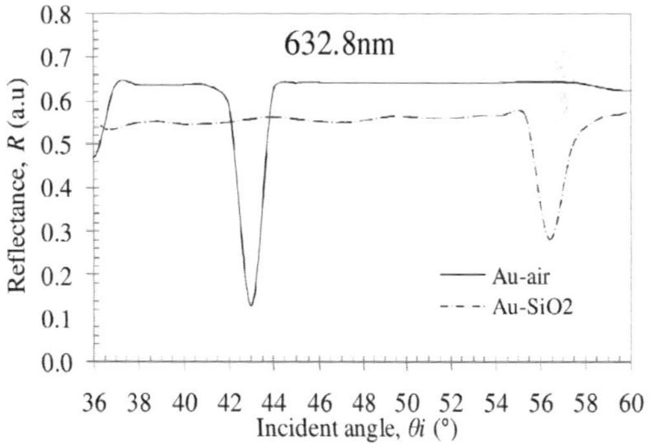

Figure 3(b). Angle shifting as the dielectric changed from air to SiO_2 using 632.8nm wavelength transmission (experimental)

For a clarification purpose, we performed a simulation using WINSPALL 3.02 simulator developed by Knoll Group, Max-Plank Institute. This simulator programme is based on a normalized Fresnel equation. Both experimental and simulated results show a good verification as proven in figure 3 and figure 4. The high Q-factors were obtained with the employment of a light source with transmission wavelength of 632.8nm. Figure 4(a) reveals the shifting of resonance angle from 47° to 71° as the refractive index of the dielectric increased from 1.00 to 1.47 with the transmission wavelength of 545nm. As the transmission wavelength changed to 632.8nm (figure 4(b)), the location of resonant dip is shifted from 44.6° to 55.4° without any effect to the reflectance.

The angles shifting which occur due to the changes in refractive index of the dielectric are an important indicator in SPR sensing application. Apart from the experimental and simulation approach, we also performed a theoretical study using equation (1) - (3) to verify the experimental results. Table 1 shows the comparison of angle shifting using three types of methods, namely theoretical, simulation and experimental. Whilst the transmission wavelength of 632.8nm used, the experimental result demonstrated a good agreement with the theoretical and simulation with small percentage difference between them. Despite that, the angle shifting for the transmission wavelength of 545nm resulted a large percentage different between experimental and both theoretical and simulation.

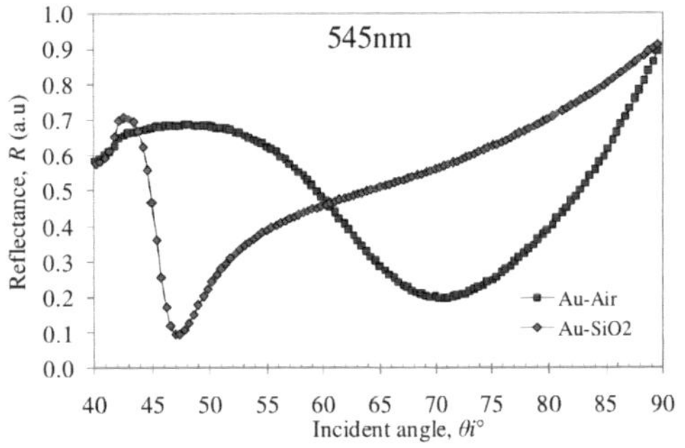

Figure 4(a). Angle shifting as the dielectric changed from air to SiO_2 using 545nm wavelength transmission (simulation)

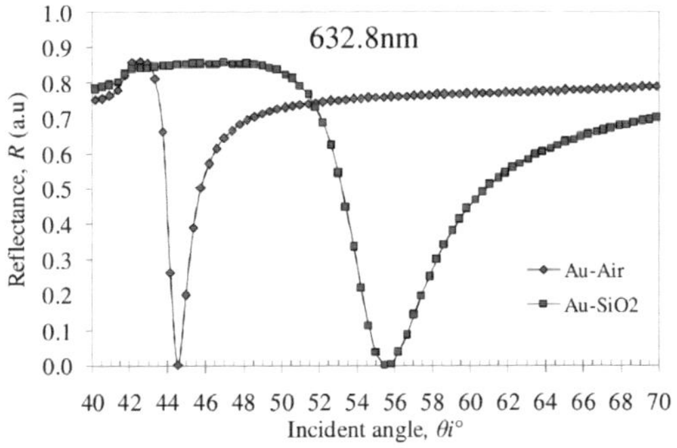

Figure 4(b). Angle shifting as the dielectric changed from air to SiO_2 using 632.8nm wavelength transmission (simulation)

Table 1: Value of angle shifting as types of dielectric change from air to SiO_2

Method	Angle shifting, Δθ (°)	
	545nm	632.8nm
Theoretical	25.9	10.0
Simulation	24.0	10.8
Experimental	10.0	13.4

4.2 Effect of wavevector value, k_x and k_{spp} in determining the SPP excitation

Value of the wavevector is the crucial part in SPR study in order to determine the occurrence of SPP excitation. By considering the wavevector, we can simply make a verification either the SPR phenomenon is occurred or not. For any given energy $\hbar\omega$, the wavevector k_x is always larger than the wavevector of light in free space, k_o [18]. Evidently, this reveals that the SPP on a planar interface cannot be excited by light at any frequency that propagates in free space. In this paper, the values of k_0 obtained are 0.00993 and 0.01153 using 632.8nm and 545nm of light wavelength transmission, respectively. As depicted in Table 2, $k_x > k_0$ which proves the potential of SPP excitation in this experiment. Table 2 lists value of k_x and k_{spp} during the SPR phenomena using the transmission wavelengths of 545nm and 632.8nm with the employment of different dielectric medium. For both wavelength transmissions under the Au-air configuration, the value of $k_x \cong k_{spp}$, which fulfil the principle of momentum conservation. Nevertheless, as the dielectric changes to SiO_2 with the transmission wavelength of 545nm, the momentum is not conserved with the fact that $k_x < k_{spp}$. This reveals the inaccuracy of the obtained result for 545nm wavelength as the dielectric was replaced with SiO_2.

Table 2. Value of wavevector along the propagation direction, k_x and SPP wavevector, k_{spp} for different transmission wavelength of laser source

Metal-dielectric	λ_0=545nm		λ_0=632.8nm	
	k_x	k_{spp}	k_x	k_{spp}
Au-air	0.012	0.015	0.011	0.011
Au-SiO2	0.016	0.042	0.023	0.013

5. Discussions

The generation of the SPR signal must satisfy the TIR condition where the excitation of SPP is above the critical angle. By considering figure 3, the SPR curves are located between 43° to 57°, meanwhile the critical angles obtained are at 38.5° and 43° for the transmission wavelength of 632.8nm and 545 nm, respectively. So, it can be concluded that the existence of absorption dips obtained are due to the SPP excitation. SPP signal strength can be measured by observing the Q-factor of the SPR curves. High *Q*-factor represents a strong signal of SPPs excitation and vice versa. Figure 3(b) illustrates a high Q-factor curve which reveals the suitable wavelength to excite strong SPR signal is 632.8nm as supported by the simulation result (figure 4(b)).

With the replacement of the dielectric medium from air to SiO_2, the angle shifting should be observed due to the change in refractive index. If we recap the calculation of k_{spp} using equation (1), the imaginary part of the gold dielectric function was omitted for the simplicity purpose. Although the imaginary part is ignored, the good verification between theoretical and simulated results has validated that this approach is acceptable as proven in Table 1. As captured in figure 3 and figure 4, the angles shifting between the experimental and simulation results show the excellent agreement with small percentage differences. However, the minimum reflectance obtained using both methods are slightly different which reveals an argument between the experimental and simulated results. The explanation for this problem is due to the value of the gold dielectric function used during the simulation. Note that the SPR phenomenon is mostly affected by metal characteristics as mentioned in the first section of this paper. Apart from the thin film thickness, the metal dielectric functions also affected the depth of the absorption dip. Since the dielectric function of gold used in this simulation is an ideal value, it is worth to conclude that the experimental value of the dielectric function is slightly different with the ideal one. This leads to the instability of results between the experimental and the simulation approaches.

Owing to the similar characteristics of the absorption dips in reflectance between the destructive interference and the SPR phenomenon, it is critical to distinguish both of them. By conducting a simple theoretical analysis based on the conservation of momentum principle, we compare the wavevector values of k_x and k_{spp} for both transmission wavelengths. This principle must be satisfied in order to generate the SPP. With the employment of a prism coupling technique, the wavevector k_x is always larger than the wavevector of light in free space, k_o. The main purpose of the prism coupling is to increase the wavevector, k_x in order to match it with k_{spp}. Value of k_{spp} is obtained by considering the dielectric functions of gold and dielectric mediums (air and SiO_2). We determine value of k_x using an angular interrogation technique by varies the incident angles until the sharp dip is observed. For the confirmation of the SPR phenomenon, the principle of momentum conservation must be obeyed which is $k_x \cong k_{spp}$. As projected in Table 2, the SPR phenomena are clearly observed with the employment of air as the dielectric medium. An excellent validation between k_x and k_{spp} proves the occurrence of the SPP excitation. Conversely, with the appointment of 545nm laser source and SiO2 layer as a dielectric medium, the wavevector for both k_x and k_{spp} does not obey the principle of momentum conservation, where $k_{spp}>k_x$. By considering the experimental and simulated results as captured in figure 3(a) and figure 4(a), the resonant angles for both Au-SiO_2 configurations are different which reveals that the SPP excitation was not occurred. We concluded that the dip observed in figure 4(a) using SiO_2 layer as the dielectric is due to the destructive interference. According to the simulated results as illustrated in figure 4(a), the SPR resonance of Au-SiO_2 thin film should be occurred at the incident angle of 71°, not at 55° as proposed experimentally.

6. Conclusions

In conclusions, we found that our custom made modified optical waveguiding assembly is capable to be used in SPR sensing. A good respond to detect changes in refractive index of the dielectric medium shows the sensing properties based on the SPP excitation. High Q-factor resulted from the transmission wavelength of 632.8nm reveals that strong SPR signal can be created using this wavelength as compare with 545nm. We prove the importance of the principle of momentum conservation for the confirmation of SPP excitation.

Acknowledgements

The authors would like to acknowledge the support of Universiti Kebangsaan Malaysia and the Malaysian Ministry of Higher Education (MOHE) for funding this work under grant ERGS/1/2012/STG02/UKM/02/2, Advanced Materials Research Centre (AMREC) Kulim, Kedah for the thin film sputtering facilities and Universiti Sains Islam Malaysia (USIM)-MOHE for the scholarship.

References

[1] Lavine B K, Westover D J, Oxenford L, Mirjankar N and Kaval N 2007, *Microchem J* **86** 147
[2] Lin C W, Chen K P, Hsiao C N, Lin S and Lee C K 2006, *Sens and Act B.* **113** 169
[3] Umar A A and Oyama M 2006, *Appl Surf Sci* **253** 2933
[4] Feng W Y, Chiu N F, Lu H L, Shih H C and Yang D 2008 *Surface Plasmon Resonance Biochip Based on ZnO Thin Film for Nitric Oxide Sensing,* 30th Annual International IEEE EMBS Conference, Canada, pp. 5757-5760
[5] Lyon L A, Musick M D, Smith P C, Reiss B D, Pena D J and Natan M J 1999 *Sens and Act B.* **54** 118
[6] Schmid A H, Stanca S E, Thakur M S, Thampi K R and Suri C R 2006 *Sens and Act B.* **113** 297
[7] Chen X, Pan M and Jiang K 2010 *Microelectron Eng* **87** 790
[8] Chowdhury M H, Ray K, Geddes C and Lakowicz J R 2008 *Chem Phys Lett.* **452** 162
[9] Maarof A I, Cortie M B and Smith G B 2005 *J Opt A-Pure Appl Op.* **7** 303
[10] Thirstup C, Zong W, Borre M, Neff H, Pedersen H C and Holzhueter G 2004 *Sens and Act B.* **100** 298
[11] Kim J B, Zou Y, Kim Y D, Kim J J and Hwangbo C K 2011 *Thin Solid Films.* **520** 1451
[12] Brahma R and Krishna M G 2008 *Nucl Instrum Meth B.* **266** 1493
[13] Neff H, Zong W, Lima A M N, Borre M and Holzhuter G 2006 *Thin Solid Films* **496** 688
[14] Islam A S and Kouzani A Z 2011 *Variable incidence angle localized surface plasmon resonance grapheme biosensor,* Proceedings of the 2011 IEEE/ICME, China, pp. 58-63.
[15] Shandilya S, Tomar M, Sreenivas K and Gupta V 2010 *J Lightwave Technol* **28** 3004
[16] Kuo C H and Moghaddam M 2008 *IEEE T Antenn Propag* **56** 1133
[17] Pun S H, Lam W W, Ho H P and Zhang Y T 2002, *A preliminary study on optical measurements of glucose concentration using a surface plasmon resonance system* Proceedings of the IEEE-EMBS Special Topic Conference on, China, pp. 68-70.
[18] Novotny L and Hecht B 2007, Principles of Nano-Optics (Cambridge: Cambridge University Press)
[19] Johnson P B and Christy R W 1972, *Phys. Rev. B.* **6** 4370
[20] Saleh B E A and Teich M C 2007, Fundamentals of Photonics (New Jersey: Wiley)
[21] Barnes W L 2006 *J. Opt. A: Pure Appl. Opt.* **8** S87

Journal of Physics: Conference Series **431** (2013) 012029 doi:10.1088/1742-6596/431/1/012029

A Theoretical Model of All-optical Switching Induced by a Soliton Pulse in Nano-waveguide Ring Resonator

I. N. M. Nawi[1,2], M. Bahadoran[2], J.Ali[2] and P. Yupapin[3]

[1]Centre of Foundation Studies, Universiti Teknologi MARA (UiTM), 42300 Puncak Alam, Malaysia
[2]Institute of Advanced Photonics Science, Universiti Teknologi Malaysia (UTM) 81310 Johor Bahru, Malaysia
[3]Advanced Research Center for Photonics, King Mongkut's Institute of Technology Ladkrabang (KMITL), Bangkok 10520 Thailand

Email: ikhwannaim@gmail.com

Abstract. We propose a theoretical model of 1×2 all-optical switching in a silicon nano-waveguide ring resonator induced by a soliton pulse. All-optical switches made by silicon fiber or silicon waveguide have attracted much attention, because the low-absorption wavelength windows of silicon material just match optical fiber communication. However, to achieve all-optical switching in silicon is challenging owing to its relatively weak nonlinear optical properties and require high switching power, which is much higher than the signal power. Such high power is inappropriate for effective on-chip integration. To overcome this limitation, we have used a highly confined nano-waveguide ring resonator structure with soliton pulse input to enhance the nonlinearity and this leads to enhance the effect of refractive index change on the transmission response. The refractive index is changed by controlling the free-carrier concentration through two-photon absorption (TPA) effect. The result indicates that a refractive index change as small as 6.4×10^{-3} can reduce the switching power to 2.38×10^{-6} W. The nano-waveguide ring resonator all-optical switching described here is achieved by using the concept of strong light confinement, and the switching power is approximately three orders of magnitude lower than the available silicon optical switches. Such controllable switch is desired for achieving high performance in nanometer-size planar structures.

1. Introduction

In recent years, efficient all-optical switching in silicon has been realized primarily in resonant devices for electro-optic modulator, wavelength conversion, ultra-high speed transmission, broadband generation, optical tweezers and chaotic signals generation [1-8]. One of the major challenges in realizing practical all-optical switching devices is the need for strong and fast material nonlinearity. All-optical switches have been demonstrated with III-IV compound semiconductors [9-10], but achieving the same in silicon is very difficult owing to its relatively weak nonlinear properties [11]. In silicon, all-optical switching has been demonstrated only by use of extremely high pumping powers and with large dimension and non-planar structure in which the pump light propagates out of the integration plane. Such silicon optical switches with high power, a large dimension and non-planar geometry are in appropriate for effective on-chip integration. The difficulty of silicon optical switches arises from the weak dependence of silicon's refractive index and absorption coefficient on the free-carrier concentration [12-13].

To overwhelm this restriction, Li and Yupapin [14] have proposed an active double-coupler all-optical switch, in which half of the fiber ring is made by an erbium-doped fibre amplifier (EDFA). The switch can accumulate the nonlinear phase shift to π, and the switching power can be reduced to miliwatt. In this paper, we propose a theoretical model of 1×2 all-optical switching in a silicon nano-waveguide ring resonator induced by a soliton pulse. We have used a highly confined nano-waveguide ring resonator structure with soliton pulse input to enhance the nonlinearity and this leads to enhance the effect of refractive index change on the transmission response. The refractive index is changed by controlling the free-carrier concentration through two-photon absorption (TPA) effect. The result indicates that a refractive index change as small as 6.4×10^{-3} can reduce the switching power to 2.38 $\times10^{-6}$ W. The nano-waveguide ring resonator all-optical switching described here is achieved by using the concept of strong light confinement, and the switching power is approximately three orders of magnitude lower than the available silicon optical switches. Such controllable switch is desired for achieving high performance in nanometer-size planar structures.

2. All-optical Switch Configuration

The configuration for all-optical switching is shown in Figure 1. This configuration is similar with the all-optical switch configuration proposed by Li [12, 13]. A silicon ring with a length of $L = L_1+L_2$ and a nonlinear refractive-index coefficient is coupled with two straight waveguides through the couplers C_1 and C_2. A continuous-wave signal beam with power P_{in} at 1550 nm and a soliton pulse pump beam with average power, P_{avg} at 400 nm are simultaneously launched into the ring through a WDM, and then the signal power passes through couplers C_1 and C_2 successively and outputs from output port 1 or 2. Here we use two other WDMs to separate the pump beam from the signal beams at the two output ports [12, 13].

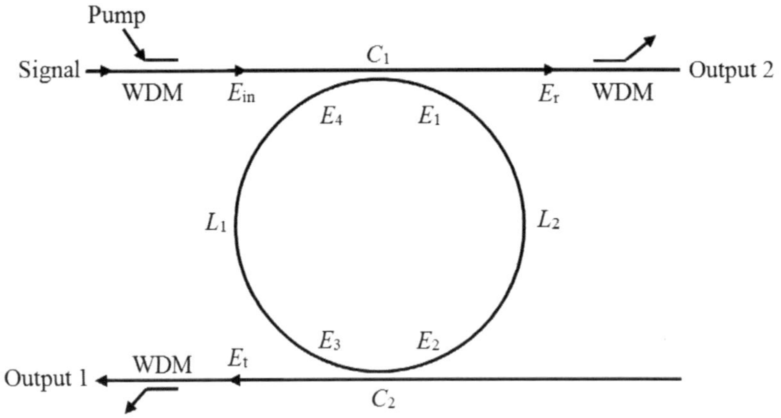

Figure 1. Configuration of 1x2 all-optical switch based on a single silicon ring resonator.

The relationships between the electrical field amplitudes E_r, E_1, E_2, E_3, E_4 and E_{in} when neglecting the losses are [9, 10,12,13]

$$E_r = r_1 E_{in} + it_1 E_4, \quad (1)$$
$$E_1 = it_1 E_{in} + r_1 E_4, \quad (2)$$
$$E_2 = \exp(i\varphi_1) E_1, \quad (3)$$
$$E_3 = r_2 E_2, \quad (4)$$
$$E_t = it_2 E_2, \quad (5)$$
$$E_4 = \exp(i\varphi_2) E_3, \quad (6)$$

where r_i and t_i (i=1,2) are reflectivity and transmission of couplers C_1 and C_2, respectively, and $r_i^2+t_i^2=1$ is satisfied, φ_1 and φ_2 are the phase shifts for two half rings. The phase shift in one cycle of

Journal of Physics: Conference Series **431** (2013) 012029 doi:10.1088/1742-6596/431/1/012029

the ring is $\varphi=\varphi_1+\varphi_2$. Letting $r_1=r_2=r$, we obtain the reflection, R and transmission, T of the nanoring all-optical switch and the power ratios of P_2 and P_4 with P_{in} [9, 10,12,13]

$$R = \frac{P_r}{P_{in}} = \frac{|E_r|^2}{|E_{in}|^2} = \frac{2r^2(1-\cos\varphi)}{1-2r^2\cos\varphi + r^4}, \tag{7}$$

$$T = \frac{P_t}{P_{in}} = \frac{|E_t|^2}{|E_{in}|^2} = \frac{(1-r^2)^2}{1-2r^2\cos\varphi + r^4}, \tag{8}$$

$$\frac{P_2}{P_{in}} = \frac{|E_2|^2}{|E_{in}|^2} = \frac{1-r^2}{1-2r^2\cos\varphi + r^4}, \tag{9}$$

$$\frac{P_4}{P_{in}} = \frac{|E_4|^2}{|E_{in}|^2} = \frac{r^2(1-r^2)}{1-2r^2\cos\varphi + r^4}. \tag{10}$$

Here r is coupler reflectivity, φ is the phase shift in one circle of the ring, which can be divided into two parts: the linear part φ_0 and the nonlinear part $\Delta\varphi$, namely $\varphi = \varphi_0 + \Delta\varphi$. We suppose $r\rightarrow 1$ in the following discussion. In the beginning without pump power and at resonance: $\varphi = \varphi_0 = 2m\pi$ (m = 0, 1, 2, 3 . . .), from Eqs. (7) and (8), we obtain $R = 0$ and $T = 1$, the signal is output from port 1. When adding a pump power to lead a nonlinear phase shift $\Delta\varphi = \pi$, so that $\varphi = \varphi_0 + \Delta\varphi = (2m + 1/2)\pi$ (m = 0, 1, 2, 3 . . .), then $R = 1$ and $T = 0$, the signal will be outputted from port 2. Therefore, as it is shown in Figure 2, the phase shift of $\Delta\varphi = \pi$ is necessary for switching in all-optical switch in nano-waveguide ring resonator.

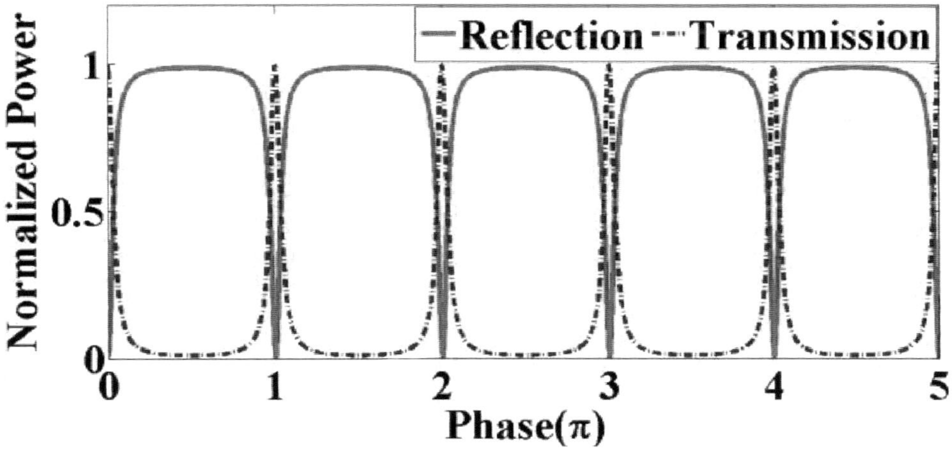

Figure 2. Reflection and transmittance response of a single silicon ring resonator with radius of 5 μm and $r_1=r_2=0.9$.

3. All-optical Switch Configuration

The soliton source for the pump is produced by a tunable mode-locked Ti:sapphire femtosecond laser at a 90-MHz repetition rate, which induces TPA effect in the silicon material. This makes changes in the free carrier concentration including a change of electron concentration ΔN_e and a change of hole concentration ΔN_h. By using Kramers-Kronig relationship, the refractive index changes Δn can be obtained from the experimental spectrum of nonlinear absorption. The refractive index change at wavelength of 1.55 μm is given by [10-11]

$$\Delta n = \Delta n_e + \Delta n_h = 8.8 \times 10^{-22}\Delta N_e + 8.5 \times 10^{-18}(\Delta N_h)^{0.8}. \tag{11}$$

The nonlinear refractive-index change in the free carrier concentration is $\Delta N = \Delta N_e = \Delta N_h$, then Eq.(11) becomes

$$\Delta n = 8.8 \times 10^{-22} \Delta N + 8.5 \times 10^{-18} (\Delta N)^{0.8}. \quad (12)$$

The free-carrier concentration N is generated predominantly by TPA [12-13],

$$\frac{dI}{dz} = \left(\frac{dN}{dt}\right) 2hv = -\beta I^2, \quad (13)$$

and the generation rate of the free-carrier concentration is given by,

$$\frac{dN}{dt} = \frac{\beta I^2}{2hv}. \quad (14)$$

Here I is light intensity, hv is the photon energy, β is the TPA coefficient.

Presuppose that the light intensity of the incident temporal soliton is [15],

$$I = \frac{P_0}{S} \mathrm{sech}^2(\tau), \quad (15)$$

where $\tau = \frac{t}{T_0}$. t is the time and T_0 is the pulse width. The Eq.(15) becomes,

$$I = \frac{P_0}{S} \mathrm{sech}^2 \left(\frac{t}{T_0}\right). \quad (16)$$

By substituting Eq.(16) into Eq.(14), then, the free-carrier-concentration change created by a single soliton pulse is given by

$$\Delta N = \frac{\beta}{2hv} \int_{-\infty}^{\infty} \left(\frac{P_0}{S}\right)^2 \mathrm{sech}^4 \left(\frac{t}{T_0}\right) dt = \frac{\beta}{2hv} \frac{{P_0}^2}{S^2} \frac{\left[T_0 \sinh\left(\frac{t}{T_0}\right)\right]\left[2\cosh^2\left(\frac{t}{T_0}\right) + 1\right]}{3\cosh^3\left(\frac{t}{T_0}\right)}. \quad (17)$$

Here P is the peak power and S is the effective cross section area. For a stream of soliton pulses, the relationship between the peak power P_0 and the average power P_{avg} is

$$\frac{P_0}{P_{avg}} = \frac{T_B}{T_0}, \quad (18)$$

where T_B is the pulse separation. Using Eq. (12), (17) and (18), the relationship between the refractive index change and the average power of the pump beam can be obtained as

$$\Delta n = 8.8 \times 10^{-22} \frac{\beta}{2hv} \frac{{P_{avg}}^2 {T_B}^2}{S^2 T_0} \left(\frac{\left[\sinh\left(\frac{t}{T_0}\right)\right]\left[2\cosh^2\left(\frac{t}{T_0}\right) + 1\right]}{3\cosh^3\left(\frac{t}{T_0}\right)} \right)$$

$$+ 8.5 \times 10^{-18} \left(\frac{\beta}{2hv} \frac{{P_{\text{avg}}}^2 {T_B}^2}{S^2 T_0} \frac{\left[\sinh\left(\frac{t}{T_0}\right)\right]\left[2\cosh^2\left(\frac{t}{T_0}\right)+1\right]}{3\cosh^3\left(\frac{t}{T_0}\right)} \right)^{0.8} . \tag{19}$$

Substitute the data for practical devices [10-13], the wavelength of pump light, λ=400 nm, hv=4.9696x10^{-19} J, β=7.9x10^{-12} m/W, T_0=100 fs, T_B=12.5 ns, $t=T_0/2$, $r_1=r_2$=0.9, φ_0=0, D=10 μm, S=450x250 nm^2, and $L=\pi D$ into phase shift equation,

$$\Delta\varphi = \frac{2\pi L}{\lambda} \Delta n. \tag{20}$$

Figure 3. represents the variation of effective phase shift against the wavelength area of pump light. The phase shift for the reflection and transmission has identical response as it enhanced interval resonance peaks.

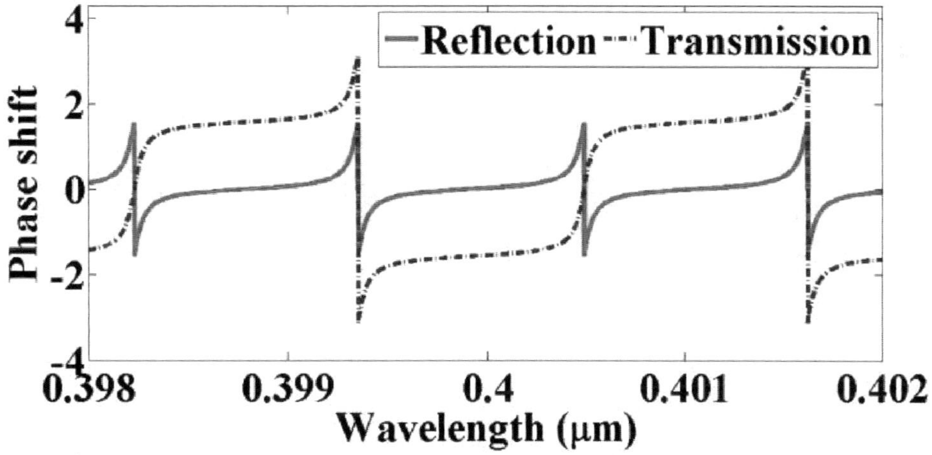

Figure 3. Phase shift in the ring resonator as a function of wavelength.

We obtain the curves of the phase shift and reflection/transmission as a function of the average pump power as shown in Fig.4(a) and 4(b), respectively. Figure 4(a) indicates that the phase shift is a parabolic function of the average pump power. When the phase shift reaches π, the average pump power or switching power is 2.38 x 10^{-6} W. Refractive index change of 6.4×10^{-3} is needed to achieve phase shift of π. Figure 4(b) illustrates when the switching power reaches 2.38 x 10^{-6} W, the output light will be switched from output port 1 to the output port 2.

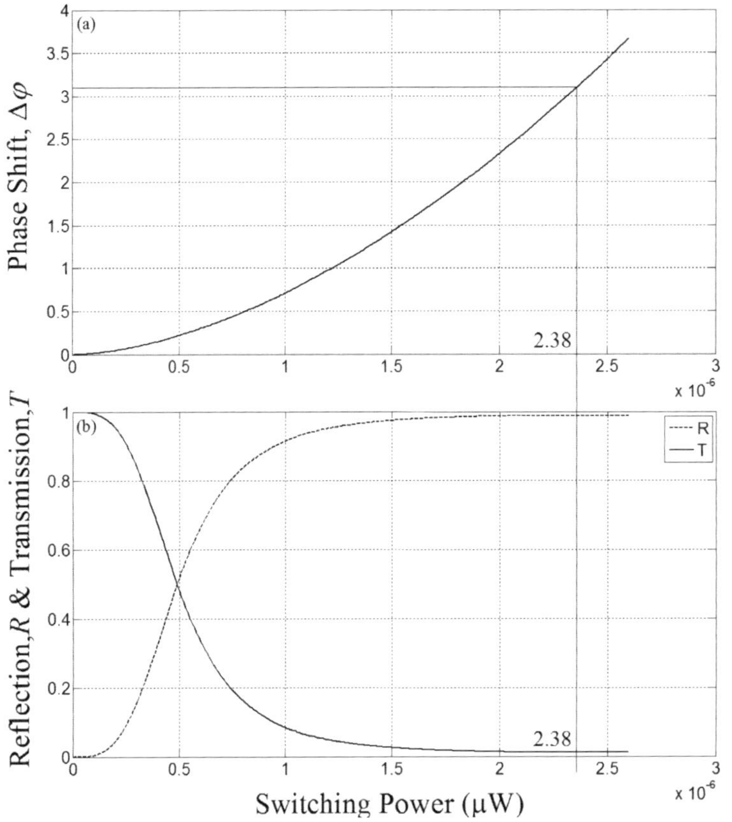

Figure 4. (a) Phase shift in the ring resonator as a function of switching power under TPA effect induced by a soliton pulse. (b) Reflection and transmission in the ring resonator as a function of switching power under TPA effect induced by a soliton pulse.

From the result, the calculated switching power is approximately three orders of magnitude lower than the Gaussian pulse pump beam obtained by Li group as shown in Figure 5 [12-13]. The calculated switching power is also approximately two orders of magnitude lower than the theoretical model proposed by Yupapin. His active double-coupler all-optical switch, in which half of the fiber ring is made by an erbium-doped fibre amplifier (EDFA) can switch the input signal by using the power range of 15.0-28.6 mW [14].

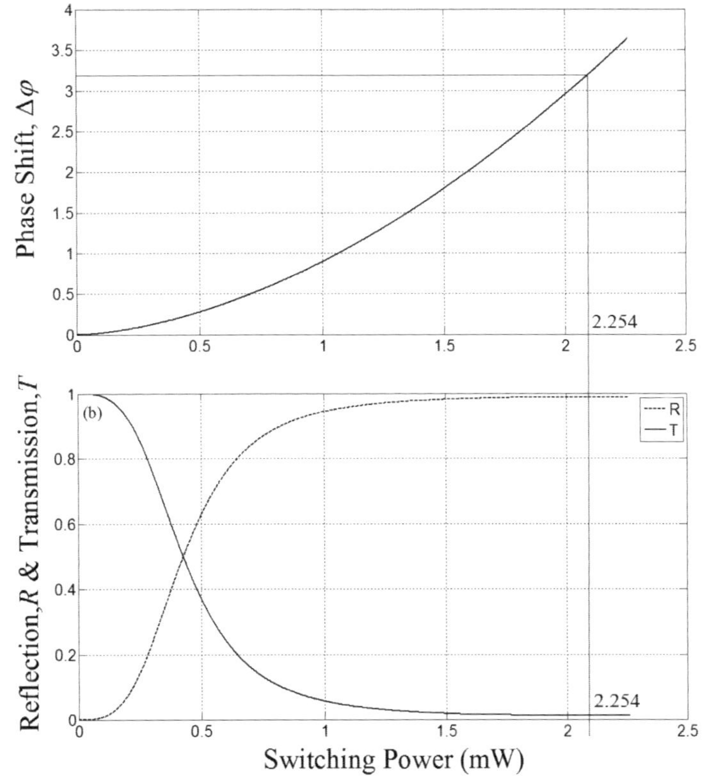

Figure 5. (a) Phase shift in the ring resonator as a function of switching power under cross pumping and TPA effect by a Gaussian pulse. (b) Reflection and transmission in the ring resonator as a function of switching power under TPA effect by a Gaussian pulse [12, 13].

4. Conclusion

We have analyzed silicon nano-waveguide ring resonator all-optical switch based on silicon nano-waveguide ring resonator based on two-photon absorption (TPA) effect induced by a soliton pulse. The result indicates that a refractive index change as small as 6.4×10^{-3} can reduce the switching power to 2.38×10^{-6} W.

References

We would like to thank the Centre of Foundation Studies, Universiti Teknologi MARA (UiTM), the Institute of Advanced Photonics Science, Universiti Teknologi Malaysia (UTM) and the Advanced Research Center for Photonics, King Mongkut's Institute of Technology Ladkrabang (KMITL) for support in research facilities.

References

[1] Guha B, Preston K and Lipson M 2012 *Opt. Lett.* **37** 2253
[2] Ophir N, Lau R K W, Menard M, Zhu X, Padmaraju K, Okawachi Y and Bergman K 2012 *Opt. Exp.* **20** 6488
[3] Lin X, Wenjia Z, Qi L, Chan J, Lira H L R, Lipson M and Bergman K 2012 *IEEE Photo. Tech Lett.* **24** 473
[4] Zhu X, Li Q, Chan J, Ahsan A, Lira H L R, Lipson M and Bergman K 2012 *IEEE Photo. Tech Lett.* **24** 1555
[5] Yupapin P, Jalil M, Amiri I, Naim I and Ali J 2010 *Circuits and Systems* **1** 71

[6] Suwanpayak N, Songmuang S, Jalil M A, Amiri I S, Naim I, Ali J and Yupapin P P 2011 *AIP Conf. Proc.* **1341** 289
[7] Nawi I N, Hairi H, Daud S, Aziz M S, Ali J and Yupapin P P 2011 *Procedia Engineering* **8** 366
[8] Lin X, Chan J, Biberman A, Lira H L R, Lipson M and Bergman K 2011 *IEEE Photo. Tech Lett.* **23** 1103
[9] Van V, Ibrahim T A, Ritter K, Absil P P, Johnson F G, Grover R and Ho P T 2002 *IEEE Photo. Tech Lett.* **14** 74
[10] Ibrahim T A, Cao W, Kim Y, Li J, Goldhar J, Ho P T and Lee C H 2003 *IEEE Photo. Tech Lett.* **15** 36
[11] Almeida V R, Barrios C A, Panepucci R R and Lipson M 2004 *Nature* **431** 1081
[12] Li C F and Dou N 2009 *Chin. Phys. Lett.*, **26** 54203
[13] Li C F and Dou N 2009 *Front. Opto. Chin.* **2** 31
[14] Li C F, Dou N and Yupapin P P 2006 *J. Opt. A: Pure Appl. Opt.* **8** 728
[15] Kivshar Y S and Agrawal G 2003 Optical Soliton: From Fibers to Photonics Crystal 2003 (California USA, Academic Press)

AUTHOR INDEX